Florida Citrus: A Comprehensive Guide

Florida Citrus: A Comprehensive Guide

D. P. H. Tucker, J. S. Rogers, E. W. Stover, and M. R. Ziegler

IFAS Communication Services, University of Florida
GAINESVILLE, FLORIDA

Produced at the University of Florida, IFAS Communication Services

Chana J. Bird, editor
Mariana Wallig, graphic designer

Cooperative Extension Service, University Of Florida, Institute Of Food And Agricultural Sciences, Larry R. Arrington, Director, in cooperation with the United States Department of Agriculture, publishes this information to further the purpose of the May 8 and June 30, 1914 Acts of Congress; and is authorized to provide research, educational information and other services only to individuals and institutions that function with non-discrimination with respect to race, creed, color, religion, age, disability, sex, sexual orientation, marital status, national origin, political opinions or affiliations. The information in this publication is available in alternate formats. Information on copies for purchase is available from IFAS-Extension Bookstore, University of Florida, PO Box 110011, Gainesville, FL 32611-0011 or visit our Web site at: ifasbooks.ufl.edu. Information about alternate formats is available from IFAS Communication Services, University of Florida, PO Box 110810, Gainesville, FL 32611-0810. This information was published December 2006 as SP-278, Florida Cooperative Extension Service.

To order contact:
IFAS Extension Bookstore
Building 440, Mowry Road
PO Box 110011
Gainesville, FL 32611
352-392-1764   352-392-2628 fax
800-226-1764   VISA/MasterCard order only
http://IFASbooks.ufl.edu

ISBN 908-0-916287-34-4

Epigraph

He who has much food has many worries; he who has no food has but one worry.

—Anonymous

Science never appears so beautiful as when applied to the uses of human life, nor any use of it so engaging as agriculture and domestic economy.

—Thomas Jefferson, 1798

Agriculture is the lynchpin of the social chariot.

—Br. of Ancient Indian Wisdom

The greatest service that can be rendered to any country is to add a useful plant to its culture.

—Thomas Jefferson

Economy consists not in saving but in selection.

—Edward Burke

Pay attention to wisdom of the past, for all wisdom is not new wisdom.

—Sir Winston Churchill

Nothing is so powerful as the silent influence of a good example.

—James Kent

Contents

Preface

The objective of commercial citrus production in Florida is to harvest high-quality, marketable fruit for juice products and fresh markets. This is accomplished using management systems with the flexibility to meet production, market, and other economic challenges in order to achieve a sustained and acceptable return on investment. While this statement may appear straightforward, the citrus business has become an increasingly complex operation. Competitive conditions, environmental concerns, labor, and the new world of canker and greening present but a few of the challenges that bring about continuous changes that effect how groves are managed and how crops are harvested and marketed. Environmental, legal, and social issues dictate that citrus management in the United States must comply with an increasing regulatory burden with added costs not equally borne by producers in developing countries. Ultimately, the production, utilization, and delivery system (including retail) should ensure for the consumer an uninterrupted supply of affordable, high quality products. Producers must play a more active role in monitoring their product through the fresh fruit packing and juice production channels to retail outlets to ensure that growers receive an equitable portion of the retail price. All of these factors are moving today's citrus enterprises toward more management intensive systems that can generate their own particular set of problems.

The Florida citrus industry must be recognized as a consistent source of high quality citrus juice products and fresh fruit to maintain and increase its customer base in order to sustain the citrus business and support local and state economies. In doing so, it must remain cost-competitive and compatible with the environment in which it functions.

Perhaps the greatest perceived threat to the sustainability of modern fruit production is its extensive reliance on synthetic pesticides and fertilizers. Fruit growing often serves as a convenient scapegoat for chemically intensive, high input agriculture, which is portrayed as exploiting and endangering farmworkers, wildlife, natural resources, and urban neighbors. Modern fruit production systems, particularly some citrus cultivars for the fresh fruit market, can involve relatively high inputs of pesticides per unit of production. Market and regulatory forces, however, increasingly discourage agrichemical usage while also mandating blemish-free fruit.

Increasing productivity of crops has usually required intensification of inputs with citrus being no exception. The perception of citrus production and farming in general by a sector of the public appears to be one of excessive consumption of finite resources, pollution, loss of wildlife habitat and misuse of sensitive land areas. There is little public understanding of the realities of citrus production and its contribution to the local and statewide economies. The value of agricultural land as opposed to urban and industrial development for water recharge purposes and the benefits awarded consumers are often overlooked.

To help producers meet these challenges, we introduce the concept of Integrated Crop Management (ICM), a whole grove management policy that enables efficient and profitable production that is sustainable and environmentally responsible. ICM is the theme throughout this book, and it expresses the view that while selecting among various management practices, producers and managers must consider the broad picture. Each practice must be evaluated not only for its direct effects, but also with regard to its influence on the overall citrus enterprise and surrounding environment.

ICM attempts to integrate climate, phenological and production factors, intensive monitoring procedures, and beneficial natural processes to ensure efficient use of off-grove inputs such as fertilizers and crop protection pesticides. Informed site selection and superior scion/rootstock selections established at appropriate planting densities are combined with pest and disease exclusion procedures and sanitary practices. The targeted use of nutrients, irrigation, and crop protection products is essential. Management decisions in an ICM system also take into account such diverse factors as economics, law, human resources, land stewardship, and air and water quality.

While management philosophies and practices vary, the basic guidelines are well documented and should be followed. Local conditions, financial limitations, and market demands influence decisions. This publication strives to provide some of the principles and basic information required to help producers develop management options, and understand how they interact to affect the vigor and productivity of the tree and the type and quality of products offered to the consumer.

As this publication goes to print (2006), Florida confronts a globalized citrus business led by sophisticated foreign fresh and processed fruit industries that compete in U.S. and overseas markets. The Florida industry is no longer the dominant force in the juice products market, and must compete with the giant Brazilian orange juice industry, which shares the U.S. market and dominates the European scene. The Florida industry must also deal with existing and new domestic regulations such as the Federal Food Quality Protection Act, as well as quality assurance standards being implemented by European countries within the European economic community, standards with which our competitors in the fresh fruit and juice product sectors must also comply.

The same independent spirit that made the Florida industry a world player may now restrict its competitive ability if certain realities are not accepted. In recent years, there have been a startling number of acquisitions, mergers, and partnership arrangements leading to increased foreign ownership of the U.S. processing industry, the full consequences of which are yet to be determined. The availability and cost of labor, most of it composed of foreign nationals, particularly for harvesting, continues to dominate the list of industry concerns. With changing government offshore labor policies that threaten a guaranteed future labor force, and more competitive wages in other industries, the total harvest of future crops is no longer a certainty. This new century will introduce new challenges and heighten existing ones in many areas, many of them brought on by global competition and domestic policies.

While the industry must defend its interests, open dialogue on controversial issues–both domestic and global–offers the best hope for future prosperity. A dialogue focused on the benefits of Integrated Crop Management should also occur.

Dialogues must engage not only those directly involved in the citrus industry, but also those in decision-making who would impact its operation and all those affected by agriculture. These issues are discussed in the sections entitled Network for Implementation of ICM Policies and Civic Action and the Citrus Industry. Readers should seek out new and updated information as it becomes available.

Acknowledgements

The editors would like to acknowledge the contribution of numerous individuals to the development of this book, including suggestions and ideas by many who are not listed as actual authors, and to those that provided photos, figures, graphs, etc., to provide information contained in their contribution of this book. In utilizing a wealth of academic and industry expertise, the publication should prove of more value to its readership. Special thanks to Kathy Davis for extensive developmental editing of original text files. We would also like to thank Kathy Snyder, UF/IFAS/CREC Word Processing Department, for the formatting of the original file submissions.

Disclaimer

The data and information contained in this publication are based on that believed to be up-to-date at the time of publication. Most of the information, figures, charts, color plates, etc., contained herein are drawn from existing research-based publications generated by, including but not limited to, the listed contributors, research papers, University of Florida Institute of Food and Agricultural Sciences Extension publications, and various local, state, and federal agencies that have been funded primarily through tax revenues, grants, and industry tax dollars, and publicly available industry documentation. Where considered necessary, references are included to recognize specific sources. Such information is made available to all sectors of the general public, sometimes at a nominal price to cover the cost of publication.

Because of the enormous variability of factors affecting the outcome of agricultural operations, including but not limited to the biological and physical environments, weather conditions, action of any governmental authority, economic markets, human factors or any other act of God, the authors make no warranties, either express or implied, as to the information provided in this book, including but not limited to warranties in fitness for a particular purpose or for merchantability.

Introduction

This publication promotes the adoption of technologically sound citrus production, harvesting, and postharvest practices designed to: (1) ensure a supply of safe, wholesome and affordable citrus fruit products, (2) provide an acceptable return on investment for producers, and (3) ensure agricultural sustainability with safeguards for the environment. It guides readers through information on all aspects of citrus production, emphasizing cultural and crop protection practices. Where possible, the guide explains the relationships among these practices and how they may be integrated into the overall citrus business, including marketing and consumer demand in the context of today's agriculture.

This guide is intended as an information resource for anyone interested in the production, processing, packing, and marketing of Florida citrus. The audience envisioned for this book consists of growers, processors, packers, distributors, retailers, educators, students, investors, and regulators. As the guide's title suggests, its primary purpose is to integrate information that will help those with management responsibilities to make sound, future-oriented decisions that provide individual success, growth, and sustainability of the industry.

Chapters bring together economic, crop development, pest management, and environmental information in a way that reveals the relationships and dependencies that exist within the various segments of the industry and the production cycle. For example, pest control involves more than identifying a pest and knowing what product(s) will control it. Managers need to understand the pest's life cycle and how chemical, physical, and biological programs influence the control of pest populations to limit crop damage. They need to know the costs of alternative management options and be aware of environmental and regulatory issues affecting each pest control strategy. This guide attempts to provide a broad approach to many decision making elements by outlining the significant criteria that influence sound management practices that are based on research and historical data, including economic returns on investment.

This guide is a companion manual to the many publications that support standards for Florida citrus production and serves to provide a format for the assembly of information that is critical for ICM. While limited in its ability to fit all needs, the intent of this guide is to support constructive decision-making in the management of Florida citrus production, utilization, and marketing. This guide is organized through the use of narratives, flow diagrams, and displays that are intended to clarify information in a concise and succinct manner. It provides a compendium of information that highlights significant historical data, presents cogent research data, and reviews viable management techniques that illustrate the interconnection of the many decisions concerning sound practices and economic returns.

The guide begins by providing basic information about the tasks of ICM, and the marketing options for fresh and processed citrus products. This introductory information is followed by discussions of tree development and the various components of citrus production, from site and tree selection through product marketing.

Special topics that are beyond classical production management, delve into regulation, labor, crop forecasting, and application technology. Special topics also include economics, risk management, and sustainability. Within each section are references to other information within or outside the guide that may have an impact on the issues discussed. In this way, the guide attempts to provide the links and connections necessary to develop an integrated view of crop management for citrus.

In spite of its broad scope, this publication cannot be all encompassing. Numerous sources of information, including printed publications from academic institutions and government agencies, industry organizations, and computer information services are available to provide more detailed and updated information about specific topics.

Each chapter begins with an introduction summarizing the content of the respective sections. The format for each section is standardized for ease of use. To emphasize critical points, summary statements are presented in call-outs to clarify important concepts.

1

Industry Overview

D.P.H. Tucker

INTRODUCTION

The Florida citrus industry contributes significantly to the state and local economies. For example, the Florida citrus industry includes citrus growers and their full-time and seasonal employees, nurserymen, fresh fruit packers, juice processors, suppliers of agricultural equipment and chemicals, and professionals in many fields. Further, it has been estimated that the Florida citrus industry consists of at least 10,000 property owners. Most of Florida's citrus is processed and sold as juice or juice products. Citrus processors separate the juice from the fruit and sell the juice in a variety of forms, including frozen concentrated juice, reconstituted juice, not-from-concentrate juice, or unpasteurized fresh juice. As of 2005, more than 150,000 people supported the Florida citrus industry by providing a wide range of agricultural services and products. Many federal, state, and county government agencies support, regulate, and oversee the Florida citrus industry also. There are several industry-related tax programs designed to assist growers with marketing and research. The Florida Department of Agriculture and Consumer Services (FDACS), in cooperation with the USDA, is responsible for the inspection of fresh and processed fruit to ensure that quality standards are maintained. There are other component groups in the citrus industry to help sustain the long-term future of the growers. This chapter provides an overview of these components and describes how they interrelate.

Peninsula Florida can be divided into four main production regions: the Central Ridge, Indian River, Peace River, and Gulf Coast. The Central Ridge can be characterized as having well-drained sandy soils, whereas the coastal areas and the flat-woods and marsh soils typical of Central and Southern Florida are poorly drained. In the latter areas, citrus trees are planted on raised beds with drainage ditches to improve aeration of the root zone. These areas each have some unique production problems but together they account for a significant percentage of the citrus production.

Almost 9 million acres are currently in citrus production worldwide. Florida citrus production accounts for about 700,000 of these acres, or 9 percent of total acreage, but represents more than 15 percent of the citrus production in the world (Plate 1.1). Florida produces 20 percent of the world's oranges on about 10 percent of the world's orange acreage, and approximately half of the world's grapefruit on about half of the total acreage planted to grapefruit. Florida also produces about 2 percent of the world's tangerines, lemons, and hybrids on about 3 percent of the world acreage planted to these cultivars. Currently, about 150 million boxes of citrus are produced in Florida each year.

The Florida citrus industry contributes significantly to the state and local economies. The on-tree value of Florida citrus is estimated at close to $1 billion, while the citrus value when leaving the processor or packinghouse is estimated at $4 billion. Gross sales activity value is estimated at about $9 billion, of which $4.1 billion is direct and $5.0 billion is indirect and induced. Economic impact earnings and income account for about $4.2 billion, which includes $1 billion indirect citrus and $3.2 billion in direct and induced value.

Tax statistics show costs involved in delivering public services to the urban sector and agribusiness interests. National statistics indicate that for every dollar raised from residential land use, as little as $1.02 and as much as $1.67 are required to provide public services for residents. Conversely, for every dollar that is raised from farmland use, $0.21 to $0.77 was required to provide public services.

Ratio of Revenue to Expenses:	
Farm and Forest Land	$1.00:$0.49
Commercial and Industrial Land	$1.00:$0.20
Residential Land	$1.00:$1.35

It is also well documented that growth in population and tax base result in higher, not lower tax bills on median-value homes.

In Central Florida, it is estimated that for every dollar generated by residential revenues, local counties spend $1.89 in services, a rating of $1.00:$1.89. Dollar for dollar, therefore, residential development creates a deficit of $0.89 compared to agricultural-related land use, like citrus production, which creates a surplus of $0.92 (rating of $1.00:$0.08).

A recent analysis of all this information shows that:

1. Increases in residential uses—without accompanying increases in revenues from other sources—will likely create an operating deficit.
2. Even though agricultural and open land uses generated a much smaller amount of revenue than do residential land uses, every time agricultural or open land is replaced with a housing development, there's the potential for a small surplus to be replaced with a large deficit.
3. It takes a concerted effort to account for all factors that contribute to the economics of a specific land use. However, with proper planning and attention to the factors that contribute to the economics of each land use; it may be possible to create a reliable process to "balance" or "mix" land uses, so the deficits caused by one land use can be offset with surpluses generated by another use.

Controversial issues often surround agricultural land use, including citrus production. However, there are also many public benefits associated with citrus production. The extensive green zones provided by forestry and agricultural landscape are aesthetically pleasing, favorably impressive to visitors, and helpful in promoting tourism. Areas planted to citrus in the highly permeable soils of Central Florida provide essential groundwater recharge with ultimate percolation to the Floridian aquifer. More than 250,000 acres of citrus in these areas contribute to rainfall generated recharge, as opposed to evaporation and runoff from urban receptors such as streets and parking lots. Citrus plantings in the poorly drained flatwoods also retain rainfall by providing retention areas to hold and recycle water and may contribute to flood control of nearby urban areas. In addition, large volumes of recycled urban water are now distributed in citrus groves as irrigation. Citrus plantings provide wildlife habitat, or may serve as wildlife corridors. Further, through ICM, citrus producers continue to refine their cultural practices, especially their use of chemical inputs, to minimize unintended and undesirable environmental effects. The citrus industry is highly regulated and operates under the constant scrutiny of government agencies, environmental organizations, and the media.

Many people with many different backgrounds and levels of training and skill are employed in citrus production. The Florida citrus industry includes citrus growers and their full-time and seasonal employees, nurserymen, fresh fruit packers, juice processors, suppliers of agricultural equipment and chemicals, and professionals in many fields (Figure 1.1).

CITRUS GROWERS

It has been estimated that the Florida citrus industry consists of more than 10,000 property owners. Individuals, companies, partnerships, or corporations may own groves. Individuals may own as little as 10 acres, and large corporations may own more than 30,000. The total number of growers has declined slightly in the last several years due to development and plant-disease pressures, but there has been an increase in corporate control of very large citrus operations. Although many groves are still family businesses or locally owned and operated, it is increasingly common for Florida citrus operations to be owned by investors who may not live in Florida, or even in the United States. These absentee owners employ professional citrus managers who provide or arrange for all grove maintenance. Much of the labor for harvesting is seasonal and may be provided by a contractor. Growers may choose to sell their fruit fresh or for processing and may select from a variety of marketing options.

FRESH FRUIT PACKERS

About 4 percent to 6 percent of the Florida orange crop and as much as 35 percent to 40 percent of the Florida grapefruit crop is sold as fresh fruit annually. In addition, about half of the specialty crop is shipped fresh each year. Fruit to be sold fresh must meet external grade standards. A relatively small but significant amount of fresh citrus is sold as gift fruit; the remainder is generally sold through fruit packinghouses. In recent years, the number of citrus packinghouses in the state has decreased, and the volume of fruit packed by a typical house has increased. Gift fruit shippers sell small, often decorative packages of fresh fruit directly to consumers or ship fruit out of the state. Packinghouses provide fruit to retailers, especially supermarkets.

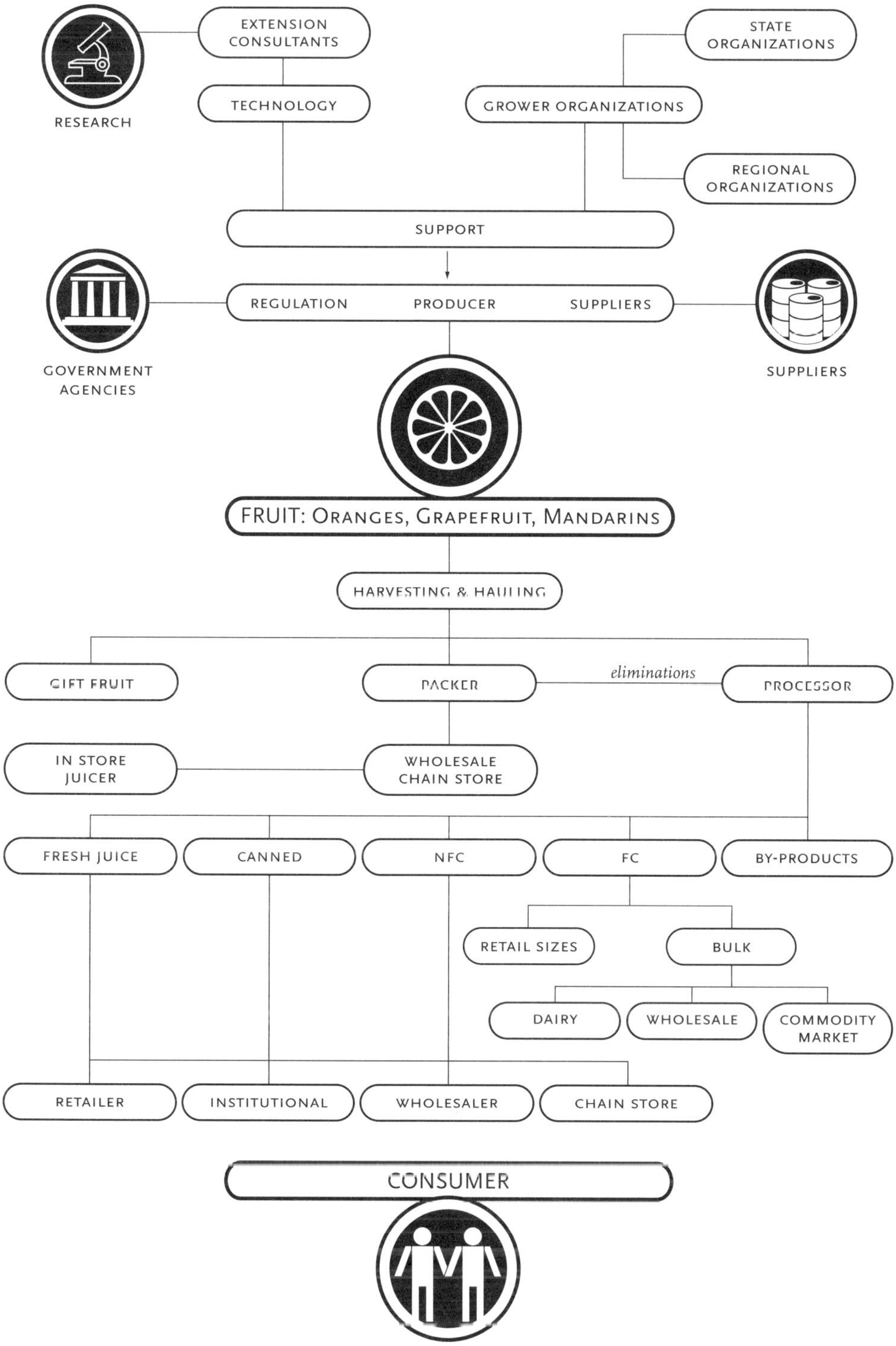

Figure 1.1 Organizational flowchart of the Florida citrus industry showing a complex, multifaceted system of production, utilization, and marketing supported by industry organizations, academic institutions, and government agencies.

FRUIT PROCESSORS

Most of Florida's citrus is processed and sold as juice or juice products. Citrus processors extract the juice from the fruit and sell the juice in a variety of forms, including frozen concentrated juice, reconstituted juice, not-from-concentrate juice, or unpasteurized fresh juice. The citrus peel, pulp, and seeds are generally dried and sold as livestock feed. There are currently about 15 citrus-processing facilities in Florida. These vary greatly in size with the largest plants processing more than 1 million boxes (90 million pounds) of fruit per week. Fruit sold fresh is usually of higher value than fruit intended for processing. However, fruit that are harvested for the fresh fruit market but do not meet external grade standards may be diverted to processing. These eliminated fruit are usually sold to processors at a slight discount over fruit that was intended for processing. In some years, eliminations may represent a net loss to the grower.

CITRUS NURSERIES

Citrus nurseries provide growers with young citrus trees to develop new plantings or to replace unhealthy or unproductive trees in existing groves. The purchaser specifies scion and rootstock variety, whether the trees are to be registered or validated, delivered bare-root or containerized, and the desired size and budding height. From 1996 to 2004, 53 commercial citrus nursery sites in Florida produced approximately 4.0 to 6.0 million trees each year.

Over the last several years, however, the citrus nursery industry has been profoundly affected by citrus canker and greening. Currently, the industry is undergoing a transition to producing trees entirely within screenhouses. This practice is to help prevent the transmission of diseases such as greening by insect vectors. It is expected that it could be early 2008 before screenhouse-produced nursery trees become commonly available.

Table 1.1, "Outlook on the Future," offers a general outlook on the possible areas of impact of citrus canker and greening diseases on each production topic presented throughout this book. Citrus canker and greening disease are both considered too widespread to eradicate. Consequently, some of the chapters in this book authored before 2006 discuss practices that may be affected by the eventual spread of these diseases throughout the state. The information contained in this table is based on the best information available in Spring 2006. However, there is substantial disagreement among experts as to what will be the likely result of living with these new exotic diseases. Nevertheless, this information is useful for framing the issues for further thought and discussion.

Table 1.1 Outlook on the Future

Title	Description
Preface & Introduction	Citrus greening (also called Huanglongbing, or HLB disease), canker and other exotic diseases become more important grower considerations beginning in January 2006. Of the approximately 800,000 acres currently in production, most experts agree that there will be a decline in producing acreage over the next 5 years or so. Estimates of future acreage range from about 500,000 to 750,000 acres. Fresh fruit acreage will likely be lower than processed unless varieties highly resistant to citrus canker are introduced soon.
The Citrus Manager	Citrus production managers will have to be trained in the recognition, management and decontamination against exotic new diseases, including canker, greening and CVC. This training may become required by the Florida Department of Agriculture and Consumer Services.
Targeting the Marketplace: Fresh vs. Processed	Exotic diseases can exclude citrus from certain markets. Fresh fruit is most affected by canker, while greening can impact yields on fresh and processed varieties.
Tree Growth, Development, and Productivity	Early control of greening and canker become very important, since early establishment of these diseases accelerates decline and death.
Influence of Environmental Stress on Tree Growth and Productivity	Environmental stress may predispose trees to infection by disease.
Cold Protection	It is not known for certain how cold protection and exotic diseases might be related. However, some bacteria related to the citrus canker causal organism are ice nucleators, so it has been speculated that trees with high canker infection may be predisposed to increase freezing injury.
Variety Development, Clonal Protection, and Tree Selection	The search for new citrus cultivars resistant to greening, canker and other exotic diseases gets higher priority.
Site Selection	Healthy trees located near infected citrus groves are more likely to become infected themselves with those same diseases. Locating nurseries at increased distances from commercial groves becomes standard.

Site Preparation and Engineering	Windbreaks play a role is reducing the rate and distance of spread of citruc canker in other countries. Despite certain logistic and economic constraints, windbreaks may nevertheless become important means of disease control in the future.
Establishing the Irrigation System	Irrigation water contaminated with bacteria and fungal diseases can easily contaminate other areas. Prevention of disease transmission, especially in nursery environments, will become more important.
Grove Design	Citrus groves will have to be designed in such a way that the likelihood of disease transmission through natural and manmade means is reduced. This will be especially important when chemo-robotic harvesting becomes more prevalent.
Grove Planting and Early Care	Early control of greening and canker become important, since early establishment of these diseases accelerates decline and death.
Introduction to Citrus Grove Management	All citrus workers will have to be trained in the recognition and management of citrus greening, canker and other exotic diseases.
Water Management	Irrigation water contaminated with bacteria and fungal diseases can easily contaminate other areas. Prevention of disease transmission, especially in nursery environments, will become more important.
Nutrient Management	Nutrition may affect disease development, but specifics are not known at this time. It is possible that managing flush timing can help reduce the incidence and severity of diseases such as citrus canker.
Grove Floor Vegetation Management-	Grove floor vegetation will factor into chemo-robotic harvesting, but specific relationships to greening and canker are not currently clear.
Integrated Plant Protection	Greening and canker diseases become much more important grower considerations beginning in January 2006. Field decontamination is more important, but common sense should be applied when practicing decontamination. More frequent applications of materials such as foliar copper for canker and imidoclprid materials for psyllids (greening) will be important management tools.

Management of Tree Size, Shape, and Cropload	Canopy size, shape and timing of flush may affect the incidence and severity of greening and canker. Experience will likely be the best teacher of how to exploit this to advantage in management of these dieases.
Plant Growth Regulators	Specific relationships to the severity of greening and canker epidemics are not clear at this time
Fruit Quality	Exotic diseases can exclude citrus from certain markets. Fresh fruit is most affected by canker, while greening can impact yields on fresh and processed varieties.
Fruit Harvesting, Handling, and Transport	Sanitation by harvesting crews becomes much more important as growers look for ways to prevent the human spread of disease.
Fresh Citrus Utilization	Fresh fruit is most affected by canker. Some experts believe that the survival of the fresh grapefruit industry will depend on the development of new highly-resistance varieties.
Citrus Processing	Canker and greening can both affect yield and quality, but some late season orange crops will probably do rather well in the presence of canker, especially during dry years.
Marketing	Exotic diseases can be issues in trade negotiations. Exceptional attention to management of these diseases will have to be demonstrated if certain markets are to remain open.
Industry Overview	Citrus greening (also called Huanglongbing, or HLB disease), canker and other exotic diseases become more important grower considerations beginning in January 2006. Of the approximately 800,000 acres currently in production, most experts agree that there will be a decline in producing acreage over the next 5 years or so. Estimates of future acreage range from about 500,000 to 750,000 acres. Fresh fruit acreage will likely be lower than processed unless varieties highly resistant to citrus canker are introduced soon.
Civic Action and the Industry	Being active in the legislative process influences when and how specific rules and regulations are adopted for canker and greening management.

Regulatory Agencies and Environmental Assessment	Canker and greening regulation compliance activities will probably be monitored by the Florida Department of Agriculture and Consumer Services. This information can be used to help maintain open markets.
Labor Requirements	Chemo-robotic harvesting will become more important as labor supplies become tighter and it is increasingly difficult to find workers willing and able to work in the agricultural industries in Florida.
Energy Utilization	Highly-refined petroleum oils are used in insect and disease control. Prices for these materials are increasing. Indirectly, conservation of energy and fuel may help keep these prices lower. Alternative fuels (e.g., ethanol) may become both produced and used by agricultural organizations in Florida in the future.
Forecasting Citrus Production	Increased prevelence of greening and canker will result in lower crop forecasts over the next 5 years or so. It is also possible that the erratic development of these diseases may introduce a confounding factor in citrus crops forecasts.
Meteorological Data Collection and Application	Weather data will be used to forecast disease outbreaks, which will provide important information for the management of disease.
Application Technology	More frequent applications of materials such as copper will be needed for canker management. Other materials will be necessary for psyllid control. New technologies for the precision application and tracking of effectiveness of these materials will have to be developed for growers to obtain the greatest efficiency in control.
Precision Agriculture and Crop Management	Geospatial tracking of the development of disease outbreaks will probably become important tools for managing disease in the field. This information can also be used for market purposes.
Grove Maintenance: Renovation, Replant, and Abandonment Options	Canker and greening can be problems in abandoned citrus groves. Infected, abandoned groves adjacent to otherwise healthy groves will have to be removed as a means to eliminate a potential source of inoculum.

Purchasing and Leasing Grove Property	There are valuation issues in normally healthy citrus groves that are found infected with greening or canker. Real estate brokers will develop methods to factor these issues into sales. The results will probably depend to a large extent on whether the intention is to maintain the grove in commercial production or to convert the grove into housing or commercial development property.
Crop Damage: Documenting and Recovering Economic Losses	Devastating diseases, eradication practices and other catastrophic factors contributing to catastrophic tree losses will become more common. Hurricanes and weather events can also cause dramatic losses in crops. Documenting tree and crop losses will become a more valuable tool for recovering losses from litigation and government aid programs.
Scouting/Monitoring	Geospatial tracking of the development of disease outbreaks will become important tools for managing disease in the field. An important aspect of tracking disease is the methods used for scouting and monitoring. In addition to traditional tools such as the hand lens, computers, digital cameras, cell phones and handheld computers will become very important methods for scouting the development of these diseases through time and space.
Economics	Canker and greening affect crop markets and prices
Risk Management	Rootstock and scion cultivar selections matched to soil profiles will probably improve the chances of tree survival when planting new genetic combinations in new geographical areas. Live databases, artificial intelligence and expert systems may prove to be a valuable tool in making better planting decisions.
Wildlife Needs in Citrus Culture	Wildlife species can transmit exotic diseases, but it is not very practical to try and restrict the movement of birds and mammals throughout groves.

Essentials of Nursery Management	Healthy trees located near infected citrus groves are more likely to become infected themselves with those same diseases. Locating nurseries at increased distances from commercial groves will probably be important. Producing trees in enclosed, protected structures will help reduce the physical transmission of disease by psyllids and other insects.
Sustainability of Commercial Citrus Production	Sustainability of commerical citrus production in the future will be impacted by development and construction in Florida. Many groves are currently being converted to housing and commercial development, with prices correspondingly increased. Better incentives to maintain citrus groves in commercial production will be important if the industry is to continue to prosper in the future.

INDUSTRY SUPPLIERS

More than 150,000 people support the Florida citrus industry by providing a wide range of agricultural services and products. Citrus growers require products such as agricultural equipment, chemicals, fuel, etc. Services include labor to maintain the grove, harvest, and transport the crop. The industry also relies on the professional services of researchers, extension agents and specialists, consultants, insurers, accountants, lawyers, and bankers.

INDUSTRY TRADE ORGANIZATIONS

Many trade organizations have been established to support the citrus industry. These organizations may represent a particular geographic region or aspect of the industry. Membership in any of these organizations is voluntary. Membership fees are determined by each association and are often specified as some amount per box of citrus produced by each member. Most of these organizations have a few paid staff members and rely heavily on the voluntary efforts of their members. Many are active in legislation to support the interests of their members. Some of the largest and more well-known organizations are:

1. Florida Farm Bureau, representing all farming operations in Florida;
2. Florida Fruit and Vegetable Association, representing all fruit and vegetable operations in Florida;

3. Florida Citrus Mutual, representing citrus growers throughout the state is especially active in legislation;
4. Indian River Citrus League, representing growers in the Indian River Citrus Production region;
5. Gulf Coast Growers Association, representing growers in Charlotte, Glades, Hendry, Lee, and Collier counties;
6. Peace River Valley Citrus Growers Association, representing growers in Hardee, DeSoto, Manatee, and Sarasota counties;
7. Highlands Citrus Growers Association, representing citrus growers in Highlands County;
8. Florida Citrus Packers, representing fresh fruit packers and shippers throughout the state,
9. Florida Citrus Processor Association, representing citrus processors throughout the state;
10. Florida Citrus Nurserymen's Association, representing the nursery industry;
11. Florida Fertilizer and Agrochemical Association, representing suppliers of fertilizers and pesticides.
12. Florida Gift Fruit Shippers Association, representing specialized gift fruit shippers.

GOVERNMENT AGENCIES

A wide array of federal, state, and county government agencies support, regulate, and oversee the Florida citrus industry. Most of these agencies enforce regulations governing the industry; the regulations have been established by legislation or by the growers themselves.

The United States Department of Agriculture (USDA) supervises federal marketing orders that regulate grades, sizes, and volume of citrus, and participates with state agencies to enforce other rules and regulations. The USDA also conducts citrus research through the Agricultural Research Service (ARS), which includes the Horticultural Research Lab that has recently relocated to Fort Pierce, and the Winter Haven Citrus and Subtropical Products Laboratory.

The Florida Department of Citrus (FDOC) is a state agency established by the Florida citrus growers to set policies and establish regulations for Florida citrus. The twelve-member Florida Citrus Commission, appointed by the governor, determines FDOC policy. The commission taxes the Florida citrus growers on a per box basis and does not receive direct funding from the state of Florida. The tax rate is determined annually by the FDOC to provide generic advertising and promotion services for both fresh and processed Florida citrus, to designate and fund some postharvest research, and to establish regulations governing the operations of the industry.

The Florida Department of Agriculture and Consumer Services (FDACS), in cooperation with the USDA, is responsible for the inspection of fresh and processed fruit to ensure that quality standards are maintained. These standards have been and continue to be established by other government agencies or grower associations.

The Division of Plant Industry in the FDACS cooperates with the USDA to enforce regulations established to control certain noxious pests. In addition, this agency controls the citrus budwood registration program. The budwood registration program provides high-quality, virus- and viroid-tested budwood to growers and nurserymen in Florida and other citrus-producing regions.

The Florida Agricultural Statistics Service in cooperation with the USDA provides monthly and annual crop estimates, a biannual commercial citrus inventory, and other statistical information about citrus and other crops produced in Florida.

The University of Florida's Institute of Food and Agricultural Sciences (IFAS) operates several research facilities with citrus programs and a network of Extension Service offices. IFAS supports research and provides information and education to growers through a wide variety of programs.

Three of Florida's five water management districts are located in the major citrus production areas. Those districts are the South Florida Water Management District, the Southwest Florida Water Management District, and the St. Johns River Water Management District. Each district regulates the quantity of water that growers are allowed to remove from water sources (surface and groundwater) for irrigation.

In addition, each county may establish regulations, especially building and zoning restrictions that affect local citrus growers. Growers must consult with county officials to ensure that they are in compliance.

There are many opportunities for growers to actively participate in their trade organizations and to advise and influence their governing agencies. Growers must be involved in the development and enforcement of reasonable and responsible legislation to maintain and promote a sustainable citrus industry in Florida.

Florida citriculture is a complex agricultural system. Despite competition from other producing regions, the Florida industry has managed to remain relatively competitive, particularly with processed products. A complex series of linkages integrates individual enterprises with distant environments. Figure 1.2 provides some insight as to how a specific region of citriculture can adapt to a changing economic and physical environment. Citrus spatial patterns are largely determined by producer response to a variety of limitations imposed by the physical environment and further defined by economic and other conditions, which dictate land value. When these factors or perceptions change, spatial patterns change among the different levels of integration in the Florida citriculture system.

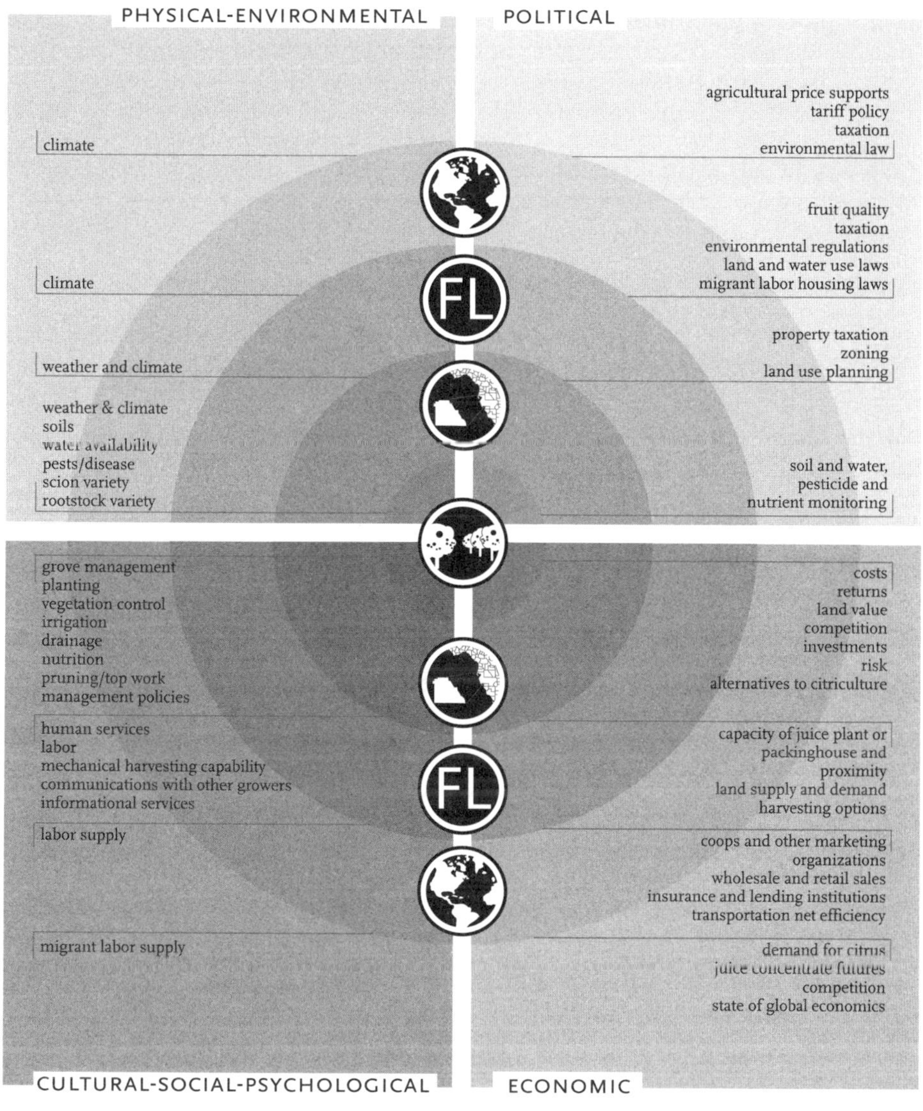

Figure 1.2 Schematic graph showing the integrated nature of the Florida citrus industry. There are different levels of integration that includes environmental, regulatory and biological components.

Growers may have a different set of interests to those of the processor. The move from the grove to the plant may follow different avenues, ranging from growers who own packinghouses, processing plants, or are a part of cooperatives that own processing facilities to those who sell their fruit on the tree. The fruit may go to cooperatives or participation pools or be sold on the cash market. For fruit going to the cash market, timing of the sale is important as very early or late harvesting is usually associated with higher cash prices. In such transactions, the financial strength of the buyer is an important consideration. For fruit going to a pool or participation plan, the harvesting date should coincide with maximum pounds solids per acre for best returns.

REFERENCES

Brey, J. A. 1985. Changing spatial patterns in Florida Citriculture, 1965-1980 (Oranges, Landtype, Land-Use Change, Agricultural Systems). Ph.D. Dissertation. University of Wisconsin, Madison, WI.

Jackson, L. K., R. M. Davis, J. M. Bulger, G. F. Fairchild, R. P. Muraro, and D. P. H. Tucker. 1990. *Introduction to the Florida Citrus Industry: Its production, harvesting and marketing practices.* University of Florida, IFAS, Cooperative Extension Service SP-4174.

2

Network for Implementation of ICM

D.P.H. TUCKER AND J.S. ROGERS

INTRODUCTION

There are four critical components to the increased adoption of Integrated Crop Management (ICM) for citrus: (1) research; (2) extension; (3) private-sector crop consulting; and (4) the field citrus manager. The development of ICM programs originates from a systematic review of existing management programs and analyses of their strengths and weaknesses. Each of the above components plays a different part in the delivery of ICM to the industry. This chapter presents an overview of the basic tenets of ICM and reviews the roles of the different areas. The citrus industry and the public ultimately benefit from the adoption of ICM practices.

Research and extension are vital to the continuing and increased adoption of Integrated Crop Management (ICM) for citrus. Research leads to the proposal of new applications that must be evaluated and adapted for production. Ultimately, researchers may produce new recommendations that should be communicated to management throughout the industry.

RESEARCH SUPPORT BASE

Humans have always searched for better ways of doing things and have therefore carried out simple research. It is an instinctive part of being human. Research helps people compare new ways of doing things with current practices.

It is important for fruit industries, like citrus, to support research. The industry must appreciate what is involved in the establishment, conduct, data gathering, and evaluation phases of experiments. The willingness of management to provide aid and support for maintenance and harvesting is key to the successful outcome of research projects. For this to occur, however, cooperators must develop and maintain an interest in such projects to completion. That coordination with researchers and harvesting managers for the collection of plot yields and other data is crucial if valuable information is to be eventually disseminated and utilized by the industry.

Citrus producers do not have access to enough information concerning factors that affect their product, and such information may not be directly applicable or freely available in the future. While public sector institutions may provide generic information, it may not be directly applicable to all cultivars, markets, and growing conditions. As public funding for research declines, it is unlikely that research will be customized enough to fit all situations, so grower funding becomes all the more essential to fill the void. Still more specific information will have to be generated by growers themselves or by consultants. While a very few corporate growers support their own research programs in the Florida citrus industry, most appear not inclined to invest in in-house research activities. However, for the future, it will be necessary for progressive growers to not only support funding for both public sector and industry supported research, but also gain the competence to conduct their own instinctive research based on subjective assessment and record keeping.

The Florida Citrus Production Research Advisory Council (FCPRAC) consists of 14 grower representatives who disburse funds to researchers working on production problems important to the Florida citrus industry. The Council works under guidelines established within the Florida Citrus Production Research Marketing Order, which was approved in a referendum of Florida citrus growers in 1991, then again in 1997 and 2004. The marketing order enacted regulations to allow growers to tax themselves up to one cent per box of fruit produced and direct those funds to help solve industry production problems by research. Funds are awarded as competitive grants selected from proposals submitted by the research community to the FCPRAC in the fall of each year.

The development of ICM programs originates from a systematic review of existing management programs and analyses of their strengths and weaknesses. Foremost in all analyses is the need for realistic assumptions and an objective appraisal of management capacities.

ICM is not intended to resolve management failures simply by leading to the adoption of new or different management strategies. Instead, ICM forces critical assessments of all systems involved in the complex array of issues imposed on citrus production regimes. Details count in the discovery of successful operations and the interactions of related decisions. For example, profitable fresh citrus marketing cannot be achieved unless timely scouting precedes pest control strategies, leading to precisely timed applications of products used in the chemical control of fungal diseases. Otherwise, the fruit surface may become blemished with lesions caused by infection of the fruit by the causal fungus. If fruit is damaged in this manner, the alternative processed market becomes the only viable opportunity for the grower, and this market likely will not generate returns equivalent to fruit marketed in fresh channels.

To implement effective ICM, it is necessary to review recommendations and examples of successful ICM practices. Thus, the purpose of this section is to lay out the basic ICM diagram with explanations on how to transform current management strategies into successful programs. Initially, managers should consider the time frame for the adoption and transition to ICM, resource requirements, information management capacities, access to critical data, and the commitment to change by those involved.

Private crop consultants provide invaluable expertise in the site-specific implementation of ICM programs. In fact, private-sector crop advisors may be thought of as the infield integrators of the diverse ICM information domain. Usually, however, the initial information resource for implementing a successful ICM program is the research, education, and extension services of the local land-grant university.

In a "shrinking" world, it is important that Florida growers and those in other sectors of the industry attend international citrus meetings to keep abreast of technological advances, and meet scientists and producers from other countries. Experience to date would indicate that the Florida industry is usually not well represented at such events. As global competition becomes more intense, the Florida producer would be wise to heed the saying—"Knowledge is Power."

RESEARCH

Most applied research requires grower cooperators in field trials under replicated and controlled conditions. These site-specific research designs permit scientists to examine critical interrelationships in the citrus management system with important practical applications for ICM. Often, projects involve educational events and field demonstrations at research sites. In all cases, data derived from such projects are evaluated for technical accuracy and application opportunities.

Applied research is conducted to resolve problems and create alternative solutions to citrus production limitations. Applied research differs from basic research that is performed under laboratory conditions for purposes of pure information discovery. It is important to reference research-based information in the development and operation of ICM because successful adoption of new and innovative intelligence is often the key to making effective decisions that achieve future economic successes. However, it is common to modify research information based on the characteristics of specific sites that may not be accommodated in the original experimental design.

Grower observations and records may indicate or suggest relationships and interactions in grove management that warrant further study. With little training in experiment design and data analysis, growers can nevertheless perform their own field experiments by trial and error and improve the effectiveness of their production practices. Regardless of the approach, the components of an effective ICM research program are generally characterized by a sequence of events:

1. awareness of the problem, such as unexplained tree attrition;
2. commitment to information discovery through a systematic approach;
3. design of a field trial;
4. establishment and execution of the field trial, including data acquisition;
5. evaluation of acquired data and information; and
6. design and implementation of changes to management practices.

EXTENSION

The principle function of the agricultural extension service is to collect, interpret, and distribute information to the agricultural community with recommendations for its effective implementation. Extension promotes the sustainable production of high-quality agricultural products and the responsible and efficient use of natural resources. Extension personnel use a variety of techniques to communicate with citrus producers, including intensive educational programs, seminars and short courses, surveys to assess industry needs and practices, meetings on emergency topics, visits to local groves, packinghouses and processing plants, field demonstrations, publications, radio and television presentations, and computer-based resources.

Private-sector industry consultants and agrochemical product representatives also use information from extension sources to advise citrus growers. However the information is obtained, it is important for citrus growers to understand that ICM recommendations are influenced by many site-specific components of the grove and its environment. Customizing production practices to a particular array of growing conditions and grower objectives is central to the concept of ICM. Climate, soil type, scion variety and rootstock, incidence of pests and diseases, market opportunities, and the availability of land, labor, and financial capital all play a role in determining an individual grower's best ICM strategies. Such strategies continually evolve as new information is acquired and circumstances change. It is in the grower's best interest to actively participate in educational opportunities in order to raise questions and assess the relevance of new information to his or her own unique situation.

The range of decisions facing management can be overwhelming, but may be resolved or reduced to manageable levels if computer-based expert systems, such as expert decision support systems, are employed. Expert systems are those based on scientific data, analytic models, and statistical evaluations, but also may include holistic or empirical approaches to problem solving. Frequently, expert systems permit the automated acquisition of weather data (Plate 2.1), market information, and cultural information followed by the automated generation of production advice that is customized to individual needs. Several such systems, provided by both the public and private sectors, are currently in commercial use in Florida. Expert information systems also permit the experiences of individual growers to be shared and documented. This leads to a better understanding of interactions in the grove, and further improvements in the application of ICM expertise.

In addition to the information services of the local university, the agricultural chemical industry and other nongovernmental groups, such as private sector crop consultants, produce high-quality, highly marketable information services that permit commercially viable ICM programs. Generally, these resources provide crop-specific information with a reasonable return on investment and minimal undesirable environmental impact.

The citrus industry and the public ultimately benefit from the adoption of ICM practices and the abandonment of overtly traditional production methods. However, citrus producers, government agencies, environmentalists, and consumers must all realize that there are costs and risks associated with ICM practices, especially in the early stages of development. In an effort to encourage the use of ICM in the commercial sector, there has been discussion of distributing the costs and risks of ICM more equitably between the industry and the public. Methods that have been suggested include providing compensation to growers in the event of crop failure due to ICM; subsidizing insurance for private ICM specialists; requiring growers to adopt certain ICM practices in order to qualify for government-approved liability exemptions; increasing support of biomonitoring through state laboratories; and accelerating the registration of low-risk, highly targeted compounds. Public support for agricultural research and extension must be maintained and increased with funding for demonstration plots and innovations in communication. Education and possibly certification programs should be provided and required for professionals in pest, vegetation, and nutrient management. University programs also need to be developed to supply qualified employees with an ICM perspective for the agrochemical industry.

Figure 2.1 illustrates the interaction of the various elements supporting ICM. While global competition has fostered the need for efficient ICM practices, public concerns, especially regarding health and environmental protection, are an important force behind the development of ICM. Citrus producers respond to the public by implementing ICM strategies. As discussed above, research, education, and extension are essential to the development and adoption of ICM practices, and developing technology furthers the practical application of ICM. Producers can be encouraged in their transition to ICM by supportive funding policies and government regulations. However, the ultimate success of ICM depends on the cooperation of government agencies and universities, industry organizations, agroproduct manufacturers, environmentalists, and other interest groups.

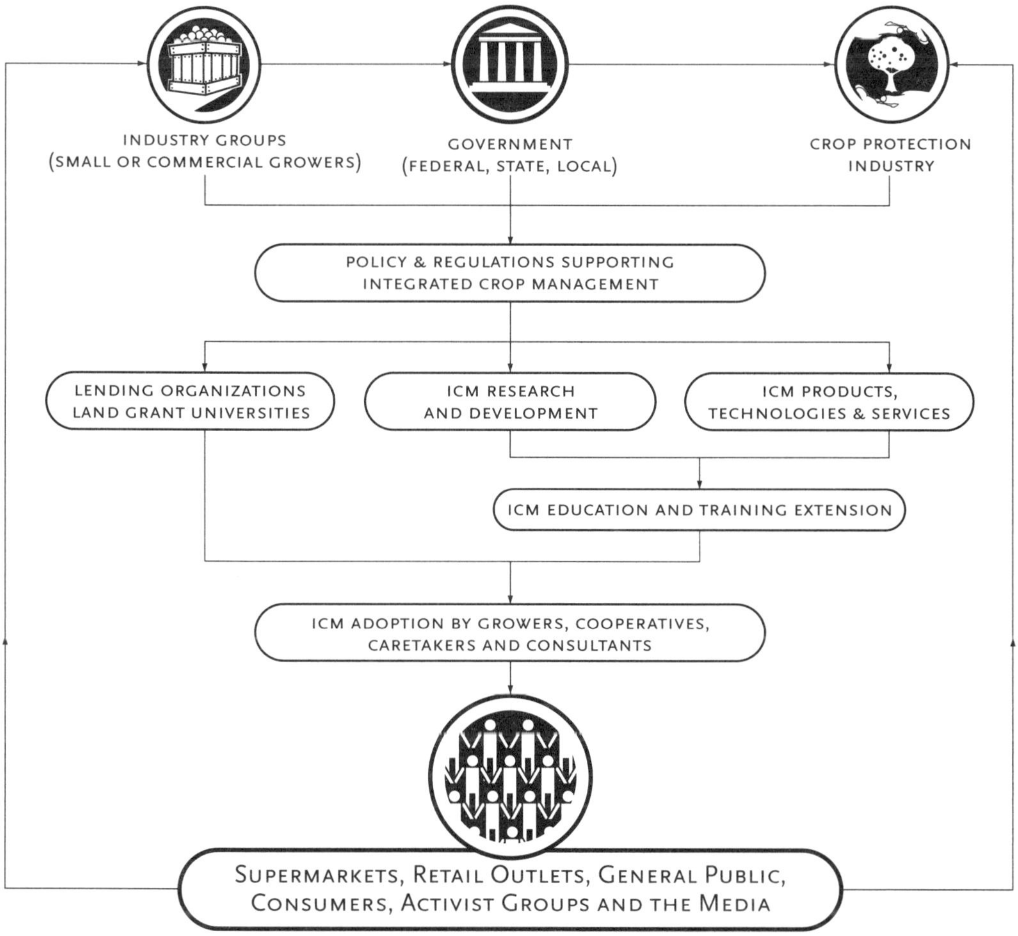

Figure 2.1 Interaction of the various elements supporting Integrated Crop Management. See text for detailed description of the components.

REFERENCES

Ferguson, J. J., C. L. Taylor, and G. D. Israel. 1995. *Citrus management surveys as tools for extension programming.* HortTechnology 5: 67-71.

Mahovic, M., and Jerry A. Bartz. 2002. *Improvements in sanitation of flumes or dump tanks in packinghouses.* Proceedings of the Florida State Horticultural Society 115: 35-37.

Rogers, S. 1998. *The citrus integrator: A decision flowchart for teaching integrated crop management.* Proceedings of the Florida State Horticultural Society 111: 151-152.

Takele, E., J. A. Menge, J. E. Pehrson, J. L. Meyer, C. W. Coggins, M. L. Arpaia, J. D. Hare, D. R. Atkin, and C. Adams. 1993. *Economic analysis of integrated crop management practices of navel oranges.* Journal of the American Society for Horticultural Science 118: 910-915.

3

The Citrus Manager

M.R. Ziegler and D.P.H. Tucker

INTRODUCTION

A citrus manager is responsible for optimizing profits and ensuring long-term sustainability of crop production. There are a multitude of unpredictable growing conditions, volatile global markets, and rapidly changing regulatory climates that can influence the manner in which ICM is carried out in the field. As the person responsible for integrating the various aspects of a citrus business into a profitable entity, citrus managers today should be aware of all aspects of the citrus business, including grove design, field operations, and fruit harvesting and marketing. Indeed, the ultimate responsibility for the success of a citrus production program rests with the citrus manager. This chapter demonstrates how the individual citrus manager can develop and implement a site-specific ICM program.

Florida citrus production today is a highly complex and competitive international business venture. The modern citrus manager is responsible for optimizing profits and ensuring sustainability of production despite many unpredictable conditions. The oversupply of citrus products in Florida during the 1990s emphasized the need for more efficient crop management. Spurred by declining profits, citrus growers sought new management strategies that minimized production costs without compromising yield or crop quality. As the person responsible for integrating the various aspects of a citrus business into a profitable entity, the citrus manager must be aware of all aspects of the business: grove design, field operations, fruit harvesting, and marketing. The production cycle can contain many complex interactions that if not fully appreciated, can ultimately limit profitable marketing of citrus products.

Citrus production management encompasses many distinct but closely related areas, including nutrient management, water management, integrated plant protection, environmental protection, water protection, marketing and economic analysis, information management, computer technology, labor management, and personnel relations. Professional citrus managers must multitask these diverse concerns in long-range planning and day-to-day decision making. Recognition of the significance of these interactions within the crop production system and the importance of coordinating grove management practices and objectives gave rise to the concept of Integrated Crop Management (ICM). ICM is a management philosophy in which the citrus grove is viewed as a dynamic, complex, networked system that includes physical, chemical, biological, and human interaction components. All of these elements are related and every change within the system affects every other component to some extent. Making decisions may seem more complicated in such a system, but the entire system can be managed more efficiently and effectively when the relationships and responses within the system are better understood.

The citrus grove manager may be thought of as a hub at the center of a very large information management wheel (Figure 3.1). The citrus manager is guided by economic considerations as the target market is identified, production objectives are established, and the seasonal management of the crop is planned. In general, management guidelines have been established for growing healthy, vigorous, and productive citrus trees, and most managers follow basic recommendations. However, the citrus manager must be prepared to respond to unpredictable and uncontrollable factors, including growing conditions, market trends, and local and global supply and demand. The citrus manager must also be prepared to modify established grove-management procedures to accommodate the changing health and maturity of the citrus trees and to satisfy new regulatory constraints or requirements that relate to tree protection, environmental protection, consumer protection, or worker protection. In addition, in order to stay competitive, citrus managers must be prepared to adopt new practices and procedures indicated by ongoing agricultural research. Thus, the citrus manager is constantly processing information concerning the developing crop, weather and grove conditions, the fresh and processed market requirements, new regulations, new research findings, etc.

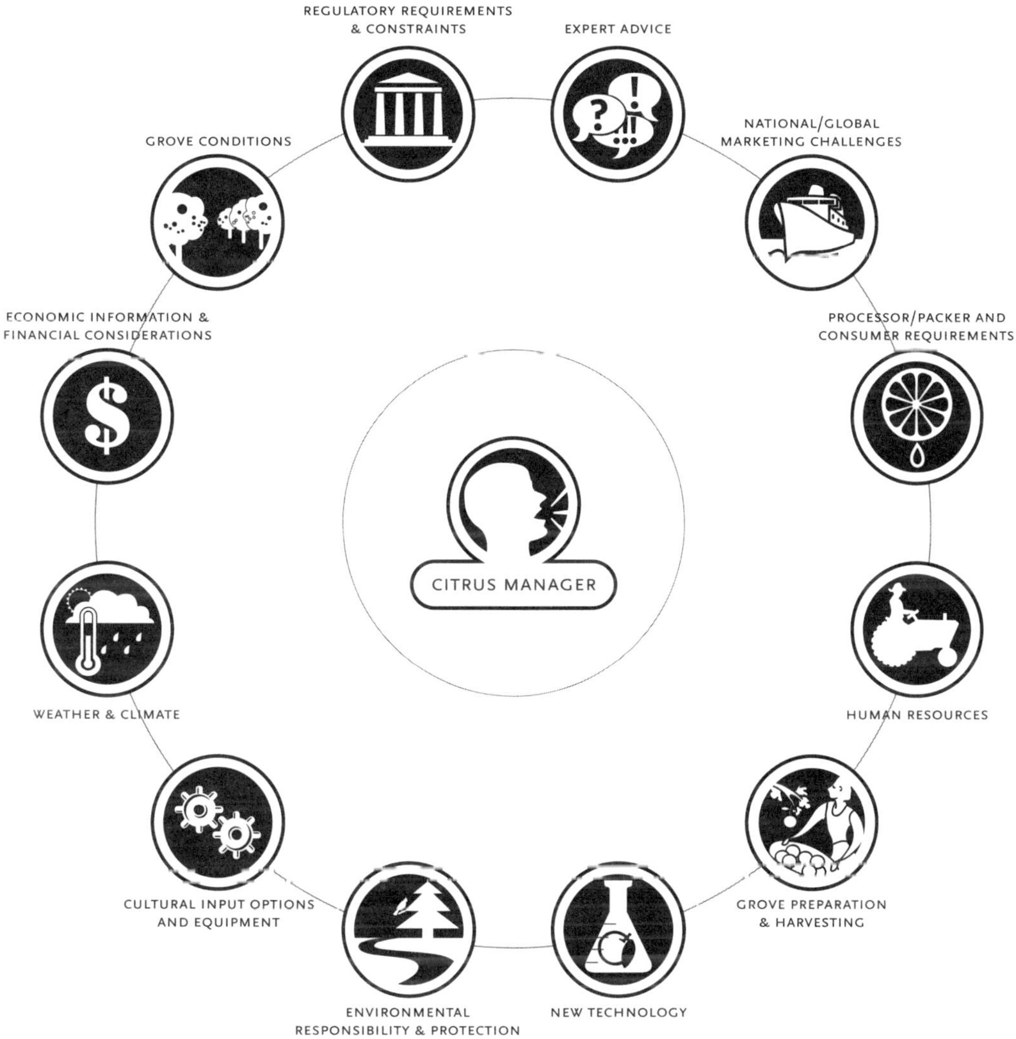

Figure 3.1 Many diverse considerations must be integrated into daily and long-term decisions by crop managers.

Citrus managers vary widely in their background and training. Their decisions are based largely on their own experience, and they often participate in continuing education programs, such as the Certified Crop Advisor program administered by the American Society of Agronomy and FLDACS. Also, a great deal of free information is available through academic and extension services. However, many citrus managers also employ professional crop and environmental consultants. Citrus managers may find it cost-effective to contract various aspects of grove maintenance or harvesting. The manager must verify the contractor's qualifications and oversee the contractor to ensure that the work is satisfactory and that the contractor is complying with applicable regulations. Nevertheless, the ultimate responsibility for the success of a citrus production program rests with the citrus manager.

The markets for fresh and processed citrus products that are most profitable for citrus growers are usually those that incorporate food safety and food quality criteria in marketing agreements. These criteria exist in both regulatory and nonregulatory incentive based programs designed to ensure that the food products marketed can be traced to the field from which they are produced and with confirmation by documentation of all production practices and crop protection materials applied. The citrus manager has direct responsibilities for compliance with the FDA, 21 CFR Part 120, Hazard Analysis and Critical Control Point (HACCP), Procedures for the Safe and Sanitary Processing and Importing of Juice, and Final Rule (published January 19, 2001). Additional compliance with the Florida Department of Citrus Chapter 20-49, "Standards for Fresh Squeezed Citrus Juices," published March 15, 2001, institute an exemplary program for food safety in Florida fresh squeezed citrus juice products. Although these two rules are examples of regulatory programs, many nonregulatory incentive based programs are available also.

Perhaps one of the most well developed examples of these is EUREGAP© (European Retailers Partnership and Good Agricultural Practice). This is a certification program established with general regulations for fruit and vegetables containing 14 chapters in a protocol of control points and compliance criteria that require verification in a checklist format. "EUREGAP is aiming to become the global standard for good agricultural practice, to reduce the need for multiple audits by different authorities, and to provide clear compliance criteria." EUREGAP certification applies in both fresh and processed marketing channels with incentives for concern for food safety, environmental protection, animal welfare, and worker health, safety and welfare.

This is an era of tremendous change in industries throughout the world. Everywhere, workers at all levels of business are experiencing painful and difficult transitions as markets shift, competition expands, traditional enterprises are abandoned, new technologies are developed and adopted, new skills are required, and objectives and responsibilities are redefined. The citrus industry is no exception to this trend. The citrus manager may be daunted by the complexity and scope of the task. However, an extensive information and support network is available to the citrus industry, and work performance and productivity can be greatly enhanced through the use of these abundant information resources. Ultimately, the production manager is the integrator to make an ICM program work.

REFERENCES

Ferguson, J. J., and G. D. Israel. 1998. *Cold protection: practices and management decisions.* Citrus Industry 79: 14-16.

Futch, S., and R. Muraro. 1993. *Budgeting citrus production costs to maintain profitability.* Citrus Industry 74: 48-51.

Lee, L. S., and D. G. Mayer. 1996. Computerized nutritional data interpretation, and preparation of fertilizer recommendations for tree crop production systems. International Citrus Congress (8th 1996: Sun City, South Africa), Sun City, South Africa 2: 794-802.

Martsolf, J. D. 1998. *Weather information dissemination and its role in decision making.* Fact Sheet: 32-34.

Rogers, J. S., and C. W. McCoy. 1996. *A laptop computer-based decision support tool for managing citrus rust mite populations.* International Citrus Congress (8th 1996: Sun City, South Africa), Sun City, South Africa 1: 625-631.

Scholberg, J. M. S., L. R. Parsons, and T. A. Wheaton. 2000. *Citrus nitrogen nutrition: Some production considerations for more efficient nitrogen use.* Citrus Industry 81: 18-19.

4

Targeting the Marketplace: Fresh vs. Processed

S.H. Futch

INTRODUCTION

The purpose of commercial citrus ICM is to bring a high quality product to market efficiently and economically. There are many paths by which Florida citrus products may travel from the grove to the consumer, but the ultimate choice of utilization for a citrus crop is a decision that must be seriously considered by the citrus manager. For example, citrus crops produced for the fresh market often incur higher production costs due to the additional spraying necessary to minimize peel blemishes. Costs and returns must continually be evaluated and reevaluated throughout the production season. However, the initial decision to produce a fresh-marketed crop occurs prior to bloom, when the previous season's crop characteristics and market performance are assessed. Whether a crop is destined for the fresh or processed market, the citrus manager has the same goal of maintaining the health and vigor of the citrus trees. The final market destination will influence the targets, timing, and type of materials that are applied in the field to help ensure a marketable crop. This chapter introduces some of the strategies and tactics that the citrus manager may draw upon to help deliver the highest quality product possible.

The choice of utilization for a citrus crop, whether to sell to fresh or processed markets, is a decision that must be seriously considered. Due to the additional spraying necessary to preclude peel blemishes, citrus crops produced for the fresh market often incur higher production costs. The grove manager must evaluate whether these costs are justified by the fresh market value of the crop. Further, a number of events beyond the manager's control may make the crop unacceptable for the fresh market. In particular, damaging weather events, such as hail, windstorms, and freezes, may force the grower to divert a crop to processing. This marketing decision (fresh vs. processing) is usually considered on a block-by-block basis prior to bloom, with continuing evaluation throughout the entire production season. Seasonal evaluations of the fruit's external condition are needed, particularly fruit size and condition of the peel, prior to the application of additional sprays. Additional sprays may not be warranted because of grade-lowering factors that already exist. Not all trees are able to produce an economically profitable fresh fruit crop because of variety, disease incidence, pest pressure, tree age, tree condition, or general environmental conditions at the specific grove site.

In addition to evaluating the condition of the fruit, the grove manager must consider economic factors that affect the prices offered for fresh and processed fruit each season. Currently, approximately 5 percent of Florida's orange crop and up to 50 percent of its grapefruit crop is marketed as fresh fruit. Tangerines and most of the specialty varieties are marketed primarily for the fresh market. Variability in fruit utilization can have a profound impact on fresh fruit prices. Blemished fruit going to processing plants as eliminations from the fresh packing operations, or in salvage conditions at harvest, typically brings lower prices than citrus originally destined for the processed market. Some juice markets may not accept additional fungicides from the drenching and waxing of fruit, which represent limitations on eliminations from fresh fruit packinghouses. There are also significant costs associated with fresh fruit handling, reducing profits if the fruit is then processed.

The initial decision to produce a fresh-marketed crop occurs prior to bloom, when the grove manager assesses the previous season's crop characteristics and market performance. At this time, price expectations play a major role in the decision. The final marketing decision occurs, of course, when the fruit is being evaluated for harvest. The external appearance and size of the fruit are evaluated to determine the anticipated or estimated percentage of the crop that will be marketed in fresh fruit channels. Any fruit that is not the desired size for the current market demand, or possesses too many peel blemishes, should be directly sent to the processing plants. In some years, when the prices offered for the eliminations are extremely low, it may be beneficial for the grower to consider spot picking the crop in an effort to match the harvested fruit with market demands for size and quality, thus leaving fruit on the tree that would have been eliminated from the fresh fruit packing operation. As the season progresses, if fruit size increases or market demands change, additional spot picking may be warranted.

Figure 4.1 is a flow chart of the decisions growers make during a growing season. As indicated in the figure, growers start the season at bloom time with the initial decision to produce a given crop for the fresh or processed market. If the fresh market is chosen for the specific crop, the grower begins applying the first spray at or shortly before bloom to protect the crop from various foliar, bloom, and fruit-infesting fungal diseases, particularly scab and melanose. As the season progresses, the grower should reevaluate the external appearance of the crop several times, applying additional sprays when necessary to maintain fresh quality appearance standards. If at any time during the year the fruit becomes blemished, or if its size does not conform to market demand, the grower evaluates current fruit prices to determine if additional sprays are practical. If the additional sprays will not produce an economic return, they are omitted and the fruit follows a processed market production program.

As a citrus crop is managed from bloom to harvest, the grove manager must take into consideration the seasonal growth cycle of the citrus tree, the development of the grove, and current environmental factors. Whether a crop is destined for the fresh or processed market, the citrus manager has the same goal of maintaining the health and vigor of the trees. The Florida citrus industry must continue to be recognized as a consistent source of high-quality citrus juice products and fresh fruit in order to maintain and increase its customer base. These issues are discussed further in subsequent chapters.

The initial decision to produce fresh crops happens prior to bloom; price expectations play a major role in this decision.

When prices for eliminations are low, it may be beneficial to spot pick the crop to match the harvested fruit to the market demands for size and quality.

If additional spraying of citrus (to protect the crop from pests and diseases) will not produce an economic return, sprays are omitted and fruit prepared for processing.

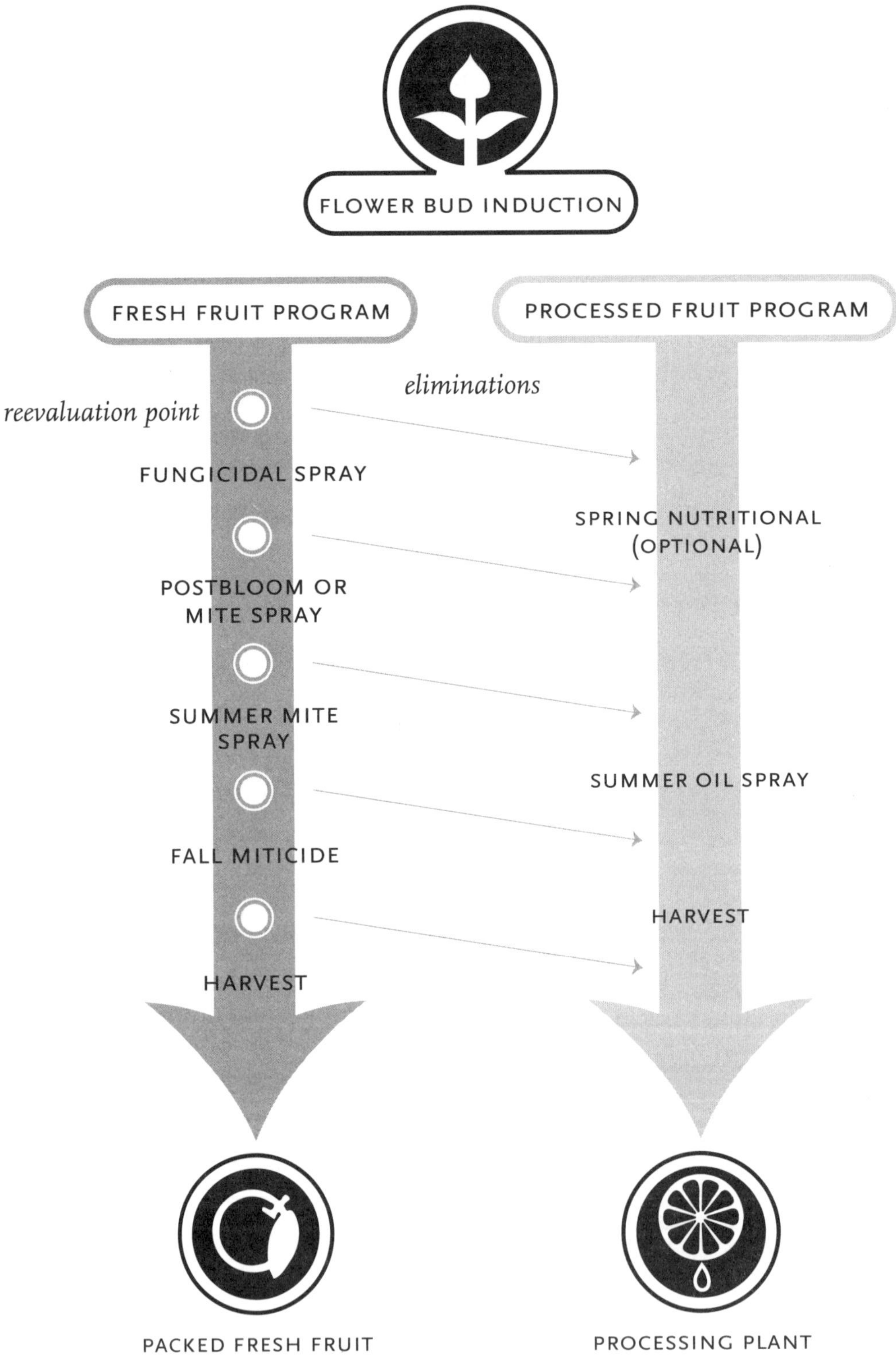

Figure 4.1 Fruit marketing decision matrix for orange production.

CONSIDERATIONS IN THE FRESH FRUIT MARKET

Generally speaking, crops targeted to the fresh market require higher production inputs than crops targeted for the processed market. This higher cost is especially evident in the additional spray applications needed to minimize crop blemishing. In addition, more specialized cultural requirements, such as in water and nutrition management programs, may be needed. In some cases, (e.g., Sunburst and other mandarins that can produce excessive yields) crop thinning may be desirable to increase fruit size and to prevent limb breakage and tree collapse. Specialty varieties also exhibit somewhat greater susceptibilities to certain fungal diseases, such as alternaria and scab, requiring additional considerations for scouting and treatment. Thus, a higher level of integrated pest management is required, especially to manage fungal diseases. Special tree spacing and pollination considerations are also required for some specialty varieties.

Special harvesting equipment is needed and more careful harvesting protocols are required in order to avoid unwanted physical damage to the fruit during the harvesting, handling, and postharvest periods. At times, there may be a conflict between early marketing, legal maturity and best price, resulting in suboptimal fruit being delivered to market. However, if harvesting is delayed too long, there may be undesirable consequences, such as fruit deterioration and fruit drop.

REFERENCES

Brown, M. G., J.-Y. Lee, and T. H. Spreen. 2001. *The decline in the domestic market for Florida fresh grapefruit.* University of Florida, IFAS, Cooperative Extension Service SP 01-15 6 pp.

Brown, M., T. Spreen, and R. Goodrich. 2000. *Trends in the NFC orange juice segment.* Citrus Industry 81: 18-19.

Finch, A. 2003. Will Florida fresh citrus varieties be acceptable tomorrow? International Society of Citriculture Congress. (9th 2000: Orlando, Florida), Orlando, Florida USA 2: 1104.

McCoy, C. W., H. W. Browning, and R. P. Muraro. 2004. 2004 Florida Citrus Pest Management Guide: Management Options for Non-Bearing Groves and Fresh or Processed Fruit, p.7. In: L. W. Timmer (ed.) 2004 Florida Citrus Pest Management Guide. SP-43. Florida Cooperative Extension Service, University of Florida, IFAS, Gainesville.

Muraro, R. P., G. F. Fairchild, and R. M. Behr. 1991. *Citrus fruit allocation: Decisions between fresh and processed markets.* Proceedings of the Florida State Horticultural Society 104: 214-217.

Muraro, R., G. Fairchild, and R. Behr. 1993. *Grower decisions between fresh and processed markets.* University of Florida, IFAS Extension Service EN-31.

Muraro, R. P. 1997. *Economic impact of fresh and processed production costs.* Citrus Flowering and Fruiting short course, University of Florida, IFAS, Citrus Research and Education Center, Lake Alfred, p.153-164.

Nelson, G. P. 2003. *A domestic and international perspective on the marketing of Florida fresh citrus.* International Society of Citriculture Congress. (9th 2000: Orlando, Florida), Orlando, Florida USA 2: 1104.

5

Variety Development, Clonal Protection, and Nursery Tree Selection

M.C. Kesinger and D.P.H. Tucker

INTRODUCTION

As seen in other chapters, there are a variety of actions that citrus managers can take to improve the health of the trees and protect them from environmental stress. However, in an ideal world, trees would be genetically impervious to damage from pests, diseases, or environmental stresses. Citrus breeding is one means of working toward this goal. Unfortunately, citrus variety improvement can be a complicated topic. With the advent of molecular genetics, the topic has gone far beyond the simple variety improvement methods of years past, such as the fortuitous identification of a branch with better color fruit, or the transfer of pollen from one plant to another. Today, citrus breeders use radiation, evaluate enzymes, DNA, and proteins in attempts to induce specific changes in citrus trees or fruits. In addition, there are regulatory authorities that help ensure the integrity of citrus cultivars. This chapter reviews these aspects of citrus variety improvement.

A cultivated citrus tree normally consists of two joined parts that originated as separate trees. The portion above ground is called the scion, cultivar, or scion cultivar. The part below ground is called the rootstock. The two connect a short distance up the trunk at the bud union, where they were originally grafted (budded) together.

Grafting a vegetative bud into the rootstock seedling produces the top (scion) of the citrus tree. The citrus tree scion is genetically identical to its parent tissue, and the process of budding it to a rootstock seedling is known as clonal propagation. The resulting combination is normally identified as scion variety/rootstock variety (that is, Murcott/Cleo). Both scion and rootstock cultivars must be carefully selected to provide the tree with an advantageous combination of traits. The continuing development and assessment of new cultivars is vital to the future of the citrus industry.

> MOST CULTIVATED CITRUS ARE COMPOUND TREES.
>
> Rootstock is selected for adaptation to soil characteristics, pest tolerance, and effects on fruit quality and yield.
>
> Scion is the primary factor of fruit characteristics, pest tolerance, and ultimate fruit use.

CITRUS VARIETY DEVELOPMENT

The Citrus Scion and Rootstock Variety Development Pipeline (Figure 5.1) illustrates the many and diverse strategies for genetic improvement of citrus. The goal is to maximize the number and variety of selections available to receive an initial screening; those that are potentially valuable to the industry are identified for further evaluation. Recently developed biotechnologies such as embryo rescue, somatic hybridization, tissue culture and somaclonal variation, and genetic transformation can be used to create genetic variation never before possible. These new techniques are used in conjunction with well-established technologies such as conventional breeding and irradiation to provide an unequaled opportunity for generating and evaluating variation.

To be successful, the entire process of development and selection must be continuous and requires fundamental support from the citrus industry. Many aspects of the process are long-term, but once the system is established and operational, solutions to chronic production problems may be discovered or developed very rapidly. It may also be expected that many completely new products will be generated, presenting new marketing opportunities for both fresh and processed products.

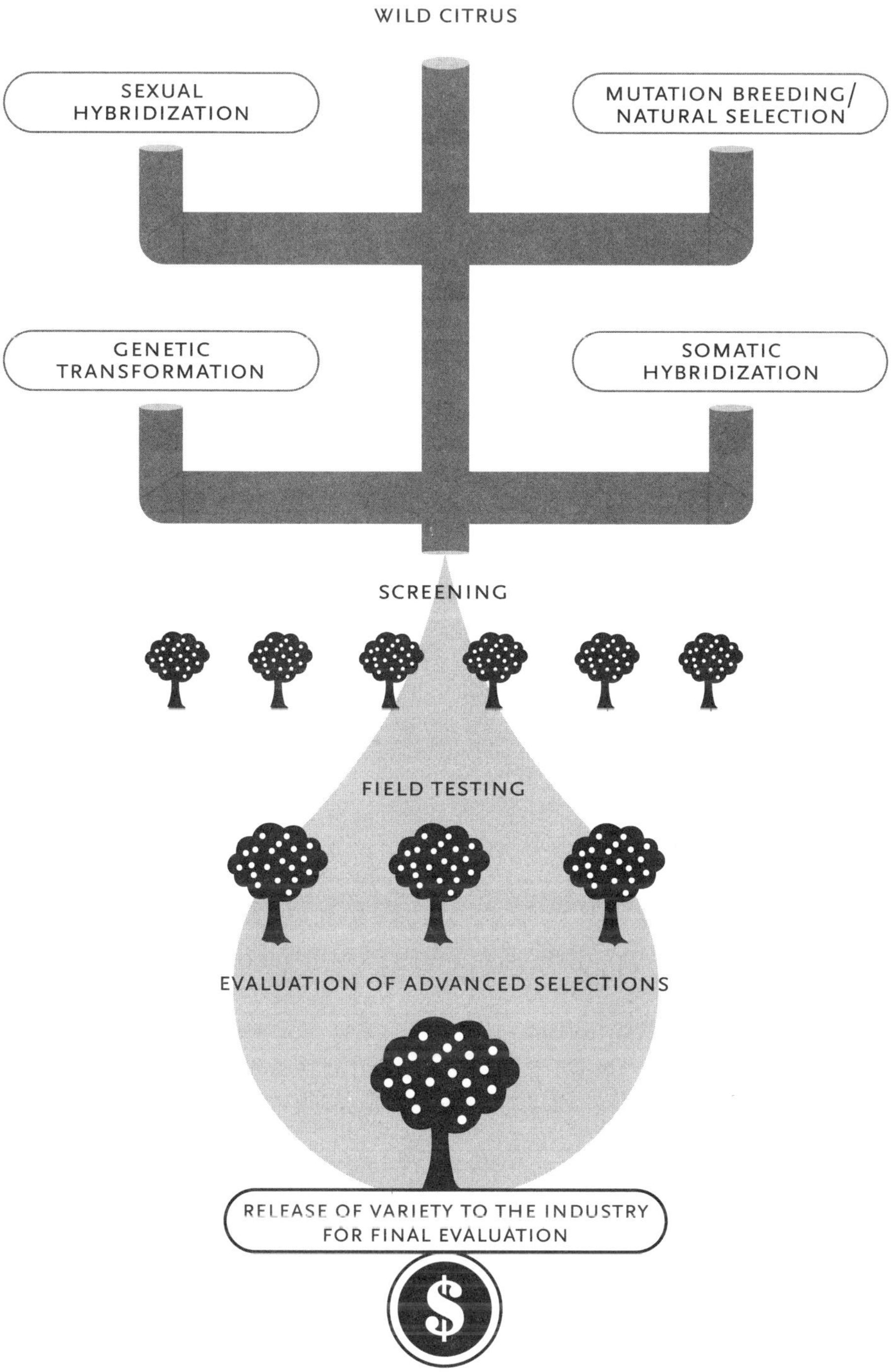

Figure 5.1 The Citrus Variety Development Pipeline illustrates approaches used in citrus breeding and the repeated process of identifying superior varieties that will eventually be released for commercial citrus production. (Concept by W.S. Castle, J.W. Grosser, and F.G. Gmitter, Jr., University of Florida, IFAS, Citrus Research and Education Center.)

The initial screening process must be followed by a period of evaluation in which additional data on field performance are collected, and varietal characteristics are studied. Newly developed cultivars may have unique susceptibilities to pests and pathogens or special nutrition and irrigation requirements. The final determination occurs after cultivars are released to the industry. Individual citrus producers evaluate the performance of the new variety under local climatic and soil conditions and management programs, and develop marketing strategies for the new variety. Producers accept some risk when they accept newly developed cultivars, since initially much may be unknown about the cultivars response to various growing and postharvest conditions. These producers are an integral part of the variety development process, and are vital to its continued success.

The United States Department of Agriculture and the University of Florida's citrus fruit-breeding programs are integrating emerging biotechnologies that facilitate the development and selection of improved fresh and processing cultivars.

OBJECTIVES OF TREE BREEDING PROGRAMS

1. Reduced seed content, greater ease of peeling, and improved internal quality, juice color, and external appearance (especially fruit color)
2. Improved storage and shipping qualities and increased shelf life
3. Reduced susceptibility of trees and fruit to pests and diseases, with a resulting reduction in dependence on agrichemicals
4. Improved cold hardiness
5. Harvest dates to satisfy consumer demand between marketing periods of existing cultivars, to capture early markets, or to permit harvest before the threat of freezing weather

Emphasis is currently on the development of high-quality, easy-to-peel, seedless mandarin types for the fresh market, and improved sweet orange selections that facilitate the processing industry's shift from frozen concentrate to not-from-concentrate (NFC) products.

Specialty cultivars may be released to the industry to fill small niche markets, provided there are no overwhelming limitations noted. For example, a cultivar that is observed as susceptible to a foliar fungal pathogen for which no effective plant protection product is available should not be considered for release, even if there is an apparently lucrative market for the variety.

CITRUS CLONAL PROTECTION

The Bureau of Citrus Budwood Registration in the Division of Plant Industry (DPI) of the Florida Department of Agriculture and Consumer Services (FDACS) administers The Citrus Budwood Protection Program. This program evaluates promising citrus cultivars for graft-transmissible diseases and horticultural characteristics, including yield and fruit quality, and makes propagation material from selected trees available to nurseries that produce trees for the industry and retail outlets. Current information on numbered selections (clones or registered budlines) of cultivars is available from the Bureau.

It is illegal to bring unauthorized citrus nursery trees or propagating material into the United States or from other states into Florida. Citrus germplasm may be legally brought into Florida only through a quarantine facility operated by the Division of Plant Industry in Gainesville. New introductions must first be approved by the Citrus Budwood Technical Advisory Committee. It is important to the vitality and growth of the Florida citrus industry that growers have safe access to competitive worldwide citrus germplasm. Our commercial groves are protected from exotic pathogens by the quarantine and indexing of all new citrus selections prior to industry release.

Florida Citrus Clonal Protection

Many diseases can be transported on budwood or plants.

It is illegal to bring citrus into Florida except through the DPI quarantine facility, which produces clean budwood.

Requests for new introductions are made through the Citrus Budwood Technology Advisory Committee.

The potential problems from the introduction of pests or pathogens cannot be overemphasized. Likewise, a supply of tested, pathogen-free propagative budwood is recognized as basic to the establishment and maintenance of a viable citrus industry. Pathogens in propagative budwood can be deleterious to tree survival and fruit production. These pathogens can be easily distributed by way of infected plants, budwood, or vectors where they become a potential hazard to existing and future plantings. Given that the principal transmission of many diseases occurs through propagation of infected stock, a comprehensive program to produce, maintain, and distribute budwood that is indexed negative for graft-transmissible pathogens is essential. The objectives of the Citrus Budwood Protection Program are twofold:

1. to provide the citrus industry with clonally propagated budwood of commercially important rootstock and scion cultivars that have tested negative for the presence of all known graft-transmissible pathogens, and is free of undesirable traits; and
2. to provide a mechanism for the safe introduction of citrus cultivars from other states and countries through extensive virus indexing and cleanup procedures.

To accomplish these objectives, the Citrus Budwood Protection Program follows a protocol for indexing and cleanup of propagative material.

A potential parent tree is identified first and is selected with great care, as many thousands of progeny trees may originate from one parent. Parent trees must have outstanding characteristics or qualities that benefit the citrus industry. Growers in every area of the state have selected parent trees after observing unique and desirable characteristics in perhaps one tree out of thousands. Many of Florida's parent trees were started as nucellar seedlings to eliminate virus or viroid infections and at the same time increase vigor and production. Other parents have originated from research breeding programs or been introduced into Florida from other growing areas worldwide. Another potentially rewarding source of parent trees is the field study of existing citrus plantings that have been significantly damaged by pests, diseases, or environmental stress, including root weevils, blight, or freezes. Once a planting is reduced to a very few trees, those remaining may be tolerant or resistant to the identified cause of tree decline or death. The trees may be saved or propagated from budwood or root material for further evaluation. Unfortunately, growers can seldom afford to maintain a planting in decline, and may be inclined to remove the entire affected block when it is no longer profitable, rather than determining if a small number of trees survive.

Potential parent trees should be at least 10 years old in order to permit evaluation of their horticultural characteristics and assessment of visible disease symptoms. Candidates are judged on a number of criteria, including yield or pounds solids per acre, fruit size, shape, color and overall quality, seedlessness, juice color and other internal characteristics such as maturity standards, consistency of bearing, and general marketability.

After candidate parent trees are selected, they must undergo indexing for graft-transmissible pathogens to ensure they are not virus or viroid infected. Buds are collected from a candidate source tree and used to propagate one or more auxiliary trees. A preliminary disease index, including tests for the important pathogens: citrus tristeza virus (CTV), psorosis virus and the cachexia and exocortis viroids, is performed on the source tree. It is important to note that many graft-transmissible pathogens are scion/rootstock specific for decline symptoms. Nonsymptomatic combinations may harbor viruses or viroids that are potential reservoirs for infection of adjacent groves. If pathogens are detected during this indexing, shoot-tip grafting (STG) is performed to eliminate the pathogen identified. This procedure takes advantage of the fact that the shoot-tips develop more rapidly than the pathogen, meaning that shoot-tips are typically free of infection.

Following shoot-tip grafting and a few months growth, the candidate STG trees are tested for previously detected graft-transmissible pathogens. If any pathogens are found, shoot-tip grafting is repeated and may be combined with thermal therapy to increase the likelihood of obtaining disease-free plants. Thermal therapy takes advantage of the fact that the plant is generally less adversely affected than the pathogen by high temperature.

Next, budwood from the candidate clean trees that has tested negative for all the known graft-transmitted pathogens is used to propagate trees that are placed in a foundation block. Foundation trees undergo yearly inspection and fruit evaluation to determine whether they are true-to-type, with extensive periodic reindexing for diseases. Trees must be tested for vectored pathogens more frequently than for those that are mechanically transmitted. After these trees have fruited, they are also registered as budwood source trees.

One important exception to this policy should be noted. Both mild and severe (quick-decline) strains of CTV are widespread in Florida citrus. Cultivars on sour orange rootstock are particularly afflicted, and sour orange rootstock is no longer propagated for this reason. The clonal protection program must now focus on excluding the stem-pitting strains of CTV that can affect most scion cultivars, but especially grapefruit, regardless of rootstock. Maintaining budwood sources free of stem-pitting CTV requires the development of routine, rapid, relatively inexpensive tests that detect all stem-pitting strains. Currently, budwood source trees must be annually tested for freedom from quick-decline and stem-pitting strains of tristeza. The budwood program requires growing all scion source trees in screenhouses to protect them from all insect-vectored diseases.

Tristeza Status of Registered Budwood

Must be free of Florida severe, quick-decline and stem-pitting strains

May be infected by mild strains that are ubiquitous in Florida

New citrus germplasm must first be tested for pathogens before it can be introduced into the Budwood Protection Program on its way to becoming a potential new variety (Figure 5.2). Budwood from registered foundation block trees is grown with protection from virus-transmitting insects in greenhouses or screenhouses (Plates 5.1a & b), and is released to citrus nurserymen and growers. Upon receiving the budwood, the nurseryman or grower may propagate certified trees directly for planting in commercial groves, or they may use the budwood to produce a registered increase block or scion grove. See Florida Department of Agriculture Rule Chapter 5B-60 and the Citrus Budwood Protection Program Procedure Manual for regulations and guidelines for growing citrus nursery trees. Propagation material (budwood, cuttings, air layers, and seeds) can only be obtained from the following source trees illustrated in the flow chart (Figure 5.2).

1. Parent tree: a tree selected for superior horticultural characteristics and indexed for citrus pathogens.
2. Foundation tree: a source tree belonging to the FDACS-DPI citrus budwood program.
3. Scion Grove tree: a designated tree located in a grove planted by nurserymen for a budwood source. Scion grove trees must originate from parent or foundation trees, and Division of Plant Industry (DPI) personnel must witness the planting.
4. Increase Block tree: a tree from a specially designated nursery block planted for quick multiplication of propagative material. The propagative material must originate from an approved source.
5. Validated tree: a source tree that has not come through the budwood evaluation and indexing program. Validated trees are tested only for vectored pathogens that exist in Florida.

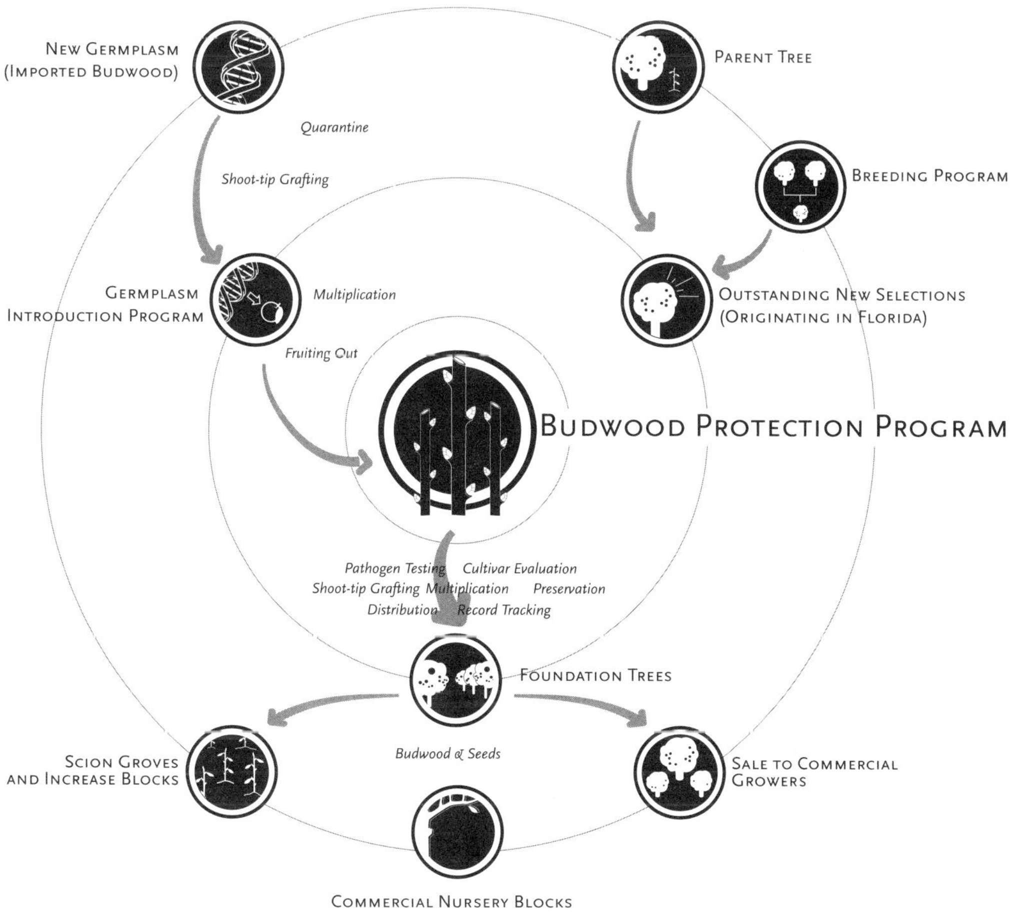

Figure 5.2 Citrus germplasm flowchart for Florida citrus budwood production.

Virus and Viroid Management

Citrus can be infected by a number of different viruses and viroids that can affect trees in different ways, causing decline, stunting, and loss of production.

Often, virus and viroid-induced diseases affect specific combinations of scion and rootstock, and may be symptomless in other combinations.

There are no treatments to cure citrus trees once they are infected; growers with virus or viroid problems must live with the problem or remove affected trees and replant.

Sterilize cutting tools in nursery and grove.

Basic control strategies are to avoid infections and to use tolerant or resistant combinations of scion and rootstock.

Quarantines are the first line of defense.

The single most effective management of viruses and viroids is prevention by propagation of new trees from registered pathogen-free budwood.

SELECTION OF A SCION CULTIVAR

The grower must consider many factors when selecting a scion/rootstock combination, especially the grove environment and the desired market niche. Typically the scion variety is selected first. This selection is based largely on the risk of freezing conditions, the pest pressure in the grove, and whether the target market is fresh or processed. The rootstock variety must be compatible with the scion and is chosen to impart a variety of advantages to the tree, including cold hardiness and disease resistance.

Although the scion is typically selected first, ultimately it must be compatible with the rootstock selection. Certain grove conditions, especially unfavorable soil conditions, can limit the grower's potential rootstock selection considerably, and only scion cultivars that are compatible with potential rootstocks can be considered.

The grove environment, especially climate, limits the selection of scion variety. In areas where freezing conditions pose a threat, a variety less vulnerable to freeze damage is essential. Fruit cultivars with a thicker rind may be less damaged by freezes. Alternatively, early-maturing cultivars may be ready for harvest before the greatest threat of late winter freezes.

When selecting a variety, growers should consider new clones that arise from mutations and are available for many modern citrus cultivars. These clones may differ widely both in productivity and in date of maturity.

New cultivars can never be fully field-tested prior to being released. Grower cooperators who plant new selections are crucial to the final evaluation of a new variety in the local environment and under current cultural practices. New releases should not be widely planted until evaluated in small-scale field trials to determine any special cultural requirements.

The most important factor in the grower's decision is the intended market niche. The grower considers whether the fruit is destined for the fresh or processed market, the target maturity date, and the characteristics of competing cultivars.

Certain varietal types are currently over produced. Those and similar cultivars may present formidable marketing challenges.

The decision to enter the fresh fruit market, particularly with specialty cultivars, is complex. The cultural requirements are significantly more intensive for the fresh market than for the processed market, and specialty cultivars present a number of unique complications. Further, the demand for specialty cultivars may be highly variable or unpredictable, and international preferences may be different from those of consumers in the U.S. Consumers may demand seedless fruit and are influenced by external fruit appearance, especially size and color. Some cultivars that are valuable in the fresh market may not represent a profit if processed, while others can be marketed either fresh or processed. In general, the grower faces more unknowns and, consequently, a greater risk of loss with specialty cultivars than with standard oranges and grapefruit.

If the fruit is intended for processing, the grower will probably select a variety that is highly productive (measured in lbs. solids/acre, or lbs. juice/acre), with good juice color and flavor. Other characteristics, such as postharvest quality, on-tree storage, and durability in processing, are also desirable.

Selecting a Scion Cultivar

Identify market for fruit or juice characteristics and season of maturity.

Observe the adaptability to grove soil, climate, and pest pressures.

Determine expected yield and price for profitability.

Consider that some cultivars are valuable in the fresh market but not profitable if processed, and vice versa.

SELECTION OF A ROOTSTOCK

The genetics of citrus are rather unusual. Whereas most plants produce monoembryonic seeds (each seed has a single embryo), many citrus cultivars produce both monoembryonic and polyembryonic seeds (each seed may have multiple embryos). Only one embryo in a polyembryonic seed, the zygotic embryo, receives the male genetic material; the other embryos in the seed are called nucellar embryos because they arise from nucellar tissue; their genetic makeup is identical to that of the mother plant. A citrus variety that produces polyembryonic seeds with a large number of nucellar embryos is called highly nucellar. This is considered a desirable trait for rootstock cultivars because such a variety produces a consistently high percentage of true-to-type seedling rootstocks from each batch of germinated seed.

Roots serve multiple functions. They anchor the tree in the ground, and absorb and transport both water and nutrients from the soil to the aboveground parts of the tree. They serve as synthesis or conversion sites for growth regulators. Finally, the roots store food reserves. It should be noted that roots also alter the rhizosphere, or soil environment, in which they grow.

A variety that is selected as a rootstock must meet certain requirements. It must be compatible with the chosen scion, adaptable to the intended edaphic (soil) and climatic environments, and capable of providing sufficient annual growth with good vertical and horizontal root distribution. Its value as a rootstock depends on its ability to enhance any aspect of citrus production that ultimately increases the profitability of the citrus operation. Rootstocks are selected for disease and pest resistance, cold tolerance, tree vigor, effects on early scion productivity, and improvements in internal and external fruit quality, yield, harvest date, and postharvest quality. The selection of rootstock has tremendous impact on the success and longevity of a citrus grove planting.

Historical rootstock use in the Florida citrus industry reflects the impact of various factors on management planting decisions. Sour orange rootstock use has declined over time and more precipitously lately due to the threat of CTV and substantial documented tree losses. Rough lemon rootstock, once one of the two most popular rootstocks, is now seldom planted because of its great susceptibility to the disorder known as citrus blight. Carrizo citrange, which largely replaced sour orange and rough lemon as a rootstock of choice, has also declined in usage somewhat, probably due to its moderate susceptibility to blight. Cleopatra mandarin rootstock, while never the most popular, has been a consistent choice throughout the industry with little variation over time. Blight and Phytophthora susceptibility and its low early production record have limited its planting. Swingle citrumelo, a replacement for some of the above rootstocks, once enjoyed great popularity, with large acreages planted. A recent decline in planting of Swingle reflects the realization in the industry of some of this rootstock's problems, including soil specificity, instances of poor tree performance, and incompatibilities with certain scion cultivars. Informed rootstock selection and rootstock diversity are key to selecting a rootstock.

Selecting a Rootstock

Must have compatibility with desired scion cultivars

Must be adaptable to soil, climate, and expected pest pressure

Observe its effects on tree vigor, yield, fruit quality, and tree longevity

COMPATIBILITY

Most commonly grown cultivars of citrus are compatible for grafting; however, the plant material of the scion and rootstock cultivars is genetically different, and incompatibilities can occur. For example, 'Pera,' 'Roble,' and 'Murcott' budded on Poncirus trifoliate or many of its hybrids are known to be incompatible combinations. The grower should limit the selection to scion/rootstock combinations that have been proven compatible. Incompatibilities can also occur when viruses or viroids are present. The grower can avoid this risk by planting only registered trees.

SOIL AND CLIMATE CONSIDERATIONS

The rootstock variety should be selected to be adapted to the grove environment, especially the soil. Of particular significance are the calcium and clay content of the soil, soil pH, drainage and irrigation, and soil salinity. For example, certain trifoliate hybrid rootstocks such as Swingle citrumelo are not adapted to grow in soils with high pH or high clay content.

RESISTANCE TO PESTS AND PATHOGENS

One of the most important benefits that the rootstock variety may impart to the tree is resistance to certain pathogens and disorders, including Phytophthora foot rot and root rot, citrus blight, CTV, and the citrus viroids exocortis and cachexia. The grower should consider the pathogen pressure in the grove, and select a rootstock that provides the required resistance. It is important to note that sour orange and citrus macrophylla are especially susceptible to Florida decline strains of CTV.

The rootstock must also be resistant to pests present in the soil, especially nematodes. Both citrus nematodes and burrowing nematodes may be of concern in Florida citrus groves, depending on soil type and location. The grower should test the soil for nematodes and choose a tolerant or resistant rootstock variety if necessary. For example, Carrizo, Kuharske, Rusk citranges, Swingle citrumelo, and trifoliate orange are all tolerant of citrus nematodes. Carrizo and Kuharske citranges are tolerant of burrowing nematodes, while Milam lemon and Ridge Pineapple sweet oranges are resistant to burrowing nematodes.

TREE SIZE AND VIGOR

The rootstock variety influences the growth rate and final size of the grafted tree. Productive rootstocks are now available that produce a wide range of tree sizes from relatively small to very large, although most dwarfing citrus rootstocks are limited by low vigor and slow growth. An appropriate combination of scion and rootstock produces a tree of adequate height with minimal need for hedging and topping. The design of the grove, including tree spacing, row spacing, tree density, irrigation system design and bedding design, and the production and harvesting system, determines the optimal tree size.

PURCHASING NURSERY TREES

After selecting scion and rootstock cultivars, most growers obtain their nursery stock from professional nurseries. Citrus nursery management is a highly complex operation, and it is generally not feasible for the grower to maintain a nursery. Often several parties are involved in the development of a grove, including the owner, caretaker, nursery, and perhaps an engineering firm; it is essential that responsibilities be clearly defined and assigned.

An agreement to purchase citrus nursery trees usually involves negotiation and a contract between the nursery and the grower. The grower may choose to purchase either registered or validated trees. Registered trees originate from sources that have undergone more comprehensive pathogen testing and horticultural evaluation. Registered trees are understood to be free of viruses and viroids. It is required that the purchaser of nursery trees be provided with a Nursery Stock Inspection Tag. The inspection tag indicates the budwood registration number, the rootstock, and whether the trees are registered or validated. Growers may select either bare-root or potted (often called containerized) trees for purchase. Field-grown bare-root trees are usually less expensive than containerized trees grown in a greenhouse. Containerized trees, however, are generally easier to handle at the planting site, requiring less effort to keep roots moist etc.

It is the responsibility of the nursery to provide the grower with trees of the quality specified in the contract. They should represent certified nursery stock of the designated scion variety/rootstock combination, and of the stated size, budding height, and perhaps heading-out height. Trees should be delivered essentially free of canopy- and rootborne pests and pathogens, including nematodes, root weevil larvae, and phytophthora root rot. It is not economically feasible or within the limits of a sound pest management program for the nursery to keep the trees free of every pest and pathogen. The presence of some canopy-borne pests and diseases, including leafminers and bacterial leaf spot, should be discussed at the time of contract signing.

The purchaser should visit the nursery at least once between placement of the order and receipt of the trees to determine if the nursery stock meets contract specifications. This is especially important in the case of large orders. When purchasing nursery stock, the purchaser should be prepared to pay an appropriate price that covers the cost of production and includes a reasonable profit to the nursery. Once trees have been grown to the purchaser's specifications and are found acceptable, they must be accepted on schedule, as stock retained by the nursery incurs costs for which the nursery should be compensated. This is particularly important for containerized material for which space is at a premium.

Once the trees are delivered by the nursery to the grove location, it is the purchaser's responsibility to inspect them, especially for proper budding height and a healthy root system. The purchaser is also responsible for appropriate preparation of the planting site. The site should be well drained and the soil should be prepared for planting and essentially free of soilborne pests and pathogens. The planting site should also be free of trash and vegetation cover (including root sprouts) in the planting strip. Preparation of the site is especially important when old citrus grove land is being replanted. The purchaser is responsible for providing a storage facility for the trees to protect them from desiccating conditions while they await planting.

The planting process is vitally important and should follow established recommendations, including size, and preparation of the planting hole, tree positioning, and final "mudding in" and watering procedures. If the nursery representative determines that the site preparation or planting process is substandard, the grower should be informed in writing. This is intended to protect the nursery against complaints should the trees fail to thrive. The site should be inspected again a few days after planting. By this time, the trees will have settled and it will be apparent if the planting height is correct and if the irrigation emitters are positioned properly. It is vital that any necessary adjustments be made as quickly as possible to minimize the stress on the young trees.

REFERENCES

Bentley, B., A. Wysocki, and P. J. van Blokland. 2002. *New citrus varieties.* Teaching and Learning Paper TLP 02-10. 19 pp.

Saunt, J. 2003. *Insights on how to test-market new citrus varieties.* International Society of Citriculture Congress. (9th 2000: Orlando, Florida), Orlando, Florida USA 2: 1105.

6

Site Selection

L.K. Jackson

INTRODUCTION

The decisions affecting the ultimate success of a citrus grove cannot simply be thought of at the moment. There is a high degree of planning that goes into the success of a citrus business. This planning extends all the way back to before and during the tree plantings. In fact, one of the first and most important decisions the citrus manager will face is where to plant the trees. There are a number of key elements that determine whether a site is appropriate for citrus production, such as cold risk, water drainage, soil quality, pest proximity, and infrastructure. This chapter reviews these considerations and explains some of the factors that must be considered when deciding upon the location for a new citrus planting.

The selection of a location for planting a citrus grove is arguably the most important decision a grower makes. Even the choice of rootstock and variety pales in comparison to site selection; the site is essentially permanent, and it is usually prohibitively expensive to remedy an unsuitable site. The selection of sites for research plantings is especially significant; soil variation and limiting environmental factors are critical considerations in experiments where tree growth and vigor, yield, and fruit quality are measured to evaluate treatment response.

When choosing a site for a citrus grove, the grower must consider many interrelated factors. These various factors (Figure 6.1) must be given the proper weight in the decision-making process to reflect their significance in the development cost and ultimate profitability of the grove. Some factors are so important that they override all others. Some are relatively minor, but may be decisive when considered in combination.

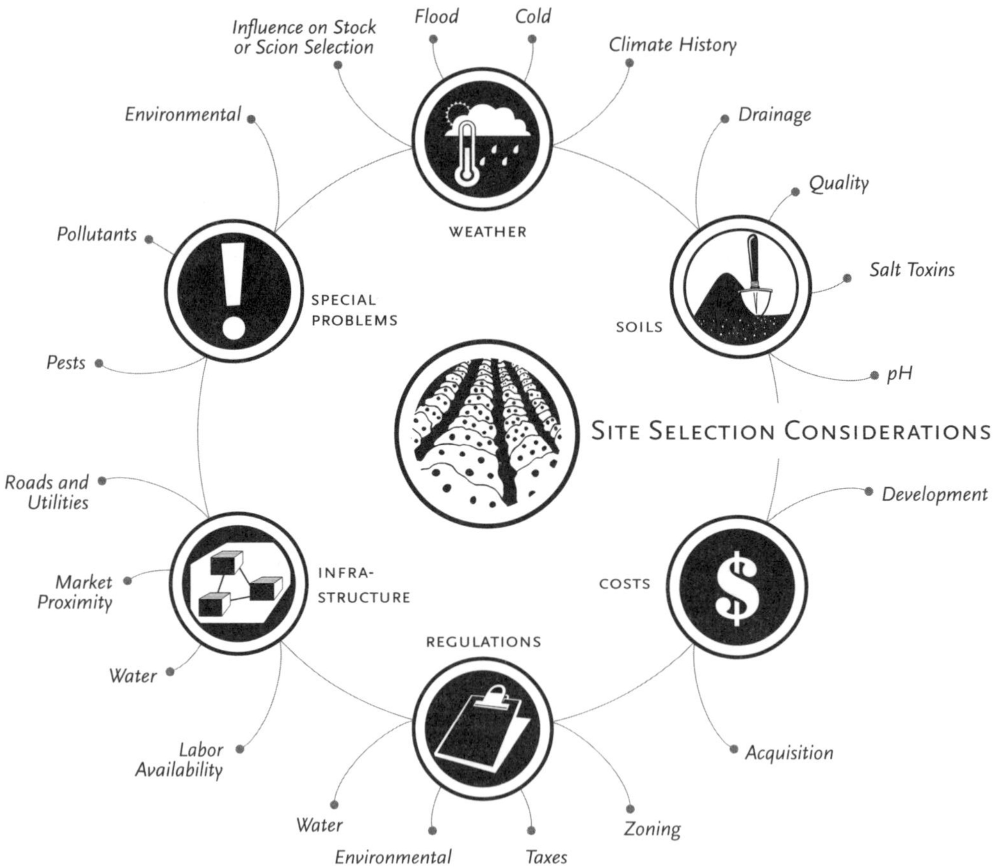

Figure 6.1 Many interrelated factors must be considered when choosing a site for a citrus grove.

There is no perfect site. Every location has some undesirable aspects that must be balanced against its more attractive features. The grower must maximize the positive attributes of a potential site and minimize the negative ones. Typically, the grower employs a process of elimination to reject inappropriate sites and finally select the best site available that represents the most favorable compromise.

> IMPORTANT SITE SELECTION FACTORS IN GROVE PLANNING
>
> There is no perfect site.
>
> Consider cold risk, water drainage, soil, pests, and infrastructure when selecting a site.
>
> Purchase price and cost of improvements are important site selection factors.
>
> To reduce the risk of cold damage, Florida citrus has moved to less horticulturally desirable areas.

SITE SELECTION FACTORS

The key elements that determine whether a site is appropriate for citrus production are cold risk, drainage, soil quality, pests, and infrastructure. In some cases, the presence of endangered species is also a determining factor. The development and maintenance cost associated with improving the site must be considered with the purchase price of the land to determine the potential profitability of a particular site.

COLD DAMAGE

Climate is one of the most influential factors in selecting a site. In Florida citrus production, the risk of damaging cold is all important. Other climate or weather concerns, including excessive rainfall, drought, excessively high temperatures, strong windy conditions, and persistent cloud cover are also important, and may be more significant than cold risk in other citrus-growing regions of the world.

All commercial citrus sites in Florida present some degree of cold risk. In general, the risk is greatest in more northerly latitudes. Arctic air masses that bring freezing temperatures to Florida are typically moderated somewhat as they traverse the length of the state. In fact, the northern limits of commercial citrus production have been established by the history of damaging freezes. Nevertheless, citrus grown in more northerly groves develop some cold hardiness by exposure to generally cooler winter temperatures, and may be less cold sensitive than citrus grown in warmer areas. It is notable that much of the prime Ridge citrus growing land, abandoned after the freezes of the 1980s, has been subjected to rampant urban development. Incidentally, no devastating freezes have impacted the northern Ridge area since the event of 1989.

Other factors also influence the risk of cold damage. A nearby heat source can keep a grove warmer when a freeze occurs, especially if the heat source is north of the site. Cold air enters Florida from the north or northwest and is warmed as it crosses over large bodies of water. Thus, groves to the south of a large lake or river receive some degree of cold protection. Sites close to the Atlantic Ocean or the Gulf of Mexico also tend to be warmer; large urban areas generate some heat as well.

The elevation and topography of the site should also be considered when assessing the cold hazard. Cold air is denser than warm air and flows and settles into low-lying areas. Thus, during periods of calm weather, without wind movement of the cold air, higher elevations are generally warmer than lower elevations. An elevation difference of as little as 4 feet can cause a temperature difference of as much as five degrees on clear nights with little wind. To prevent the formation of cold pockets, a grove should have an outlet to permit cold air to drain down and off the site into an adjacent lower area. Plantings on high slopes with adequate air drainage are typically least damaged by cold weather. However, not all freezes occur under calm conditions, and, historically, some of the most damaging freezes have been accompanied by high winds. Movement of the cold air by wind may eliminate the advantage of elevation. In some of these cases, the windy freezes actually produced the greatest damage to trees planted in exposed areas at higher elevation.

In addition to determining the latitude and elevation of a site and its proximity to a heat source, the grower should research the history of the site. Temperature records for the area should be studied and other groves in the area should be examined for signs of previous cold damage. The surrounding area should be examined for potential obstructions to air drainage from the site. It may also be helpful to interview other growers in the area.

If a site with moderate to high cold risk is selected, the grower needs to choose scion and rootstock combinations that are less vulnerable to cold damage. Scion cultivars should be chosen to mature early and be cold hardy, with rootstocks that contribute to the overall hardiness of the tree.

Cold Risk in Florida Citrus

Cold risk is the most important factor in determining a site.

Risk is generally reduced as you move further south.

Large lakes and urban areas immediately north of a site, and proximity to the Gulf or ocean, mitigate cold risk.

Cold-drainage is important in hilly Central Florida.

WATER DRAINAGE

Soils at the site should be well drained, but not excessively. At sites with some degree of elevation and adequate drainage, trees are planted at ground level and beds and drainage conveyances are not necessary. Such plantings are easy to maintain and development costs are relatively low. The rolling hills of Central Florida in the area referred to as the Ridge and a few small pockets in the flatwoods are characterized by light sandy soils, usually underlain by a sandy clay layer, and are generally well drained. These soils are also well aerated, allowing development of an extensive root system. However, much of this land is already in agricultural production or urban development and availability of these desirable sites is very limited. Further, the low nutrient-holding capacity of these soils necessitates frequent fertilization.

Sites, in the southern portion of the state referred to as the flatwoods area, tend to be quite flat and are usually poorly drained. These soils are usually white sand underlain by an organic layer or hardpan that inhibits internal drainage. The soil may be acidic at the surface, but calcareous rock can occur in the subsoil. Sites near the coasts are also poorly drained and usually are underlain with shell, marl or limestone. This soil is also typically acidic at the surface, but the subsoil may be either acidic or alkaline. Since citrus does not tolerate flooding, these poorly drained sites need to be artificially drained to provide an adequate rooting zone for citrus trees. Groves in these areas are usually planted on raised beds and use a system of water furrows and ditches to remove excess water from the rooting zone of the trees. This construction increases initial development costs, requires maintenance, and interferes with the movement of equipment in the grove. Oolitic limestone soils are found in extreme South Florida, predominately in parts of Dade, Broward, and Monroe counties. Soils may vary markedly within a short distance, so the site should be carefully surveyed.

Water Drainage

It is cheaper to develop and maintain well-drained sites than those with poor drainage.

Poorly drained soils must be bedded and require expensive infrastructure to prevent flooding.

SOIL QUALITY

The soil should also be of adequate quality to support the growth of citrus without extensive, potentially expensive modification. The overall quality of the soil depends on a number of factors, such as texture (proportion of sand, silt, and loam), soil particle sizes, amount of organic matter, available rooting depth, pH, uniformity, and the presence of toxins or harmful biological agents. A thorough evaluation of each of these elements is important. Again, it is important to note that the soil quality may vary significantly within the site. Soil variation is an important determinant of tree growth, and such variations are often visible from the air (Figure 6.2).

Figure 6.2 Aerial photography can identify soil variability and other important factors critical in making sound management decisions about land use. This site-specific technology is well-developed and is becoming more fully utilized by the industry.

Although citrus can survive in impoverished soil, areas with minimal soil fertility (such as some of the deep, white sands) should generally be avoided. Soil should contain some clay or organic material to help retain moisture and nutrients for use by the trees. Areas with soil that is predominantly sand are likely to be excessively drained and probably will require greater irrigation and fertilization to produce economic crops. Similarly, areas with a hardpan (Plate 6.1) that inhibits drainage, aeration, and satisfactory root development should be avoided or modified.

The pH of the soil is also important. Citrus performs best in a soil pH of 6.0 to 7.0. The pH of excessively acidic soils can be increased with the application of agricultural limestone, but this amendment represents an additional development cost. Soils with high calcium content, such as those containing large amounts of marl, shell, or limestone, may be excessively alkaline.

There is currently no satisfactory method for reducing pH in alkaline soils. Trees grown in high pH soils require extra applications of certain heavy metal elements (usually as sprays) to maintain nutritional health. Further, rootstock choices for high pH soils are very limited; many of the currently popular rootstocks perform poorly in high pH soils.

It is prudent to check both soil and water for excess salt content, because salts in the soil or water are deleterious to good citrus growth. The levels of copper and other potentially toxic metals should also be checked, especially at sites previously planted in citrus or vegetables. Sites with high copper levels should be avoided because tree growth will be compromised.

Soils for Citrus

Soil should contain some clay or organic matter.

The ideal pH is 6.0 to 7.0, though lime can be added to acid soils.

High salt or copper in soils can be a problem.

Pests and Diseases

A potential site and the surrounding area should be examined for the presence of soil pests and diseases, such as nematodes, root weevils, and Phytophthora root rot. Controlling or reducing pests and diseases may be difficult and expensive. Once again, interviews with other growers in the area may be helpful.

INFRASTRUCTURE

Finally, general infrastructure in the surrounding area should be considered. This includes roads, utilities, water, availability of labor, and proximity to market. Roads in the area must be able to accommodate the periodic movement of heavy equipment to and from the site, and the site should be reasonably close to either packing or processing facilities. The cost of electricity, water, and labor also affect the profitability of the grove.

LISTED SPECIES

The Central Florida Ridge and other well-drained, sandy soils in Florida support a unique flora and fauna with a high degree of indigenous, plant and animal species found only in these areas. Due to the initial relatively sparse distribution of this type of habitat, and the rapid conversion of these areas to citriculture, housing, golf courses and other development, federal and state agencies now list many of the species unique to them as threatened or endangered. Consequently, development in those areas remaining in their natural vegetative state is severely restricted. Therefore, in Central Florida or the coastal dunes, it is advisable to select land that has already been cleared. Former row-crop farming areas are especially desirable, because the site has been cleared and leveled, and the soil has often been amended and improved.

COST OF THE SITE

There are two components to the cost of any new grove site: land acquisition cost and development cost. To perform an economic assessment of a potential site, one must determine the total acquisition and development cost and the number of acres that can be planted. The grower then calculates the cost per acre of planted grove and uses this figure for comparison.

The cost to purchase the land depends almost exclusively on real estate value. Some of the most desirable sites for a grove may also be attractive for other enterprises, especially if the site is near a population center. This inflates the price, potentially making the site uneconomical for the production of citrus. A high purchase price may also result in high property taxes, and special agricultural exemptions are not always applied.

The development cost is the money that is spent to make the land suitable for the planting of citrus. Land clearing, leveling, mowing, and disking are required for most sites. Areas that experience seasonal flooding require the construction of raised beds, water furrows, and ditches to facilitate drainage. It may be necessary to construct or repair roads, pumps and wells, and have irrigation systems installed or modernized. Legal costs of purchase and development must be taken into consideration, and permitting costs vary greatly between areas. There may also be some set-aside requirements that do not allow the planting of all the purchased land area; this increases the effective cost per acre developed.

It is important for the grower to be aware of relevant regulations and to incorporate the cost of compliance into the price of the land. Regulations may make a site more expensive by placing additional development costs on the landowner, or may even completely disqualify a site from consideration. Most of these fall within the categories of site development restrictions, water rules and regulation, and environmental regulations.

The most common form of site development restriction is zoning. If zoning restrictions permit grove development, growth and zoning in the surrounding area should be examined to determine if home construction nearby is likely. Additional complications or expenses may arise in operating a grove adjacent to, or even in close proximity to, residential areas. The grower may also be burdened by ad valorem tax rates. It may be possible to obtain tax relief at some expense.

Water regulations are numerous and complex, and are generally implemented by water management districts. Permits are required for well construction, and consumptive use permits are necessary to use the water that is pumped. These permits and the water allocations specified influence the irrigation system selected and how it is used. Hazards to water quality are also regulated; backflow prevention devices are required, and the use of certain chemicals may be restricted or prohibited in areas where leaching is a problem.

Various regulations are in place to protect the environment. These regulations can restrict the discharge of drainage waters, often compromising the landowner's ability to effectively drain a given site. Environmental protection of wetlands may restrict a property owner's use of the land at a particular site.

Advantages of Choosing Cleared Sites

Undeveloped areas of native vegetation may contain threatened or endangered species, restricting development.

Farmed areas may already have undergone expensive improvements, such as leveling and the addition of soil amendments.

In general, wetlands cannot be disturbed; however, there may be a provision that permits a land exchange. In such cases, landowners create new wetlands to offset the use of others. For example, it might be acceptable to fill and plant a small wetland area in the middle of a grove and create a new wetland area of equal or greater size elsewhere on the site. However, these regulations are quite restrictive and create additional development costs that must be factored into the total site selection process.

Clearly, site selection is vitally important and extremely complex. A wide variety of factors determine the suitability of the site and its cost. Many of the issues that are considered in site selection are revisited when the grove is designed. Therefore, the issues of climate and weather, soil quality, the need for irrigation, and the requirements for compliance with water regulations and general environmental regulations are discussed in even greater detail in other chapters.

REFERENCES

Jackson, Larry K., and Frederick S. Davies. 1999. *Citrus Growing in Florida.* 4th edition, University Press of Florida, Gainesville.

7

Site Preparation and Engineering

B.J. Boman and D.P.H. Tucker

INTRODUCTION

As explained in the previous chapter, a number of factors influence the location where groves are planted. However, it is rarely possible to find the perfect site. Generally speaking, some degree of site engineering may be required to make the site more suitable for citrus production. For example, in the poorly drained soils of the South Florida flatwoods, an extensive drainage system may be needed to control the naturally high water table and provide for the rapid removal of excess surface water after rainfall. In fact, a surface water management system for citrus production in the flatwoods should be designed to remove 4 inches of water per day. A typical modern drainage system for citrus production in the flatwoods consists of beds and water furrows, lateral ditches, collector ditches, tile drains (if needed), and a perimeter ditch. The successful design and construction of such structures will involve engineering. This chapter provides an overview of the various considerations that are incorporated into engineering a new citrus grove site.

Once a site has been selected for citrus production, a grove design or planting arrangement is developed and the site is prepared. Site preparation includes clearing the land, amending the soil as necessary, providing for adequate drainage, and installing irrigation equipment. Environmental impact is a central concern, especially for water management. Water quality must be protected, and wetlands must be preserved. The grower must ensure that designs for both drainage and irrigation comply with all water management district regulations and must obtain all necessary permits before land preparation and well drilling start.

LAND CLEARING

In areas of native vegetation or previous citrus production, the first requirement for land preparation is clearing. Trees and shrubs, such as longleaf pine, slash pine, oak, palmetto, and wax myrtle or old citrus, can usually be cleared using front-end loaders fitted with root rakes. Heavier equipment, such as bulldozers or backhoes, may be needed to remove tree stumps. Some trees, especially pines, may be sold as timber. The cleared vegetation is usually piled and burned. On previously cleared land, only disking may be needed. The soil should be thoroughly analyzed for plant nutrients, pH, etc., during the site selection process, and amended as indicated by that testing. Flatwoods soils generally require pH adjustment via the application of lime or dolomite.

DRAINAGE

Drainage is usually adequate in the sandy soils of the Central Florida Ridge, and no grove preparation is needed to improve this feature. However, in the poorly drained soils of the South Florida flatwoods, an extensive drainage system may be needed to control the naturally high water table and provide for the rapid removal of excess surface water after rainfall. A surface water management system for citrus production in the flatwoods should be designed to remove 4 inches of water per day. However, most native flatwoods areas drain at rates of one inch or less per day.

Drainage

Natural water drainage is usually adequate in the sandy soils of the Central Florida Ridge.

Flatwoods soils usually require extensive drainage systems.

Drainage should permit removal of 4 inches of water per day.

Typical Central Florida citrus soils are sands with low water holding capacities, only 0.2 to 0.8 inches of water per foot. However, in the coastal flatwoods areas some of the finer-textured, sandy clay soils may hold as much as 2 inches of water per foot of depth. The actual water-holding capacity at any site depends on the amount of organic matter, clay content, and other water-retaining materials in the soil.

Several important soil factors need to be considered in a properly designed drainage system: the soil type, the thickness and continuity of strata in the soil profile, and the hydraulic conductivity and porosity of the various soil layers. Hydraulic conductivity is the rate at which water moves through soil in terms of distance per unit time (inches/hour or feet/day). It varies tremendously, and is usually specified under saturated conditions. Porosity is the volume of pore space divided by the total volume of soil. It determines how much water can drain out of the soil as the water table is lowered.

In flatwoods soils, control of the water table is usually achieved with the construction of evenly spaced ditches and elevated planting beds often with drain tiles (Plate 7.1). A typical modern drainage system for citrus production in the flatwoods consists of beds and water furrows, lateral ditches, collector ditches, a perimeter ditch, a drainage pump, and tile drains (if needed) (Plate 7.2).

BEDS AND WATER FURROWS

Planting beds are typically oriented north to south. They are constructed with vee-ploughs or motor graders to shape water furrows that are generally 48 to 55 feet apart. The water furrows are cut 2 to 3 feet deep and the soil is mounded between them to create a bed, with the crown of the bed 3 to 4 feet above the bottom of the furrow. Beds of these dimensions accommodate two rows of trees spaced 24 to 27.5 feet apart. However, some growers prefer single row beds. Wider beds accommodating 4 to 6 rows of trees are also used, particularly in areas where shallow, fractured limestone is encountered. It is helpful to laser-level the land prior to constructing the beds to facilitate rapid surface water removal by furrow drains.

Considerable skill and experience are needed to develop bedded groves that provide adequate drainage, good water-holding capacity, and acceptable soil characteristics. Skilled grove developers use their knowledge of nutrients, pH, and water holding for each soil layer in developing their strategy. Typically, the beds are constructed primarily from topsoil, but frequently lower soil strata are useful additions, enhancing water- and nutrient-holding capacity. Variable soil characteristics within a grove may encourage different approaches for each area. For example, areas with high clay content may be made productive through incorporation of sand excavated during ditch preparation.

Experienced grove developers frequently recommend greater height from bed crown to furrow bottom for soils with poor drainage. Depth to spodic layer or hardpan and the expense of importing fill dirt limits the bed height in most grove sites. Furrows should not be dug below the hardpan, because wet areas may develop that interfere with equipment movement and compromise drainage. Careful shaping and prior reshaping of furrows is necessary to maintain adequate drainage. Drainpipes should be placed so that the soil is slightly below the pipe interior bottom at the low end of the furrow. Otherwise, water will pond at the outlet inside the furrow.

Even greater care is necessary when a single-row bedded grove is renovated into a wider two-row bed system. Most grove owners report superior results from leveling the site to approximate original soil distribution prior to rebedding. Without this step, one row in the new bed typically contains a much higher portion of subsoil, and this may result in poor tree development.

Not all land successfully produces citrus, even within an area well suited to bedded grove development. There are many areas within existing groves where poor or excessive drainage makes production uneconomic. Careful assessment of potential citrus production throughout the planned grove may return profits by identifying areas that should be excluded from planting.

LATERAL DITCHES

Lateral drainage ditches should be cut at right angles to the beds and water furrows and spaced no farther apart than 1,320 feet from center to center. Topsoil spoil from the ditches can be used as fill for low areas in the adjoining fields. Subsoil spoil provides a grove road base on either side of the lateral ditch. Furrows drain into the ditches through 6- to 8-inch flexible polyethylene or rigid PVC pipe that can be installed either before or after furrow construction. A laser level is sometimes employed in this operation, but is not essential. The pipe is installed in the bottom of the water furrow and sloped to discharge approximately one foot above the bottom of the ditch.

Lateral Ditches

Ditches are prepared perpendicular to beds and furrows.

There should be no more than 1320 feet between ditches for proper drainage.

Ditches should be a minimum of 14 feet wide at the top, 4 feet wide at the bottom, and 5 feet deep.

Drainpipe from furrows should emerge at least one foot above the ditch bottom.

COLLECTOR DITCHES

Drainage water from lateral ditches flows into larger collector ditches, and is conveyed off-site (Plate 7.3). Gravity drainage is preferred, but discharge pumps are needed where there is insufficient topographic relief. The size of the collector ditches and any associated pumping facilities depends on the soil, the size of the area being drained, the design of the beds and water furrows, and the slope of both lateral and collector ditches.

TILE DRAINAGE

Drain tiles may be installed for additional control of the water table. Perforated, flexible 4-inch diameter polyethylene pipe covered with a nylon fabric sock is installed down the center of the beds. It is normally sufficient to install the pipe only in every other bed. The pipe should be installed on a slope of at least 0.10 percent at depths averaging no more than 3 to 4 feet, depending upon the location of impermeable spodic (organic matter hardpan) or clay layers in the soil. In general, drain tile should not be installed below the depth of the hardpan. In some areas where sandy soils have underlying impermeable hardpan, tile drainage may be used to drain groves without forming beds and furrows.

PERIMETER GRADING

Natural drainage from high water tables on adjacent properties may result in subsurface flow towards a grove. A perimeter ditch and external dike are needed to control the off-site water table and intercept off-site surface flow. The size of the ditch depends on the anticipated flow rate. Perimeter ditches can also serve as collector ditches. Pumps may be needed in the perimeter ditches to maintain satisfactory water table depths in the developed grove.

ENVIRONMENTAL IMPACT OF DRAINAGE

The principal environmental concerns associated with citrus production in the flatwoods area are the quality and discharge rate of runoff water and the effect on protected wetlands. A registered civil or agricultural engineer must certify the water management system in order for the developer to obtain a permit from the water management district.

> ENVIRONMENTAL IMPACT OF DRAINAGE
>
> System must be certified by registered engineer.
>
> In most areas, retention areas to reduce runoff rates are required for new groves.
>
> Off-site discharges must not exceed predevelopment rates.

DISCHARGE RATES

Properly designed surface water management systems control storm water runoff rates. Runoff rates are typically reduced by the use of retention areas, often called reservoirs, located between the grove and the ultimate off-site discharge points. Typically, these areas are diked and receive inflow from the grove area either by gravity drainage or pumped discharge. Outflow from the retention areas passes through discharge structures that are designed to restrict the flow rate to predevelopment rates. Thus, water levels rise in retention areas for a short period of time following major rainfall events.

DISCHARGE WATER QUALITY

Discharge water may contain sediments and dissolved agrichemicals, including pesticides. In order to obtain an Environmental Resource Permit from the appropriate water management district, a grove developer must either agree to provide ongoing monitoring of discharge water quality, or retain an engineer to design a system that meets the criteria for water quality treatment.

Generally, the retention/detention areas used to control runoff rates are effective in providing water quality treatment by allowing sediment to settle and providing more time for natural degradation of agrichemicals. In the Southwest Florida Water Management District (SWFWMD), the use of grassed filter strips is an accepted method of achieving these results. For small systems serving less than 40 acres, engineers in the USDA Natural Resource Conservation Service (NRCS) can provide designs to obtain a "permanent agricultural exemption" from the WMD. For larger systems, a professional engineer must be retained.

IMPACTS ON WETLANDS AND DIVERSION OF WATERSHEDS

It is important to try to eliminate impacts on wetlands from proposed designs. The WMDs have a fairly clear legislative mandate to protect wetland resources. In addition, the U.S. Army Corps of Engineers (COE) regulates direct impacts to wetlands from water discharge or deposition of fill. Both agencies are also tasked with assessing and regulating secondary and cumulative impacts to wetlands. The issues can be fairly complex, so it is usually cost-effective to retain a consultant to develop a design and prepare the permit application.

Grove design should start with a thorough investigation of predevelopment drainage patterns through and around the site, and these should be incorporated into the design. Care must be taken to eliminate or minimize diversion of natural watersheds, particularly if they are feeding off-site wetland systems. Off-site flows downstream can be routed through projects in natural flowways or constructed channels, or they can be taken around projects in constructed ditches, provided they outflow at their predevelopment location. Off-site flows are usually kept separate from project flows; otherwise, water quality monitoring and treatment must be provided for the entire volume.

Direct impacts to wetlands can be avoided by maintaining an upland (nonwetland) buffer between the wetland and grove. This also eliminates the need to obtain a permit from the Corps of Engineers. When setbacks are large enough, the indirect impact of water table drawdown can be minimized or eliminated. This approach may also satisfy new requirements designating upland buffers to accommodate wildlife in groves. Some Water Management Districts have established procedures to calculate buffer zone dimensions needed to minimize water loss from wetlands.

Incorporation of wetlands in retention areas also avoids illegal drawdown, or lowering of water levels in wetlands. However, if the retention areas are meant to reduce runoff rates, the water levels may rise as much as 4 feet after a major rainfall, which is the maximum safe depth for small impoundments. If water remains at this depth for a prolonged period of time, it may have a deleterious effect on natural wetlands and violate environmental restrictions.

Maintaining grove-associated wetlands with natural flow-ways carrying off-site water through a project can provide for a habitat mix of wetlands and uplands, and connectivity between off-site and on-site undeveloped areas (Plate 7.4). This also meets the goal of accommodating wildlife species that are sensitive to habitat fragmentation.

REFERENCES

Reitz, H. J. 1978. *Higher density plantings for Florida citrus: Introduction to symposium.* Proceedings of the Florida State Horticultural Society 91: 26-33.

Sites, J.W., L. C. Hammond, R. G. Leighty, W. O. Johnson, and K. D. Butson. 1964. *Information to consider in the use of soils of flatwoods and marshes for citrus.* Circular S-135A: 7 pp.

Whitney, J. D., A. Elezaby, W. S. Castle, T. A. Wheaton, and R. C. Littell. 1991. *Citrus Tree Spacing Effects on Soil Water Use, Root Density, and Fruit Yield.* Transactions of American Society of Agricultural Engineers 34: 129-134.

8

Essentials of Nursery Management

M.C. Kesinger

INTRODUCTION

Nursery management is a highly specialized operation requiring a significant investment of time and resources. It is generally less expensive for growers to obtain stock from a commercial nursery than to develop and maintain their own. The purchaser specifies scion and rootstock variety, whether the trees are to be registered or validated, delivered bare root or containerized, and the desired size and budding height. Commercial citrus nurseries clonally propagate large numbers of trees to grower specifications and most citrus nurseries use an inverted "T" budding technique to graft a small bud eye onto a rootstock liner (small tree of the rootstock variety) at the appropriate height. After the bud is inserted, the eye and seedling are wrapped in budding tape for several weeks. Later, the budding tape is removed and the seedling is lopped or bent over to force the bud to sprout and grow. To produce such trees, a number of factors have to be considered that relate to location, site selection of the nursery, etc. This chapter provides a brief overview of how citrus trees are produced in the nursery.

Nursery management, only discussed briefly, is a highly specialized operation requiring a significant investment of time and resources. It is generally less expensive for growers to obtain stock from a commercial nursery than to develop and maintain their own. The purchaser specifies scion and rootstock variety, whether the trees are to be registered or validated, delivered bare root or containerized, and the desired size and budding height.

Commercial citrus nurseries clonally propagate large numbers of trees to grower specifications. Most citrus nurseries use an inverted "T" budding technique to graft a small bud eye onto a rootstock liner (small tree of the rootstock variety) at the appropriate height. Budding height is important for various reasons such as disease control and having an adequate straight section of trunk below the first scaffold branches to allow mechanical harvesting. Only vigorous, healthy, true-to-type seedlings are budded. Seedlings with stem or root rot must be discarded. After the bud is inserted, the eye and seedling are wrapped in budding tape for several weeks. When the budding tape is removed the seedling is normally lopped or bent over to force the bud to sprout and grow. Once the trees reach the specified size they are delivered to the grower. Trees are expected to be healthy and delivered essentially free of canopy- and root-borne pests and pathogens.

Growing conditions at the nursery are vitally important. The site for a field nursery must be well drained, and plants should be spaced appropriately to provide good ventilation, adequate tree growth, and permit worker access (Plate 8.1). In greenhouse and shade house nurseries, containers should be located on raised structures (Plate 8.2). Containers are frequently moved to sort plants by size and provide optimal light for those trees not growing satisfactorily; it is critical to maintain tree identity during these moves. Many container nurseries tag trees with plastic labels, which may be printed directly from a personal computer.

Young nursery trees require intensive cultural care and pest management. The seedlings are very vulnerable, and they must be constantly monitored for pests and diseases and evaluated for environmental stress. Regular soil analyses are necessary to determine the nutritional needs of the seedlings, and soil and water pH must be considered (a range of pH = 5.5 to 7.0 is preferred). The stage of plant growth also affects plant nutrient requirements. For example, fertilization prior to budding may affect percentage of bud "take," with fewer buds that survive and grow when plants are overfertilized. Overfertilization can also lead to salt damage due to cation excess, ground water nitrate contamination, and unnecessary growing costs.

Irrigation must be adjusted seasonally for rainfall, temperature, soil type, and stage of plant growth. Applications of water should be timed to allow drying, as extended wet periods are conducive to fungal or bacterial invasion. Excessive watering leaches nutrients and herbicides out of the root zone.

Plants in field nurseries must be guarded against freezing conditions, and the temperature in greenhouse nurseries must be monitored and maintained by cooling fans and heaters, with an emergency system for power failures.

Good weed management is critical to a successful field nursery operation. Until the more widespread use of herbicides, managing weeds was a very expensive, labor- and machinery-intensive operation. Modern programs implement integrated weed management, incorporating good sanitation measures and a variety of weed control techniques. These include fumigation, mechanical tillage, and the application of both soil residual and postemergence herbicides. Several of the safer preemergence herbicides may be successfully used in rotation (use of different products in successive applications over several years) at lower rates to prevent development of resistant weed types and reduce risk of tree injury. Seedling liners and budded stock are very susceptible to nonselective contact and systemic herbicides. These herbicides must be applied with shielded equipment and under calm conditions to avoid plant contact.

Weed management in containerized production in closed structures such as greenhouses relies on good sanitation procedures. However, weed seed can enter the greenhouse, and weed infestations do occur on floor areas below structures supporting containerized trees. The use of herbicides in closed structures can be hazardous due to volatilization and should be approached with caution. Most herbicide products are not registered for indoor use.

Protection of the seedlings from pests and pathogens is of the utmost importance. Fungal pathogens are a particular threat to the susceptible plant tissue. For all field or enclosed nursery operations, exclusion and sanitation procedures should be the primary means of protecting the young trees. Chemical intervention should be considered a last resort. Both field and enclosed nurseries should be fenced, and access to the facility should be limited to trained nursery workers. Enclosed nurseries may use more sophisticated devices such as double-door entrances, sanitary foot dips, and hand-washing facilities, as well as protective clothing, especially for visitors. Water sources should be protected from contamination from point of origin to application. In greenhouses, hose ends and nozzles should be kept off the ground. Potting soil mixtures should be sterilized and stored under sanitary conditions to avoid infestation from pests, pathogens, and weed seed. All cutting tools should be sterilized in 20 percent household bleach solution before use, as citrus viroids such as exocortis are easily transmitted on budding knives and pruning tools. All equipment should be cleaned before it is brought to the nursery site and designated to remain on site. Screen-enclosed nursery stock and clean budwood sources are the most effective control for insect-vectored pathogens.

REFERENCES

Dekkers, M., and R. F. Lee. 2000. *Evaluation of commercial potting mixes for optimizing growth of citrus in containers.* Proceedings of the Florida State Horticultural Society 113: 130-133.

Lee, R. F. 2003. *Value of clean stock programs and quarantine measures for citrus.* International Society of Citriculture Congress. (9th 2000: Orlando, Florida), Orlando, Florida USA 2: 948-951.

Roistacher, C. N. 2003. *Essential ingredients for a successful certification program for citrus.* International Society of Citriculture Congress. (9th 2000: Orlando, Florida), Orlando, Florida USA 2: 942-947.

Roka, F. M., R. P. Muraro, and R. E. Rouse. 1998. *Economic impact of the Florida citrus nursery industry.* Proceedings of the Florida State Horticultural Society 111: 166-168.

Timmer, L. W. and L. W. Duncan. 1999. *Citrus health management.* APS Press, St. Paul, Minn.

Zekri, M. 1999. *Evaluation of citrus trees planted directly from the nursery and trees grown for two additional years in bags before transplanting in the field.* Proceedings of the Florida State Horticultural Society 112: 56-60.

9

Grove Design

D.P.H. Tucker, T.A. Wheaton and R.V. Russ

INTRODUCTION

After a site has been engineered and the water delivery systems have been installed, further consideration must be given to decisions about tree spacing and row orientation. These decisions will to some extent depend on the site's climate and soil characteristics, the vigor of the scion/rootstock combination to be planted, the expected life of the grove, and the space required for the operation of production and harvesting equipment. Windbreaks may also be desired to reduce spread of pests and diseases through the grove. Ultimately, if tree spacing is not matched to tree vigor, yield in a closely planted grove will decline when the trees reach maturity and maintenance size is reached.

Many tactics can be applied when deciding on plant spacing. For example, the bearing volume, and thus the production potential, may be increased by increasing tree height, reducing tree spread, reducing the width of middles, increasing the number of trees per acre in a young grove, or forming a continuous hedgerow in a mature grove. Canopy volume per acre increases most rapidly when trees are densely planted and the individual trees grow at their maximum rate. In a mature grove, the canopy covers most of the land area and collects a high percentage of available sunlight for growth and fruit production. It is possible to calculate the bearing volume per acre for a range of grove designs.

This chapter provides relevant background related to tree spacing and grove geometry. Regardless of which methods are applied, the new grove must be designed to provide maximum economic return averaged over the expected life of the grove.

The design of a citrus grove begins with site selection and a plan for the physical arrangement of trees in the planting. All future investment in a grove may be compromised by a poor grove design where trees fail to fill their space, are excessively crowded, or do not receive adequate pollination as required. A good design results in healthier, more productive trees and higher fruit quality with little pruning. Changes to the design of an established planting are difficult, time-consuming, and costly, so the design of a new grove merits careful consideration. The grower generally seeks to maximize the lifetime economic return of the grove and must consider both the biological impact and the management implications of a design. Like so many of the grower's decisions, selection of an appropriate design is based on a variety of factors and concerns. Decisions about tree spacing and row orientation depend on the site's climate and soil characteristics, the vigor of the scion/rootstock combination to be planted, the expected life of the grove, and the space required for the operation of production and harvesting equipment. The design must also provide appropriate pollenizer cultivars at proper intervals for some specialty cultivars. In addition, the grower should consider the value of the land and the cost of labor, whether the target market is fresh or processed, and the available resources, equipment and financing. After these decisions are made, the site is prepared, irrigation is installed, and the trees are planted. When trees grow beyond their allocated space, mechanical hedging and topping are used to control tree size.

TREE SPACING

The allocation of space/tree in a grove is determined at the time of grove design. Enough space is allocated for each tree to fill at maturity, and clear space (the drive middle) is provided between the rows for access. Tree spacing is an important factor in the profitability of a planting. Optimal tree spacing changes with the age of the trees, so any decision about tree spacing represents a compromise. Groves planted with trees closely spaced provide an earlier return, but are more expensive to establish and more difficult to maintain. If tree spacing is not matched to tree vigor, yield in a closely planted grove will decline when the trees reach maturity. In contrast, groves with more widely spaced trees require a longer time to reach their maximum yields, but may ultimately be as productive as a closely planted grove and are easier to manage at maturity. The effect of tree spacing can be discussed in terms of the following points:

1. the relationship between sunlight interception and productivity,

2. the relationship between canopy bearing volume and productivity,
3. the importance of matching tree vigor to the space allocated for each tree,
4. the compromise between the higher productivity of young trees when closely planted, and the easier management of mature trees when widely spaced.

SUNLIGHT INTERCEPTION

Sunlight is the source of energy for tree and fruit growth. The focus of grove design is the arrangement of tree canopies to capture sunlight efficiently. Sunlight falling on the ground between the trees does not produce fruit. The grove design should be based on the optimization of light interception over the life of the planting. This requires the minimization of the amount of light falling on the ground and the development of canopy structures such that a maximum number of leaves receive an optimum intensity of light. It is important to note that citrus leaves do not use full, direct sunlight efficiently; most effective photosynthesis occurs somewhere between full shade and full sunlight.

CANOPY BEARING VOLUME

The yield potential of a citrus grove is closely related to the bearing volume of the canopy. The bearing volume of a citrus tree is defined as the outer 3 feet of the canopy; fruit bearing is largely limited to this zone, and 90 percent of the solar radiation is absorbed in this outer 3 feet. Production per acre is a function of bearing volume per acre. A grove design in which a large proportion of the available space is devoted to bearing volume produces more fruit than a design with a high percentage of space devoted to structural tree support, drive "middles" (clear space between rows of trees), or other uses, such as drainage.

The bearing volume, and thus the production potential, can be increased by increasing tree height, reducing tree spread, reducing the width of middles, increasing the number of trees per acre in a young grove, or forming a continuous hedgerow in a mature grove. There are, of course, practical constraints to each of these approaches. Tree height should be limited to prevent loss of tree skirts due to shading and to aid workers and equipment. Enough space must also be left in the middles to allow for maintenance and harvesting equipment and operations. Finally, tree spacing must be in balance with natural tree growth so that trees can continue to be maintained in their allocated space without loss of productivity. Thus, the grove design should provide maximum canopy without loss of adequate access in the grove. Maximum production is obtained when about 70 percent of the land area is covered by canopy.

In practice, bearing volume per acre for mature trees, i.e., trees that have grown to fill their allocated space, is similar over a range of tree densities and arrangements. Hence, at maturity, a range of tree spacings results in similar production potential. The dimensions of the rectangular area allotted to each tree generally describe tree spacing.

Between 1960 and 1994, average tree density in new plantings almost doubled. In 1960, new plantings spaced at 25 feet by 25 feet were common, which provided about 75 trees per acre. Planting density increased during the 1960s and again during the 1980s. By 1994, the average density of new plantings was 140 trees per acre. A spacing of 12.5 feet by 25 feet is frequently used, and provides this density. However, many new groves are planted at even higher densities. For example, a 10-foot by 20-foot spacing is not unusual and provides 218 trees per acre, but usually leads to increased canopy maintenance and reduced yields.

DEVELOPING GROVE

In a young grove, tree canopies cover only a small part of the ground area and capture only a fraction of the available sunlight. Yield per acre is low because canopy volume per acre is low. Canopy volume per acre increases most rapidly when trees are densely planted and the individual trees are growing at their maximum rate. The combination of higher tree density and improved management results in substantially increased early production. There are, however, increases in cost when tree density and management inputs are increased. Careful analysis is required to determine the tree density and management strategy that will provide optimal returns.

It is possible to predict the annual yield for trees planted at different densities if the rate of canopy development and individual tree yield are known. Although yield per tree varies with variety, climate, soil, and other factors, the prediction is useful when enlarging or replanting an existing grove where growth rate and yield for an individual tree may be known. This approach was used to compare annual production during the first fifteen years for trees planted at three different spacing. The study with Hamlin orange trees growing at a good site with excellent management, resulted in the highest density, 10 by 20, reaching its maximum yield in eight to ten years, while the 15-foot by 25-foot spacing approached its maximum in 15 years, and trees at a 25-foot by 30-foot spacing did not approach maximum production within fifteen years. Grove experience indicates the closest spacing will probably have less yields at twelve years and beyond because of overcrowding and required hedging.

MATURE GROVE

In a mature grove, the canopy covers most of the land area and collects a high percentage of available sunlight for growth and fruit production. Trees have filled their allocated spaces and are maintained at that size by hedging and topping. Most leaves are located in the outer 3 feet of the tree where there is adequate light for fruit production, and production per acre is closely related to the bearing volume. It is possible to calculate the bearing volume per acre for a range of grove designs (Table 9.1). Such calculations demonstrate that grove design and tree spacing have little effect on bearing volume in a mature grove. Researchers have observed that maximum production is similar for a range of grove designs and tree densities. Thus, the advantage of the higher-density planting exists only during the early years of the planting.

Table 9.1 Canopy bearing volume per acre of a mature grove is similar for a variety of grove designs and tree spacings. Therefore, production of a mature grove is similar over a range of designs and densities.

Spacing (feet)	Bearing volume/acre (1000 cubic feet)
Individual trees	
20 x 25	210
25 x 25	213
25 x 30	213
30 x 30	213
Hedgerow design	
20	248
25	224
30	205

The distribution of fruit on the tree is affected by tree spacing. In widely spaced trees, a high percentage of the fruit is located close to the ground. As row spacing decreases, however, the fruit tends to concentrate in the top center section of the tree. This change in fruit distribution affects harvesting efficiency. In general, it is more difficult to harvest fruit that is high in the tree, although platform harvest aids may be helpful. A high-density planting of small trees produces a wall of easily accessible fruit and provides attractive harvesting. However, the placement of containers for conventional harvesting operations is difficult if solid hedgerows are spaced less than 20 feet apart.

HEDGEROWS VS. INDIVIDUAL TREES

Citrus trees planted closely together in the row grow rapidly to form a continuous hedgerow. The close spacing increases tree density so that canopy bearing volume and yield increase rapidly in the young grove. It was previously believed that citrus was most productive when all sides of the tree were exposed to sunlight, and the suitability of a hedgerow planting was considered questionable. Hence, citrus was traditionally grown and maintained as individual trees in an open planting. However, the hedgerow has been found to be an optimal production system for Florida citrus. Attempts to maintain or create spaces in the row between trees or the removal of every second, fourth, or fifth tree down the row generally reduce yield.

The hedgerow offers several advantages in addition to increased yield. A dense, solid hedgerow shades out weeds and reduces the need for herbicide applications. It also allows tubing used for microirrigation to run along the tree row on top of the ground. Movement across the row for harvesting or other operations and the placement of containers may be difficult in a close planting; however, the advantages of the hedgerow greatly compensate for these obstacles. Virtually all new citrus plantings in Florida are designed to rapidly form a hedgerow.

EXCESSIVE VIGOR

In some plantings, a decline in yield has been observed after trees reached containment size. These groves grow rapidly and produce very well while the trees are young. However, the groves experience a decline in yield when the mature trees begin to require hedging and topping, even though the trees appear to be healthy. This problem is typically observed in groves planted at average or above-average tree densities on a vigorous rootstock. Such trees receive high levels of nutrients and irrigation after planting in order to promote early canopy development and increase fruiting potential. The problem may be more severe for trees on better soil types. The trees appear to be in a vegetative state and produce excessive regrowth following hedging and topping.

The probable cause of this declining yield is an imbalance between the natural growth habit of the tree and space it has been allocated. Trees in this situation may benefit from a reduction in nutrient and irrigation inputs. Hedging or topping two times per year may also improve fruiting. In severe cases, it may be necessary to remove every other or every third tree to provide more canopy volume per tree. The problem can be avoided at the grove design stage by selecting tree spacing appropriate to the inherent vigor of the variety and rootstock while taking into consideration of the soil and climate at the site.

FRUIT QUALITY

Tree spacing appears to have minimal effect on fruit quality, provided sufficient space is allocated for the natural growth habit of the tree. However, a grove design that does not match allocated space to tree vigor may ultimately result in poor-quality fruit on mature trees. The fruit quality of very young trees is always inferior. The fruit tends to be larger and to have poor external color, a coarser peel texture, and lower juice-soluble solids concentrations. Mature trees that are too confined by hedging and topping revert to a more vegetative or juvenile state. This may result in low production of poor-quality fruit characteristic of young trees. Hedging or topping that substantially reduces crop load (or any other factor significantly reducing crop load) also compromises fruit quality. Fruit size increases as the crop load decreases, resulting in a decline in juice-soluble solids concentration.

No consistent effects of tree spacing on internal or external fruit quality have been observed in close plantings where the allocated space was appropriate for the vigor of the trees. Concerns about fruit color or other quality factors in higher density plantings are unfounded. Therefore, fruit from higher density plantings is suitable for the fresh market as well as for processing.

COLD PROTECTION

Several observations in Florida indicate that freeze damage is reduced in higher-density plantings in both calm and windy freezes. For calm (radiation) freezes, the continuous canopy traps more of the heat that is radiated from the ground. During windy (advecting) freezes, hedgerows may act as an effective windbreak, reducing heat loss from the site. Thus, higher density plantings may be better for colder locations. However, grove design in such cases should also ensure adequate air drainage.

SPACING TIPS

Spacing of 20 feet to 25 feet between rows is generally recommended in a citrus grove. Vigorous cultivars such as grapefruit, lemons, and some tangelos require more space than oranges and mandarins, and should be planted at the upper end of this range. The spacing of 20 feet to 25 feet between rows leaves room for a clear middle of 7 feet to 8 feet, which is adequate for current production and harvesting equipment. Some commercial and experimental plantings have been successful at 15-foot row spacing if rootstock vigor is moderately low, and the advent of new rootstocks may make this row spacing more attractive. It is difficult, however, to perform conventional harvesting operations with row spacing of less than 20 feet. The miles of travel per acre also increase as spacing between rows decreases.

Within rows, it is desirable to space trees about 10 feet to 15 feet apart. Inside this range, the influence of climate, soil, variety, and rootstock on tree vigor must be considered. With modern production practices in a moderate and subtropical climates such as Florida, trees at these spacings grow together in the row and form a hedgerow within a few years. There is little advantage to a spacing of less than 10 feet in the row because trees start competing for space before significant production begins. Spacing trees at regular intervals in a single row is preferable to planting trees in groups of more closely spaced trees. For example, a consistent 10-foot spacing between trees is preferable to groupings of two trees spaced 5 feet apart with 15 feet between the groups, even though the number of trees per acre is the same. Trees planted at regular intervals develop the desired hedgerow more rapidly.

Spacing at 15 feet by 25 feet provides a tree density of 116 trees per acre and spacing at 10 feet by 20 feet gives 218 trees per acre. Table 9.2 shows the tree spacing and number of trees per acre for several commonly used grove designs. There may be justification for tree densities outside this range in some situations. Some variety and rootstock combinations can be grown successfully at up to 350 trees per acre, but under Florida growing conditions there appears to be little or no economic advantage to this practice.

In older established groves with trees spaced 20 to 30 feet apart in the row, it is common to interset additional trees in the row. This can improve production if trees have not already grown together in the row. However, unless there is 10 feet or more of open space, competition from adjacent trees may be so severe that young interset trees never develop adequately. Removal of mature trees in a hedgerow due to disease also leaves space for intersets. Intersets should be planted only when there is adequate space for development, and should be evenly spaced. The practice of resetting with two trees spaced less than 8 to 10 feet apart is not advisable.

Citrus is adaptable to a range of planting densities. A final decision on tree spacing requires careful economic analysis. Complete financial analysis includes appropriate discounting of money over time and must include additional factors such as loss and replacement of trees and the expected life of the grove.

Table 9.2 Tree spacing and trees per acre in common grove designs.

Spacing between rows (feet)	Spacing within rows (feet)	Trees per acre
20	10	218
20	15	145
22	11	180
22	22	90
25	10	174
25	12.5	139
25	15	116
25	20	87
25	25	70
30	25	58
30	30	48

TREE HEIGHT

The designed tree height (topping height) should be specified when space is allocated. The height of the tree affects the level of light that reaches the lower tree skirt. A tree height to middle width ratio of 2:1 provides a conservative guideline for maximum tree height for adequate sunlight in the lower canopy. For the common 8-foot clear drive middle, a tree height of 16 to 18 feet is ideal.

ROW ORIENTATION

Row direction or orientation is not critical for citrus. Although light distribution is more uniform on both sides of a tree in a row running north to south, the effect of row orientation decreases closer to the equator. Rows should run in the direction most suited to a particular site. This may depend on the row length, air or water drainage at the site, or the need for erosion control.

POLLINATION

Most citrus cultivars require pollination for fruit production. Self-pollination, which is sufficient for most cultivars, occurs naturally when the anthers contact the stigma, as pollen falls or when pollen is transported from the anthers to the stigma by insects, especially bees. However, some cultivars are self-incompatible, or produce sterile pollen. These require pollen from some other source and a means of pollen transport, typically bees. Such cultivars include the Orlando tangelo and the Minneola tangelo, as well as the Sunburst tangerine and many other newer cultivars.

In designing a grove, it is important to include pollen source trees (trees with compatible pollen and overlapping bloom, known as pollenizer trees) if the variety being planted is self-incompatible. A tree should be no farther than two trees away from a pollen source, which can be accomplished with a variety of arrangements for interspersing pollenizers. A pollenizer may be planted in every third tree space in every third row. However, this may be inconvenient at harvesting, since the pollenizer may be harvested at a different time or for a different market than the variety of interest. Another effective arrangement that may be more convenient for harvest is alternating one or two complete rows of pollenizer trees with four rows of the variety of interest.

Citrus pollen is heavy, sticky, and not effectively transported by wind, so insects are required to move the pollen. Honeybees are the best pollinators available. The grove design should facilitate the pollen transport, exploiting the bees' natural tendency to make efficient use of convenient nectar sources and to select the easiest and most direct flight path between the nectar source and the hive. Rows should be oriented with the prevailing winds, space between rows should be suitable for pollinator flight, and open spaces at each block should be shaped and maintained for placement and removal of beehives (Plate 9.1).

Generally, one colony of bees for every two acres of grove is adequate. The hives should be introduced and located appropriately within the grove after the target variety bloom is at about 10 percent. This encourages the bees to identify the citrus blossoms as an attractive nectar source. Once a nectar source is located, the bees continue to harvest that source until the scouts locate a better source. Water should be conveniently available, and disturbance in the grove should be minimized. Insecticides that are hazardous to bees must not be used during the bloom period. Grove management activity such as disking, spraying, and picking can disrupt the colony, or cause the bees to become aggressive. Communication and cooperative negotiation between the citrus manager and the beekeeper are necessary to ensure successful pollination.

The selection of the pollenizer variety is important. The pollenizer must have viable pollen and a flowering period corresponding to that of the target variety. Cultivars that are especially susceptible to pests or diseases should be avoided, and a market should exist for the fruit of the pollenizer. Temple and Orlando are suitable pollenizer cultivars for Minneola and Sunburst; Temple is the usual choice for Orlando. Fallglo does not require a pollenizer for satisfactory production, and Murcott also produces well on its own. There is evidence to indicate that Ambersweet's poor early production is better in a mixed plantings.

In summary, if a mandarin or mandarin hybrid requires a pollenizer variety, the grove design must include a suitable ratio and arrangement of pollenizer trees. Spacing and arrangement of both cultivars should facilitate bee movement and pollen transfer between the two cultivars and provide open spaces and water sources for the bee colonies.

PLANNING FOR THE FUTURE

The life of a citrus grove extends over many years, so the design of a new grove must incorporate the latest information, technology and materials available, and anticipate changes during the life of the grove. The new grove must be designed to provide maximum economic return averaged over the expected life of the grove.

The citrus variety should be selected based on predicted long-term or future market potential and demand. Citrus juice has become an international commodity, meaning that market potential in any one area is affected by current and future production throughout the world. Fresh fruit is also marketed both locally and globally. In many cases, the popularity of a particular variety may result in over planting of that variety and a subsequent reduction in value.

New genetic material should be considered in the design of a new planting. In particular, the incorporation of new rootstocks that provide some degree of tree size control may be advantageous. With proper combinations of variety, rootstock, and tree spacing, it should be possible to minimize future requirements for hedging and topping.

Mechanical harvesting for processing fruit is expected to become important in some citrus-producing areas. Trunk shaker and catch frame systems may be used extensively for citrus harvest. This requires a trunk that is free of branches to a height of 30 inches or more. Thus, it appears prudent to prepare for this by growing trees in the nursery headed (cut to stimulate branching) at a height of 30 inches or more.

The successful design of a new grove requires consideration of the site characteristics, new variety development, varietal growth characteristics and requirements, modern grove management practices, and global trends in citrus production and marketing. The new grove should be designed to provide maximum economic return averaged over the expected life of the grove.

REFERENCES

Futch, S. H. and L. K. Jackson. 1993. *Cross-pollination planting plans.* Citrus Industry. 74: 9.

Whitney, J. D., T. A. Wheaton, W. S. Castle, and D. P. H. Tucker. 1994. *Optimizing orange grove factors for fruit production and harvesting.* Transactions of American Society of Agricultural Engineers 37: 365-371.

10

Establishing the Irrigation System

B.J. Boman

INTRODUCTION

All citrus planting sites must be engineered to some extent, and each site will be different. Due to annual rainfall distribution patterns, almost all Florida citrus plantings have an irrigation system. If operated properly, irrigation systems provide sufficient moisture for good tree growth and production and minimize percolation of water and nutrients below the root zone of the tree. Many considerations factor into the design of an irrigation system: materials, water quality, water source, etc. Irrigation systems also can deliver fertilizers and pesticides to citrus plantings. Growers can obtain yield increases and minimize leaching losses (and pollution) by injecting nutrients and other chemicals through the irrigation systems. The Environmental Protection Agency (EPA) requires that all pesticide products be labeled to clearly state whether injection into irrigation systems is permitted.

Most irrigation in Florida today is carried out through microirrigation, the control of which may be manual or automatic. A well-designed microirrigation system provides good root zone coverage at efficient application rates but also provides freeze protection at higher application rates when required. However, irrigation with a small-diameter spray pattern for one hour or more can drive water below the main root zone. By varying the emitter output, spray diameter, and irrigation duration, citrus managers adapt microirrigation systems to meet the water needs of the trees efficiently with minimal leaching and provide cold protection when needed.

Research since the 1960s has consistently shown that irrigation can significantly increase production for most commercial citrus cultivars in Florida. The research shows that soil moisture level is critical from the spring flush until the young fruit is at least one inch in diameter. The soil moisture should be maintained at no less than two-thirds field capacity during this time. Field capacity is defined as that amount of water remaining in the soil after the excess has been eliminated. Soil moisture levels during the remainder of the year can be allowed to fall to about one-quarter field capacity without adversely affecting production. Generally, irrigation can increase tree size, fruit yield, fruit size, and juice content. The source of irrigation water should be identified, and the irrigation system should be designed and installed during site preparation, prior to planting.

> IRRIGATION
>
> Irrigation significantly increases productivity.
>
> Water use is highly regulated in Florida.
>
> Various irrigation methods are used; almost all new groves are microirrigated.

PERMITS AND REGULATIONS

Five water management districts (WMDs) regulate water use in Florida. The districts issue water use permits (WUP) to allocate water for reasonable beneficial uses, including citrus production. Water is generally allocated from lower-quality sources for agricultural uses. A consumptive use permit is granted for a fixed period of time not to exceed twenty years. Permits are revocable only for false statements made regarding water use, for willful violation of permit conditions, and for nonuse of the water supply. Thus, except during times of water shortage or emergency, a grower has full rights to the water specified in the permit.

Applicants proposing a new water use are required to establish that the proposed use is a reasonable beneficial use, that it will not interfere with a presently existing legal water use, and that it is consistent with the public interest. Reasonable beneficial use is defined as the use of water in such quantity as is necessary for economic and efficient utilization for a purpose and in a manner that is both reasonable and consistent with the public interest. Applicants must present evidence to support the quantity of water requested, the need, purpose, and value of the use, and the suitability of the use to the source. The method and efficiency of use, the use of water conservation measures, and the practicality of reuse or the use of lower-quality water are also considered.

Applicants must show the proposed use will not interfere with any presently existing legal use of water. Any adverse effects from the water use, such as water quality degradation or increased flood damage, are also considered. In the case of groundwater withdrawals, extensive modeling (and sometimes testing) of the aquifers involved may be necessary.

METHODS OF IRRIGATION

A variety of irrigation methods have been used on Florida citrus, with advantages and disadvantages to each. Installation and maintenance costs, water use and efficiency, and secondary benefits, such as freeze protection, will determine the best irrigation system.

Seepage and crown flood irrigation are practiced in some Flatwoods areas in the Indian River area and in Southwest Florida. These types of irrigation can be quite efficient for appropriate soils and topography, managing irrigation and drainage together with suitable water control structures and ditches. These systems usually require large pumps to move large volumes of water. It is important not to allow the water to remain in the root zone for extended periods of time, as damage to the tree root system may occur. These systems also offer good cold protection, as growers can flood the furrows and perimeter ditches during freeze periods. However, since large volumes of water are required, the grower must be able to justify this use to the Water Management Districts, even though a high percentage of the water is returned to the shallow aquifer and retention ponds.

Permanent overhead sprinkler systems and pop-up tree sprinklers have declined in popularity over the last twenty years. Permanent overhead sprinklers are expensive and inefficient in high winds. Pop-up tree sprinklers may be damaged by grove equipment, and their riser movement may be impaired by sand contamination. High-volume under-tree sprinklers continue to be used on limited acreage in Florida, but low-hanging branches interfere with the water distribution pattern of these under-tree systems.

Most of the newer groves have microirrigation systems installed. These systems are less expensive to install and generally require less water than other irrigation systems. They operate at lower pressures, reduce energy consumption, and are easily automated. Since lower volumes of water are required, these types of systems can utilize less productive water sources. If operated properly, these systems can minimize water loss due to evaporation and deep percolation below the root zone of the tree. In addition, the spray-jet systems offer some degree of cold protection during radiation-type freezes. However, microirrigation systems require frequent maintenance. Algae, ants or other insects, and minerals can clog emitters.

MICROIRRIGATION:
is less expensive to install and run,
is an efficient use of water,
can be used for cold protection,
requires frequent maintenance.

IRRIGATION EQUIPMENT

Pump Units

If freeze protection is desired the pumping unit must have enough capacity to irrigate all zones during the most extreme conditions. In zoned systems, all zones should be about the same size and have about the same pressure requirements, because an irrigation pump operates most efficiently at a single flow rate and pressure. The system design should meet all local and state standards, and should be approved by a professional engineer.

Centrifugal pumps are the most economical option for water supplies at the surface or in wells where the water level is above 20 feet. Turbine pumps must be used to pump water from depths greater than 20 feet. Turbine pumps have an advantage over conventional centrifugal pumps in that they do not require priming for the pump to operate. For large systems, deep-well turbines with power units on the surface are commonly used. For smaller systems, submersible turbines are a less expensive option. When submersible turbines are used, electric motors are connected directly to the pumps and lowered into the well.

For automatic operation, electric motors are recommended as power units for microirrigation systems. They have lower initial costs than internal combustion engines, especially for smaller sizes. Local power company policies will determine actual costs, and these policies should be investigated before a decision is made.

Internal combustion engines are recommended when irrigation systems will be used for cold protection because of the possibility of electric power interruption and loss of pumping capability on cold nights. Diesel power units are the most efficient and the most popular choice for citrus irrigation in Florida.

The Pump Station

The pump station should be protected from the elements, as rain can cause moisture in motor windings and direct sun can significantly raise motor operating temperatures. Safety issues must also be considered in the design of the pump station, since high-voltage power lines, fuel tanks, and various toxic materials (chlorine, pesticides) may be present. Generally, a sturdy shed that can be locked should be constructed with a concrete floor and concrete mounting for pumps and power units (motors or engines). For diesel engine installations, the mounting should be insulated from the floor to minimize vibration.

Filters

Filters to remove small particles are nearly always required to prevent clogging of microirrigation systems. The type of filtration system required depends on the type of emitter used and the source, volume, and quality of the irrigation water; filters should be selected based on emitter manufacturer's recommendations.

Organic matter may be a problem when water is pumped from a surface source such as a pond or canal. The intake should be positioned below the water surface to avoid floating debris, and above the bottom to avoid sediment. A strainer should be used on the pump intake to exclude as much organic matter as possible. A media (sand) filter should be used as the primary filter.

Screen filters are normally adequate when water is pumped from a well, unless large amounts of sand are being pumped. Hydrocyclone sand separators or settling basins may be used to remove large amounts of sand, but settling basins can create problems if organic matter such as algae is present in the basins.

Valves

Valves are required to control the filling of irrigation systems at pump startup to control flows to the desired subunits of a system and to allow flushing of irrigation pipes. Only properly pressure rated irrigation valves should be used to avoid failures due to system pressure and water hammer problems. Valve materials and components must be resistant to corrosion by the irrigation water and by any chemicals injected during irrigation. Automatically controlled irrigation systems will require the use of automatic valves. These may be controlled by electric solenoids or hydraulic pressure, depending upon the type of timer or controller used.

Backflow Prevention

Florida law requires the installation of a backflow prevention system on most irrigation systems. Backflow prevention systems are always required when chemicals are injected into an irrigation system.

When chemical toxicity Category I pesticides are injected into irrigation systems, a double check valve, low pressure drain, and vacuum relief valve assembly are required. These pesticides are marked with the keywords "Danger" or "Poison" on the label. The Environmental Protection Agency (EPA) requires that all pesticide products be labeled to clearly state whether injection into irrigation systems is permitted. Pesticide labels must also list the backflow prevention equipment requirements and application instructions.

Chemical Injection Equipment

Pressurized irrigation systems are often used to apply chemicals, especially fertilizers. Growers can obtain yield increases and minimize leaching losses (and pollution) by injecting nutrients and other chemicals through the irrigation systems. Many growers currently inject fertilizer through sprinkler and microirrigation systems. Chemical injection equipment is required to add the correct amount and rate of chemical. Several types of chemical injectors are commercially available, ranging from the low-cost Venturi devices to high-precision, high-cost positive displacement pumps. If a high degree of precision is required (e.g., for pesticides), more precise injection methods must be used. Microirrigation systems may also require high-precision chemical injection pumps to precisely control biocides and water amendments that are used to prevent emitter plugging.

CHEMICAL INJECTOR EQUIPMENT

Injecting nutrients in an irrigation system is called fertigation.

Fertigation can increase yields and minimize leaching, especially in young trees.

Venturi injectors are cheaper but less accurate than positive displacement pumps.

High accuracy is important for injecting pesticides and cleaning agents in microirrigation systems.

Irrigation Controllers

Irrigation controllers are devices that automatically turn the irrigation system and associated equipment, such as chemical injection pumps, on and off. Controllers are not mandatory for system operation, but they are time- and labor-saving devices. They are especially economical and efficient for management of microirrigation systems on Florida's sandy soils because of the need for frequent irrigation. Controllers range in capabilities from simple timers, which can turn a single valve on and off at preset times, to computer controllers. Computer controllers are very complex systems that are programmable and have microcomputer capabilities. They can collect data from sensors, make calculations, and adjust water and chemical application schedules in response to plant needs.

Emitters

The choice of emitter always represents a compromise. A microsprinkler should provide good root zone coverage at a low application rate to minimize leaching, while adequate freeze protection may require a higher application rate. To select an emitter, the grower should consider the cost, uniformity of wetting, wind effects on spray pattern, system constraints, and maintenance of the emitter, as well as the particular characteristics of the site. Emitters with distinct spoke patterns may not be as efficient as those with more uniform patterns, especially in very sandy soils or where root systems are shallow. Although the system should be installed before planting, the grower must anticipate the water requirements of the trees at maturity. Higher density plantings with smaller trees will require less water per tree than more widely spaced trees. Larger wetting patterns may be desirable for more widely spaced trees, and various microsprinkler emitters provide different wetting patterns (Figure 10.1).

The water requirements of the trees should be met with reasonable operating times and without percolation of nutrients and water below the root zone. Irrigation with a small-diameter spray pattern for one hour or more can drive water below the main root zone. A high application rate can also lead to leaching of nutrients and pesticides. However, irrigation at a higher application rate is needed to provide cold protection. Approximately 2000 gallons per acre per hour is the minimum required for reasonable cold protection. Even at higher application rates, increasing emitter output and decreasing the spray pattern size from a 360° spray to a 180° or 90° spray can usually increase freeze protection capability. By varying the emitter output, spray diameter, and carefully regulating duration of irrigation, the grower can usually manage microirrigation systems to efficiently meet the water needs of the trees with minimal leaching, while also providing cold protection when needed.

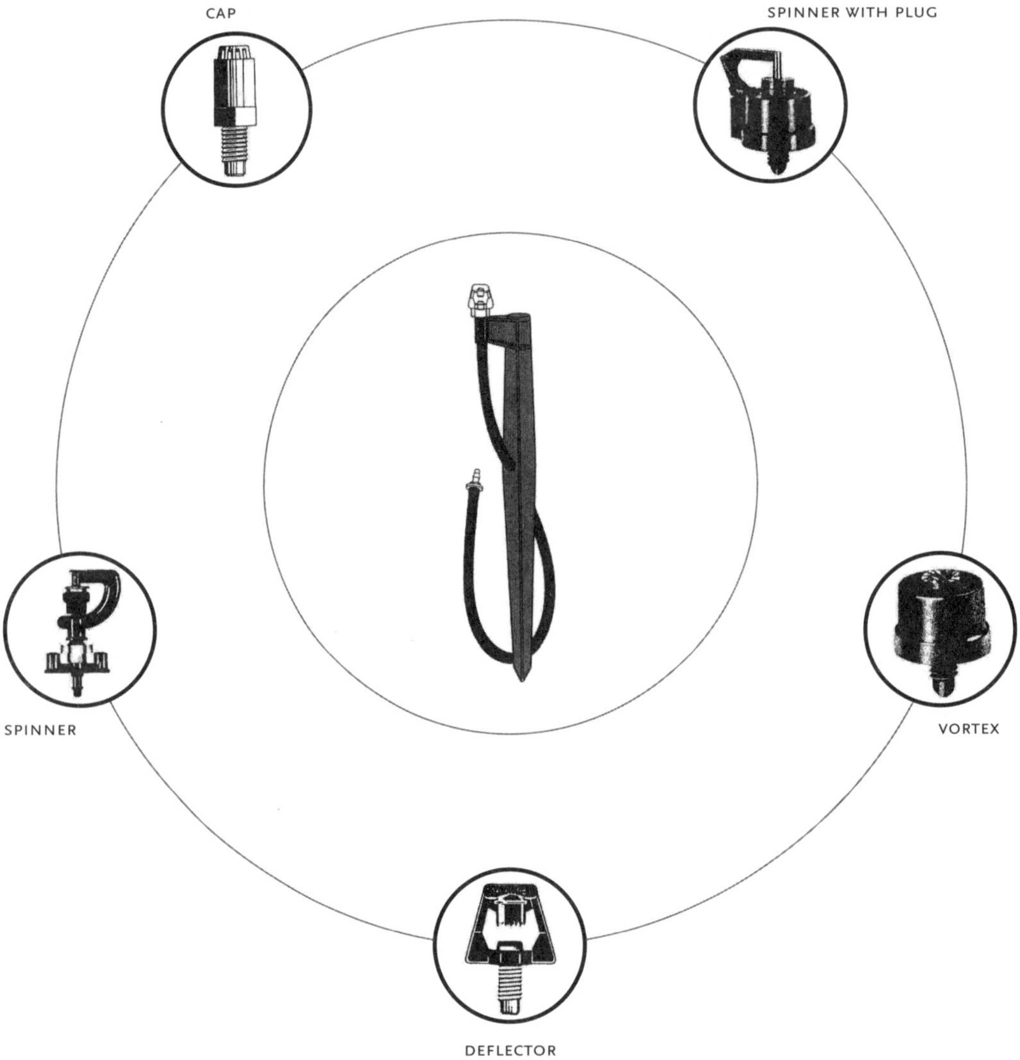

Figure 10.1 A variety of microsprinkler emitters are used in citrus irrigation and are fitted to the microjet assembly. Each emitter type controls a different rate of water flow, and various attachments permit the proper positioning of the emitters in the ground or tree canopy.

REFERENCES

Boman, B. J. 1994. *Evapotranspiration by young Florida flatwoods citrus trees.* Journal of Irrigation and Drainage Engineering 120: 80-88.

Boman, B. J. 1997. *Soil conditioner applications to salinity-stressed citrus.* Proceedings of the Soil and Crop Science Society of Florida 57: 7-11.

Fares, A., A. K. Alva, P. Nkedi-Kizza, and M. A. Elrashidi. 2000. *Estimation of soil hydraulic properties of a sandy soil using capacitance probes and guelph permeameter.* Soil Science. 165: 768-777.

Fares, A., L. R. Parsons, T. A. Wheaton, K. Morgan, and J. Šimunek. 2001. *Simulated drip irrigation with different soil types.* Proceedings of the Florida State Horticultural Society. 114: 22-24.

Ferguson, J. J., and G. D. Israel. 1999. *Citrus irrigation methods and water sources: A 1996 survey.* Soil and Crop Science Society of Florida Proceedings. 58: 103-106.

Obreza, T. 2000. *How uniform is your microirrigation system?* Citrus and Vegetable Magazine. 64: 14.

II

Grove Planting and Early Care

D.P.H. Tucker

INTRODUCTION

As discussed in previous chapters, there are a multitude of considerations that must go into the design, engineering, and planting of new citrus groves. All of these activities lead up to preparing the site for the actual planting of young citrus trees. One- to two-year-old, grafted, nursery-grown trees eventually will be planted at the new site. Generally, there are two options growers have in selecting their trees: bare root and container grown. The trunk caliper is generally greater in bare-root, field-grown trees, and bare-root trees have been found to grow more quickly than container-grown trees in some trials. Bare-root trees are supplied for planting with all soil removed from the root system. It is important to plant soon after digging bare-root trees, and to keep roots wet until the trees are safely in the ground. Further, bare-root nursery trees need to be pruned back at planting so that the canopy and root system remain in balance after the loss of roots during digging. Container-grown trees, often grown in screenhouse structures, are smaller but may be easier to transport and plant. Young citrus trees can be successfully planted in the grove anytime of the year but there are a number of safety factors that must be considered when working within and around the new trees. For example, the application rates of some herbicides need to be adjusted; rates that are suitable for older, bearing trees may damage young trees with tender trunks and shallow root systems. This chapter covers the essentials of selecting, planting, and taking care of young citrus trees.

A newly planted grove will not bear fruit for several years and requires intensive management during those years (Plate 11.1). Young trees have an especially high demand for water and nutrients and are more vulnerable to freeze injury, diseases, and insects (e.g., citrus leaf miners). It may be tempting to minimize financial outlay during the nonproductive years, but appropriate and timely investments in a new grove are likely to be returned during the economic life of that grove.

TREE SELECTION

After the site is prepared, one- to two-year-old, grafted, nursery-grown trees are planted at the grove site. Selection of well-formed (i.e., straight), pest- and disease-free registered trees is critical for proper grove development. Container trees can be produced more quickly than field-grown trees, which can be an important advantage when they are custom budded. In addition, container-grown trees have been purported to suffer less root damage in the planting process. However, the trunk caliper is generally greater in bare-root, field-grown trees, and bare-root trees have been found to grow more quickly in some trials. The relative merits of these two types of nursery stock are still debated by growers.

PLANTING

The bare roots of field grown nursery trees must be kept wet from the time of digging until the trees are planted in order to avoid desiccation and consequent poor tree survival. Container trees are generally grown in greenhouses in small plastic pots. The pots or plastic bags should be removed immediately prior to planting and the outer soil loosened from the roots. Roots that are pot-bound should be pruned, but proper aged container trees should not have pot-bound roots (Plate 11.2). Trees with irregular root systems should not be planted, since they often grow more slowly and have substandard productivity. After being readied for planting, trees should be placed in holes, which are slightly larger in diameter than the original root biomass. Several gallons of water should be added as the soil is replaced, and the trees should then be "mudded in," meaning slurry of soil and water is worked around the roots. Trees must be supplied with water frequently during the first few weeks to ensure survival.

PRUNING

Bare-root nursery trees need to be pruned back at planting so that the canopy and root system remain in balance after the loss of roots during digging. Generally, removal of some of the canopy is appropriate. If the intent is to mechanically harvest the grove, the requirements of the harvesting system should be considered when pruning. For example, harvesting with trunk shakers often requires that the trunk be free of branches up to 30 inches above the soil. Rootstock suckers and shoots below the desired branch height should be pruned away regularly to avoid loss of resources to unwanted growth. Trunk wraps or sprout inhibitors can reduce the need for hand or mechanical sprout removal. Minor selective pruning may be beneficial during the first several years to help develop tree form. Vigorous branches that extend beyond the rest of the canopy may be cut off, but there is no evidence that this is advantageous.

TIME OF PLANTING

Young citrus trees can be successfully planted in the grove at any time of the year. Optimum first year growth is obtained, however, when trees are planted early in the spring or at least before the summer rains. In regions where there is a potential for freeze damage, planting should be delayed until the danger of frost has passed. In regions with no risk of cold damage, planting in the winter, when trees are dormant, is recommended.

IRRIGATION

The primary objective during the first years in the grove is the rapid development of tree canopy. Substantial fruit production is not possible until the canopy is well developed. Fruit production and fruit quality become considerations only after trees have reached sufficient size to support a substantial crop. Irrigation and nutrition are critical factors in the growth of young trees. (This is discussed in more detail under the respective chapters on these subjects.) Vegetative growth is particularly susceptible to water stress, so it is important to maintain adequate soil moisture at all times to ensure maximum tree growth. Irrigation needs will depend on a number of factors, including rainfall, temperature, the size of the tree, and soil and irrigation system characteristics. For example, during the cool days of winter in Florida, irrigation may be required only every ten days to two weeks. On the other hand, irrigation may be required daily during May to maintain optimal soil moisture for young tree growth.

NUTRITION

Young trees need to be supplied with fertilizer on a regular basis, and this is best achieved with numerous applications in small doses. The addition of fertilizer at planting provides little benefit and is likely to burn sensitive roots. The first fertilizer application should be delayed until soil is well settled around the roots, typically a few weeks after planting. Fertilizing between November and February in areas where there is a danger of freeze injury is not advisable. Fertilizer may be applied as a dry material (four to six times per year) or as an injection into the irrigation system, a process known as fertigation. Fertigation is a common practice in Florida. Some growers inject small amounts of fertilizer every time they irrigate, while others inject on a weekly or monthly basis. Initially, fertilizer is applied on a per-tree basis, meaning that more fertilizer is used per acre in higher-density plantings. As trees get older and begin to fill their allocated space, per-acre fertilizer requirements become more similar over the range of tree spacings and should be based on fruit production and leaf analysis. At that point, hedging and topping become important grove management practices.

PEST CONTROL

Monitoring for insect pests and diseases is critical in newly planted trees. In an established planting, the grower may choose to tolerate some damage to new growth in order to encourage the establishment of biological control agents. However, young trees are especially vulnerable to pest damage, and relatively low pest levels warrant pesticide application in a young planting. Fire ants feeding on trunks of young trees can kill or greatly slow their development. Regular inspection for ants, especially under trunk wraps, is very important. Canopy-invading pests such as aphids, leaf miners, orange-dog, caterpillars, grasshoppers, and katydids can also significantly damage young trees and slow canopy development. Chemical control for leaf miners is seldom needed for the spring flush. However, it may be helpful to treat young trees during the summer and fall flushes.

TRUNK PROTECTION

During the first years in the field, wraps may be placed around tree trunks. These are typically either insulating water-jacketed Styrofoam wraps or simple corrugated plastic wraps. The wraps serve several functions. Insulating wraps are often used in colder locations to provide some freeze protection. All types of wraps provide protection from herbicide applications and mechanical or rodent damage. Wraps also inhibit sprout formation, helping to maintain a trunk without branches up to the heading or branching height developed in the nursery. However, wraps may become detrimental if they harbor insects. They also retain moisture around the trunk, potentially increasing the risk of disease. In some cases, trunk sprays or brushing with sprout inhibitors may be viable alternatives. Some growers use sprout inhibitors as a spot treatment where sprouts are recurring.

CROPPING

Some trees produce a small amount of fruit during their first two years in the grove. However, the fruit quality of very young trees is usually inferior. It tends to be larger than fruit borne on mature trees, have poor external color, a coarser peel texture, and lower juice-soluble solids. A very heavy crop may compete for resources and inhibit tree growth; in such cases, the fruit may be removed from the young trees. In general, however, fruit removal from young trees prior to commercial production is seldom economically feasible. In addition, there may be significant tax advantages to producing early crops in order to begin depreciation of planting costs.

WEED MANAGEMENT

Weed management is especially important in newly established groves to reduce weed competition and enhance young tree growth. The application rates of some herbicides will need to be adjusted; rates that are suitable for older, bearing trees may damage young trees with tender trunks and shallow root systems.

RESETTING TREES

For a variety of reasons, a grower may decide to reset young trees even in a young grove if trees are stunted (Plate 11.3). The needs of bearing trees differ from those of developing trees, so special care must be taken to meet the needs of the young reset trees. In general, reset trees in an older planting should receive care appropriate to a newly planted grove. This ultimately results in enhanced tree development and increased productivity.

RESETTING OPTIONS

The average annual tree loss in the Florida citrus industry is about 3 percent to 4 percent. Such losses result from a number of factors including diseases and disorders such as CTV, blight and footrot, and insects such as root weevils. Localized drainage conditions in wet years and major freeze events also contribute to tree mortality. The extent of losses among individual groves can vary from 0 percent to 10 percent or more, thereby having substantially different impacts on grove revenues and equity. The cost of tree removal, resetting, and other expenses associated with mixed-aged groves can add 10 percent to 13 percent to the annual expenditures for a bearing grove. It is therefore important to evaluate the sustainability of regular resetting versus replanting after grove productivity declines to some unprofitable level. The question then becomes at what intervals of tree loss should replacement take place: every year, every three years or not until the block has become unproductive or nonprofitable. A number of scenarios were compared with a hypothetical zero tree-loss situation, and the analysis indicated that resetting the block would result in positive returns over no resetting. If a block is experiencing a rapid decline in yield and returns, removal, and replanting of the entire block should be considered.

As a citrus grove is a highly capitalized, long-term investment, evaluating the economic feasibility of the replacement is an integral part of the management decision-making process. Grove managers should therefore evaluate the grove site, soil type, scion/rootstock combination along with the market outlook for the variety in question before initiating an aggressive removal and replanting program.

REFERENCES

Jackson, L. K., J. M. Bulger, and R. M. Davis. 1985. *Planting and care of young citrus trees.* Citrus Industry. 66: 19, 21-23, 26.

Jackson, L. K., and D. P. H. Tucker. 1988. *Citrus tree planting and establishment practices.* University of Florida Press, Gainesville, pp. 1-14.

Zekri, M. 2000. *Evaluation of orange trees budded on several rootstocks and planted at high density on flatwoods soil.* Proceedings of the Florida State Horticultural Society. 113: 119-123.

12

Introduction to Citrus Grove Management

D.P.H. Tucker

INTRODUCTION

The ultimate profitability of a grove depends on the yield and market value of the fruit, offset by site acquisition and preparation costs, labor and equipment costs, tree purchase costs, agrichemical application costs, and grove management costs. For many years, Florida citrus production was highly profitable, and conventional high-input grove management practices received little scrutiny. However, current conditions have forced citrus growers to re-evaluate all factors affecting the efficiency of their citrus operations. Such factors include integrated management, crop quality management, exotic diseases, tree health, environmental concerns, operational factors, and more. This chapter provides an overview of these factors and serves as an introduction to the remainder of the book, which will cover each of these topics in greater detail.

The first phase of establishing a commercial citrus operation is the acquisition and preparation of a suitable site and the selection of one or more variety/rootstock combinations. These initial decisions are extremely important, and the grower generally considers a wide array of options.

Once the grove is designed and planted, a management program is implemented. Routine management of groves includes nutrient and water management, grove floor vegetation management, plant and crop protection (pest and disease management), tree size management, and freeze protection. Each of these aspects of grove management is addressed elsewhere in this publication, and in greater detail in other publications.

INTEGRATED MANAGEMENT

The ultimate profitability of a grove depends on the yield and market value of the fruit, offset by costs of site acquisition and preparation, labor and equipment, tree purchase, agrichemical application, and grove management. For many years, Florida citrus production was highly profitable, and conventional high-input grove management practices received little scrutiny. But as the profitability of citrus production declined, growers sought methods to minimize the costs of grove management and crop inputs. In particular, the relationship between various crop inputs and crop management practices began to be studied extensively in an effort to learn how to maintain sustainable levels of production with minimal cultivation and agrichemical inputs. Understanding and managing these relationships is central to the concept of ICM, and ultimately will lead to more efficient and sustainable production.

For example, excessive fertilization is an inherently wasteful grove management practice. Worse, it may contribute to increased pest and disease pressure, and stimulate the growth of weeds, further increasing grove management costs. In addition, excessive fertilization increases the possibility of pollution of surface or groundwater and possibly undermining the public and political position of the industry. Such a chain reaction is possible in the context of inappropriate grove management practices. Through ICM, the citrus manager can maximize the efficiency of grove management practices and minimize unintended and undesirable consequences. This concept is emphasized throughout this book, as specific grove management practices are discussed and recommendations presented.

Appropriate grove management depends on a number of variables, especially the stages of tree growth and development (see Table 12.1). Other significant factors may include the intended market, production equipment and technology, variety and rootstock, soil type, and local climatic conditions. Management strategies and the level of management inputs are inevitably controlled by the value of the fruit.

Table 12.1 Stages of Tree Growth and Development.

Stage of Tree Development	Management Goals and Input Levels
Nonbearing	Rapid vegetative growth for development of canopy fruit bearing surface. High resource inputs per small canopy volume, protection of the tree canopy and root system.
Young, bearing	Continued high growth rate with fruit production. Continued high resource inputs per canopy volume. Improved internal and external fruit quality for processing and fresh outlets, respectively.
Mature, bearing	Reduced growth rate as trees reach containment size. Reduced or stabilized inputs for enhanced fruit quality with maintenance of fruit production and moderate tree vigor.

The following list outlines the various stages in citrus production management that are involved in maximizing returns from processed fruit, referred to as Pounds Solids Management. This outline can be adapted for fresh fruit production through modifications in variety/rootstock selection, cultural and pest management practices, harvesting and handling procedures, and consideration of fresh fruit traits influencing the price of fruit. The important steps in the production cycle are to maximize returns from fruit for the juice market, a process referred to as "pounds solids management." Each step is discussed in the context of processing considerations.

Evaluate existing research-based information on production, exclusion/sanitation practices, harvesting, utilization and marketing.

Select suitable planting site compatible with selected scion/rootstock and other factors.

Acquire quality trees propagated from horticulturally-tested, pathogen-indexed numbered selections with high solids/acre, color and flavor potential and appropriate harvesting dates.

Use correct planting practices and tree spacing/ architecture compatible with scion/rootstock vigor, soil type and harvesting systems.

Follow correct sampling, monitoring and analytical procedures and evaluate information.

Adopt and monitor recommended, cost-effective and environmentally-compatible production practices specific for variety and market.

Optimize harvest period for fruit maturity and specific market based on fruit testing and/or customer specifications.

Prepare grove trees for manual/mechanical harvesting, contract with harvesting crews and coordinate

Minimize transportation costs and in-transit damage.

Deliver sound, debris-free fruit of acceptable ratio with high solids and good juice color and flavor.

Maximum opportunity for best returns per load of fruit and optimum profitability.

MAXIMIZING RETURNS

Citrus production has evolved from small-scale agriculture into sophisticated, well-managed, cost-driven operations. Recent industry discussions indicate that in the 1990s, the industry's focus shifted from short-term to long-term profits. This section outlines some of the considerations involved in achieving the latter with respect to fruit grown for processing.

The concept of Integrated Pounds Solids Management (IPSM) is considered throughout this publication, and is essential to analyzing methods for achieving maximum profit. IPSM is based on the premise that efficiency is increased and value is preserved when all of the stages in a citrus production system are closely linked. This linkage enhances long-term grower returns by optimizing processing value through quality of the fruit produced. Traditionally, fruit production, harvesting and fruit processing in Florida have been separated from one another, so linking provides for an enhanced understanding of the entire system.

Prior to the planning and establishment of groves, existing information relevant to the individual situation must be evaluated. Numerous sources are available, including the University of Florida's Institute of Food and Agricultural Sciences, the Cooperative Extension Service, USDA, consultants, product vendors, and information from other citrus-producing regions. It may also be appropriate to establish contacts with fruit procurement managers at processing plants to determine fruit specifications and to learn what constitutes desirable processing fruit. Evaluation of consumer trends in fruit, beverage, and juice consumption may also be important for long-term planning. This information can be obtained from trade and industry journals and private organizations and associations (such as Florida Citrus Mutual and the Florida Citrus Processors Association) as well as from governmental agencies such as the Florida Department of Citrus, the USDA, and the Florida Department of Agriculture and Consumer Services.

Next, a suitable planting site must be selected and prepared. Data exist regarding the individual characteristics of the four major citrus growing areas in Florida: the Ridge, the Indian River area, the Peace River, and the Gulf Coast area. Location will dictate much of the site preparation effort.

While keeping other planting site constraints in mind, it is wise to at least consider the logistics involved in moving fruit from a proposed site to the processing plant(s) that will most likely utilize the fruit. For both quality and economic reasons, fruit transportation distances should be limited. Although citrus fruit is structurally tough compared to other produce, the Florida climate and the large trailers used for hauling fruit can cause significant yield and quality loss in processing fruit.

Virus-free registered trees should be sourced from reputable citrus nurseries, which should provide certification regarding pests and diseases. Current recommendations regarding nursery practices should be in place. It is important to obtain healthy nursery trees with proper scion/rootstock identification and accurately account for where these trees are planted. When they come into production and are expected to provide a financial return, this tracking may be useful. Marketing rules often require accurate identification also.

Young trees should be planted correctly with due consideration for tree spacing and grove layout. While only about 5,000 acres of juice oranges were mechanically harvested during the 1998-1999 season (about 1 percent of the total juice orange acreage), future labor and other cost issues suggest that much of the fruit for processing will be mechanically harvested in the future. Preliminary evidence indicates a quality impetus, since some processors have reported that mechanically harvested fruit is cleaner than traditionally harvested fruit. Some mechanical harvesting systems require specific tree spacing and training for effective harvesting, while others have specific row dimension requirements. In all cases, these issues need to be addressed prior to tree purchase and actual planting.

Soil and plant analyses of plant materials and soil allow effective and early intervention in the case of nutrition or pest problems. Appropriate monitoring of fruit quality parameters provides important data over time that a grower can use to more accurately predict overall returns.

Fruit for processing generally requires less input than fruit destined for the fresh market. The priority should be on internal quality. Excessive nutrient and pesticide treatments are wasteful and unnecessary. In some cases, especially for processors emphasizing by-product sales (such as essential oils) and for processors selling to some international markets, residues may be forbidden. It is important to understand the effect of all production practices on fruit quality at harvest before embarking on any particular management program.

While pounds solids are important to the grower, ratio and color are also key factors for the processor, once the fruit is above a certain minimum maturity. Vulnerability to freezes may impact the price offered to the grower, with discounts applied to fruit from vulnerable areas and premiums for more protected fruit.

The location of fruit production is a factor for the entire industry and offers enormous opportunity for cost savings and improved efficiencies. Processing plants depend on a steady supply of fruit, and difficulties in obtaining contracted fruit at the appropriate time lead to losses for both the plant and the grower.

Together with harvesting, the transportation of fruit is a critical issue. Harvesting and hauling are highly labor-intensive and involve considerable liability and expense. Transportation issues affecting fruit prior to receipt at the processing plant have a direct impact on the fruit's profitability to the grower. Damage to fruit during transport, juice loss and leakage, and poor handling can result in the delivery of lower total pounds solids.

To obtain optimum returns from fruit, sound fruit of acceptable ratio and high juice solids, good color and flavor must be delivered to the processor. State test extractors must be properly set up and operated in order for the grower to receive fair compensation. The procedure for the handling of culls (fruit not acceptable to the processor and for which the grower is not paid) should be defined and agreed upon by both the processor and the producer prior to fruit harvesting. Ideally, fruit exceeding processors' quality standards, or delivery of fruit that has a specific attribute that is highly regarded or urgently needed should result in premium spot payments to the grower. These performance standards can be written into contracts and can provide better returns to the grower by incorporating efficient pounds solids management into the long-term production plan.

A partnership between producers, packers, and processors integrates many aspects of citrus production management discussed above. Such a partnership has been promoted by the industry and is supported by the consolidations underway in various sectors of the citrus industry. Potentially, it allows for greater integration, greater efficiency, better quality, and shared risk with the various parties improving performance and profitability.

MANAGEMENT OVERVIEW

The Management Timing Guide (Figure 12.1a) illustrates stages of vegetative development, pest and disease pressures along with monthly averages of temperature and rainfall. The calendar is provided only as a reference since these ICM management decisions should be based on the environmentally controlled timing of tree growth, pests, and diseases. Figure 12.1b is the second part of the Management Timing Guide that illustrates stages of flowering and fruit development and their relationship to timing of ICM decisions in citrus production and fruit marketing. Again, the calendar and monthly averages of temperature and rainfall are provided only as a reference since fruit growth and maturity characteristics begin with the timing of flowering and fruit set. Specific management decisions should be based on crop development, specific varietal requirements, variations in seasonal environmental conditions, market outlet, and economic considerations.

VEGETATIVE GROWTH, PESTS, AND DISEASES

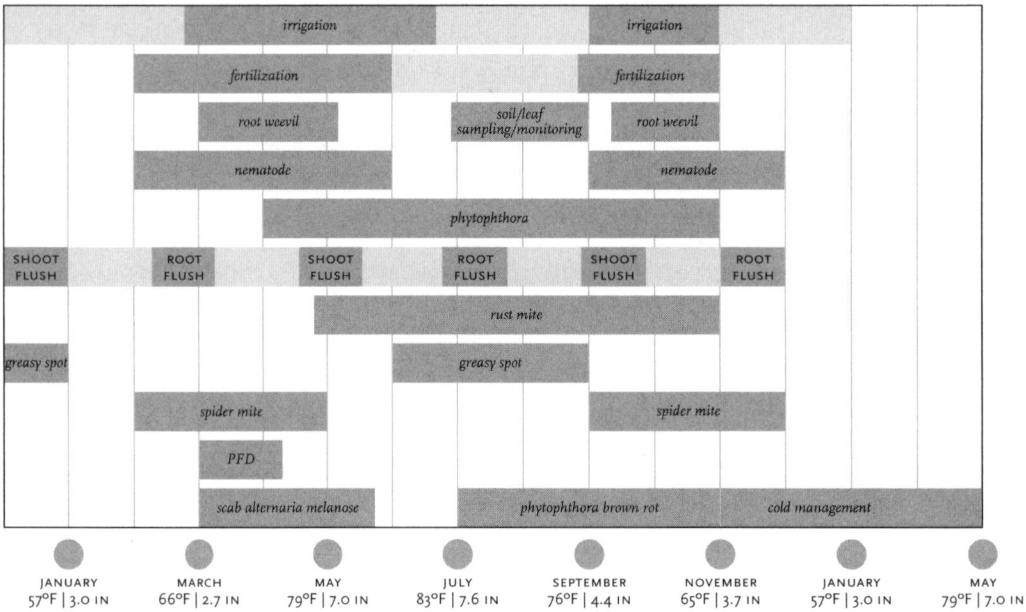

Figure 12.1a Approximate annual timing of vegetative development, pest and disease pressures for making ICM decisions about production practices, pest and disease management. A monthly calendar and averages of temperature and rainfall are provided as a reference. (PFD = post bloom fruit drop.)

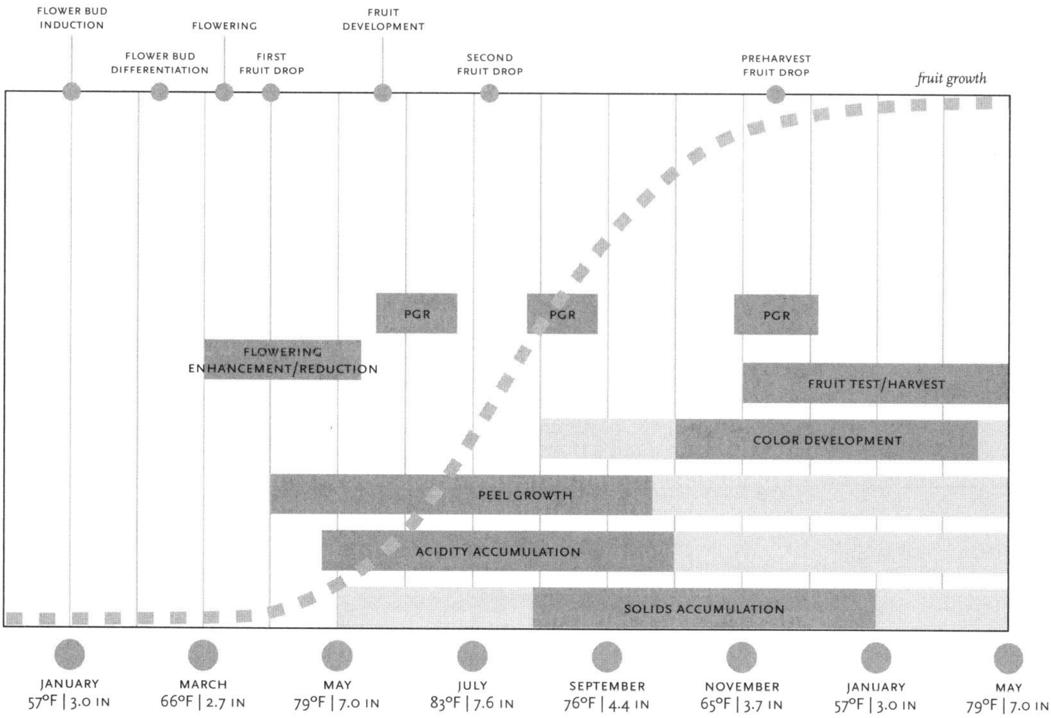

Figure 12.1b Approximate annual timing of flowering, fruit growth and maturation characteristics and potential timing for applications of plant growth regulators (PGR). A monthly calendar and averages of temperature and rainfall are provided as a reference.

13

Tree Growth, Development, and Productivity

J.P. Syvertsen and L.G. Albrigo

INTRODUCTION

Once a citrus manager has decided on a general strategy for the marketing of a crop, the question of how to maintain healthy trees must be addressed. Healthy citrus trees are more likely to provide higher returns in the long term. Maintenance of citrus tree health and vigor requires a basic understanding of tree requirements and seasonal growth cycles along with the impacts of pests, diseases, nutrition, irrigation practices, and environmental factors on tree and fruit growth. For example, poor soil water drainage and cold temperatures are two very important environmental factors that can limit the successful growth of citrus.

Best evaluation and management of these factors occurs when they are based on the changing condition of the trees and crop throughout the season. Basically, the seasonal growth cycle of a citrus tree can be discussed in terms of root development, leaf and flower development, pollination, and fruit development and maturation. The first leaf growth flush appears on Florida citrus during February and March each year. From this time forward, the citrus manager should continually evaluate tree condition and crop development and combine this information with market information to optimize production practices.

BASIC TREE REQUIREMENTS

About 70 percent of a living citrus tree is water. Water and dissolved nutrients enter the tree system from the soil through the roots. The soil should have good water-holding capacity or be provided with irrigation as needed, and also should be well drained to allow air to reach roots below ground. Roots require oxygen for respiration, which must diffuse through the spaces between soil particles. Roots accumulate water and nutrients at concentrations above those in the soil. These are then transported to the canopy. Soil water drainage and cold temperatures are the two most important factors that can limit the successful growth of citrus throughout the world.

If water were removed from a tree, the remaining dry weight would be over 50 percent carbon, about 5 percent calcium, and about 2 percent nitrogen, along with other mineral nutrients. Carbon (C) gets into the tree system through the process of photosynthesis, which takes C from carbon dioxide (CO_2) in the air into stomatal pores in leaves, fixing the C into sugars that ultimately are used to fuel tree functions and to make carbohydrates used to construct roots, leaves, and fruit. Photosynthesis requires light energy from the sun and occurs in all green tissues, especially in the leaf chloroplasts. Sugars are translocated to fruit and most accumulate in the juice. Sugars are also transformed within the tree system to become structural carbohydrates, such as cellulose and lignin.

Integrated crop management (ICM) strives to supply adequate amounts of water, nutrients, and light to the tree. Trees must be planted and pruned in a way that allows adequate absorption of light to permit economically optimal cropping. ICM must also control weeds that compete with citrus trees for resources. Pests and diseases must be controlled as well, since they also reduce the resources available for growth and fruit production.

> Maintenance of citrus tree health and vigor requires a basic understanding of the seasonal growth cycle, the specific requirements of each variety, and the impact of pests, disease, and cultural/environmental factors.
>
> Soil drainage and cold temperatures are the two most important factors that can limit citrus growth in many areas.

TREE GROWTH AND DEVELOPMENT

The seasonal growth cycle of a citrus tree can be discussed in terms of leaf and flower development, pollination, and fruit development and maturation.

LEAVES

An important leaf growth flush appears on Florida citrus during February and March each year. Other flushes (usually two flushes) occur during the warm rainy summer periods followed by a weak flush in the fall. In Florida, the summer flushes usually produce about twice the leaf area of the spring flush, whereas fall flushes seldom produce more than 1 percent to 2 percent of the total leaf area on a mature tree. Sporadic winter flushes can occur in the southern growing regions of Florida, particularly during warm, wet winters. Typically, little growth occurs during the shorter days of winter.

Individual leaves remain on the tree for one to two years, with a few leaves surviving three years. Leaf drop is usually greatest during the spring (February and March) at the time of the new flush of leaves and flowers. Factors such as freezing temperatures, desiccating wind, low soil moisture, nutrient deficiencies, high soil salinity, foliar diseases (e.g., greasy spot), and mites can influence the time and intensity of leaf drop.

> Main leaf growth flushes appear on Florida citrus during February/March and July/August.
>
> Other, less prolific flushes can occur during the fall, and even winter.

FLOWERS

Flower induction (or flowering) refers to the process by which some vegetative buds begin to develop into reproductive buds that then produce flowers. Flowering in citrus can be repetitive under tropical conditions of dry-wet cycles. Citrus is relatively insensitive to day length, and induction of flowering occurs during quiescent periods when conditions are not conducive to growth. Initiation of differentiation of flower buds and flowering then occurs during a period of more favorable conditions for growth. Nongrowth periods are generally associated with drought stress or cool winter temperatures. Cool winter temperatures in Florida and California provide the needed induction and can result in a single major bloom event each year particularly in cooler winters as in California. Cool, nonfreezing winter temperatures that stop growth and that are followed by a warm, wet spring usually produce the most intense and uniform bloom. Under Florida conditions, several minor cool induction periods can occur during the winter with interspersed warm periods. This frequently leads to multiple blooms or a protracted bloom during the February-to-April bloom period. In addition, a poor winter induction in Florida, disease outbreaks or a prolonged spring dry period without irrigation or rainfall, can trigger an irregular or off-bloom (June bloom). Under more tropical conditions, there may be several drought induction periods resulting in flowering more than once a year. In such a situation, two or more crops commonly exist on the tree at one time.

Flowers are borne on new flush called inflorescences, which may bear one or more flowers. They generally form from buds in leaf axils on wood that is one year old or less, either as a mixed sprout (leafy bloom) or without leaves (leafless or generative bloom). A bouquet bloom can occur if the bloom is mostly leafless and well synchronized. The proportion of leafy to leafless bloom varies from year to year and by variety. Although leafless bloom is often more abundant in cooler subtropical climates, a higher percentage of fruit set occurs from leafy bloom, which is more prevalent under Florida conditions. Young trees and trees with excessive vigor frequently do not flower well. Excessive leaf drop due to diseases or freeze damage may also reduce flowering. Small variations in flowering and fruit set may translate into large variations in subsequent yield. There is evidence that excessive flowering may reduce sequential cropping in some citrus cultivars; this can be managed using plant growth regulators.

> Climates for best citrus production must provide good cool temperature induction of flowering, with minimal limitation of wintertime photosynthesis.
>
> Cool, nonfreezing winter temperatures followed by a warm, wet spring usually produce the most intense blooms.
>
> Although leafless bloom can be abundant, a higher percentage of leafy blooms usually result in higher fruit set.

POLLINATION

The citrus flower has multiple male structures (stamens) and a single female structure (pistil). Most cultivars produce viable pollen, although some important commercial cultivars, such as navel oranges, do not. Some citrus cultivars, such as Hamlin, Pineapple, Midsweet, and Duncan, are self-compatible and can be fertilized by their own pollen. Some hybrid cultivars, however, are self-incompatible and require pollination by another cultivar in order to set fruit. Many citrus cultivars can produce fruit without fertilization and seed formation (for example, Navel, Marsh, and Clementine). A few nearly seedless cultivars, most notably Star Ruby grapefruit, require viable pollen to achieve a commercially acceptable level of fruit set.

Pollen grains, when moved by insects onto the stigma, germinate to produce pollen tubes that grow through the style into the ovules in the ovary. Each pollen tube contains a sperm cell, which may fertilize an egg cell in the ovule to produce a zygotic embryo. After fertilization, the ovules develop into seeds. Seeds are important in fruit retention and fruit growth of some cultivars since they signal the tree to send resources into the fruit. Plant growth regulators (PGRs) can be used to increase the number of seedless fruit retained.

In most citrus cultivars, one or more asexual embryos arise from nucellar tissue of a fertilized ovule, producing seedlings genetically identical to the mother plant. The nucellar embryos often crowd out the sexual embryo. Nucellar embryos provide uniform rootstock material and can also be used in some cultivars for scion material to avoid virus-infected material, since most virus diseases aren't transmitted through seeds. Several seedlings may grow from a polyembryonic seed but usually one to three seedlings dominate and survive.

FRUIT DEVELOPMENT

After bloom, the young ovary (now a fruitlet) continues to undergo rapid cell division for 4 to 6 weeks. Growth during this time is characterized by increasing peel thickness. After this stage, the cells enter a period of rapid enlargement. The peel, including the outer colored flavedo and inner white albedo layers, reaches maximum thickness early in this period, becoming thinner as the fruit juice segments with juice vesicles continue to increase in size over several months. A maturation period follows in which fruit continue to grow at a slower rate for most of the time they are left on the tree.

During the early maturation stage, acids (mainly citric acid) accumulate and reach their maximum levels. As maturity progresses, sugar and juice content increase while acid levels decline. Unlike most deciduous fruit, citrus fruit has no clearly identifiable point of maturity. Color may serve as an approximate guide for early maturing cultivars, but generally is not reliable, since color development depends largely on temperature (especially low night temperature) and the mineral nutrition of the tree. For oranges, mandarins, and grapefruit the ratio of soluble solids (mainly sugar) to acids is more important than color for determining legal maturity. The fruit usually continue to improve in flavor and taste for several weeks or even months after it is considered to be legally mature. There are early cultivars that are harvested at peak maturity in October and November, midseason cultivars that are harvested November through February, and late-season cultivars (for example, Valencia) that are harvested March through June. Many cultivars of citrus fruit can remain on the tree for extended periods of time with minimal adverse effects, but premature drying out of some juice sacs and fruit drop can occur, depending on variety and rootstock. Certain hormonal

PGR sprays can prolong fruit retention on the tree, but do not delay internal quality changes.

> Most cultivars are self-compatible and can be fertilized by their own pollen.
>
> However, some hybrids require pollination from another cultivar to set fruit.
>
> In most cultivars, one or more asexual embryos arise from nucellar tissue of a fertilized ovule. Nucellar embryos provide uniform rootstock material and can be used in some cultivars for scion material to avoid virus-infected material.

Late developing fruit produced by very late off-bloom seldom have adequate maturity for harvest with the main crop and may complicate the harvest. In some citrus-growing areas, off-bloom fruit are harvested late and receive high prices, but they are seldom numerous enough for harvest in Florida. In addition, late-bloom or off-bloom fruit may be misshapen and have rough peels because high temperatures during early development tend to alter fruit growth.

FRUIT DROP

Citrus usually blooms abundantly, but most flowers and many young fruit drop from the tree before harvest. Typically less than 5 percent and, in some cases, as little as 0.2 percent (California navels) of the flowers produce the final crop. A combination of environmental and physiological factors seems to determine which flowers develop into fruit that persist to harvest. Any factor that results in faster-growing fruit increases the likelihood that the individual fruit will survive to harvest, as slower-growing fruit appear to be less competitive and drop off the tree at an early stage of development.

> During early maturation, citric acids accumulate and reach maximum levels.
>
> As maturation advances, sugar and juice content increase while acid levels decrease.
>
> For most cultivars, color is not a good indicator of maturity, as it is greatly influenced by climatic conditions (accumulated cool temperatures); the ratio of soluble solids to acids is a more important monitor of edibility for most cultivars.

The fruit drop reduction in crop load occurs in several steps. The largest drop occurs during and shortly after flowering. This is a natural drop that may be increased by water stress or nutrient deficiency. During this period, drop of very young fruit occurs by separation of the pedicel from the twig.

Another drop typically occurs in May through June in Florida when fruit range from pea-size to the size of a large marble. This is often called "June" drop because it parallels the well-known June drop of deciduous fruits. For citrus, this is the final major drop and largely determines the eventual crop load. This drop involves separation of the pedicel from the young fruit. Water stress at this time increases drop and reduces the final crop. In Florida, this is a critical time for irrigation, since May is usually hot and dry. PGRs may be used to enhance this physiological drop, ultimately producing larger and more valuable fruit in small-fruited, primarily fresh-marketed cultivars.

Additional fruit drop, called preharvest drop, varies from year to year and among cultivars. It usually ranges from 5 percent to 15 percent, but in the case of Murcotts can be as high as 50 percent. At later stages of development, fruit drops for a variety of reasons related to formation of an abscission zone in the button of the fruit. Much of the physiological basis for excessive drop at this time is not known. Fruit splitting, which can begin in late summer or early fall, causes some loss of fruit but is difficult to manage other than by good (uniform) irrigation practices. Fruit also may be lost due to disease or from mechanical damage caused by equipment traveling through the grove.

Some cultivars experience a physiological drop with little visible sign of the cause. Navel oranges go through several periods of drop during the summer and early fall; in some cases this results from the abscission of the secondary fruit or navel. The Pineapple orange variety frequently begins a substantial natural drop when fruit reaches maturity in January or February. Drop of both Navel and Pineapple oranges can be reduced using PGRs. Hamlin orange frequently drop when harvest is delayed into late February or March.

After flower intensity is determined, the second most critical time for obtaining adequate fruit set and optimal crop load is during the early flowering and June drop periods. Good soil moisture (above one-quarter depletion) should be maintained during this period to avoid water stress.

> The largest drop occurs shortly after flowering; it is a natural drop that can be increased by water stress or nutrient deficiency.
>
> May/June drop is the final major drop, is smaller than the first drop, but may be important to determining final crop loads.
>
> Additional fruit drop (preharvest drop) varies from year to year, ranging from 5 percent to 15

percent (up to 50 percent in some cases with Murcott).

Fruit losses also occur due to fruit splitting, mechanical damage, and physiological damage.

The critical period for obtaining adequate fruit set and maximum crop load is during early flowering and the May-June drop periods.

ALTERNATE BEARING

Alternate or biennial bearing is the tendency of a citrus fruit tree to produce a heavy crop one year (on year), followed by a light crop the next year (off year). It may occur only on part of a tree, on a whole tree, in a block of trees, or over an entire citrus-growing region. It varies in severity with variety and location, but has been observed in all citrus-growing regions of the world. In Florida, alternate bearing is a significant problem in some mandarin cultivars. However, it is usually a minor problem in oranges, though it may be more severe in Pineapple and Valencia oranges. It is less noticeable in Hamlin.

The economic impact of alternate bearing depends on its severity and duration. In extreme cases with the Murcott variety, it can be so severe that the tree dies during the on year. Heavy crop loads usually result in smaller fruit and delayed maturity, both of which are problems in the fresh fruit market. Alternate bearing tends to cycle on a regional basis and reduces not only marketable fruit but also total fruit production in an off year. There are significant economic benefits from regular cropping. A regular cash flow is preferred over one that fluctuates widely from year to year, and regular bearing often increases the cumulative crop value by producing a higher cumulative yield of marketable fruit sizes and avoidance of years with supply exceeding demand.

The causes of alternate bearing are not completely understood, but a number of internal, climatic, and management factors may be important. Tree nitrogen and carbohydrates are typically higher in off years and lower in on years. Either an especially heavy or light crop year may induce alternate bearing. Environmental stress, such as a significant freeze or an unusually hot and dry period during June drop, can result in an unusually light crop and can trigger alternate bearing. Pests and diseases, such as postbloom fruit drop, can also alter crop load substantially. Unusually favorable flowering or growing conditions may also result in an especially heavy crop and trigger alternate bearing.

Management practices clearly affect alternate bearing. Practices that reduce flowering or fruit set during the on year reduce alternate bearing. Several practices are useful in reducing the crop load during an on year. Fruit may be thinned by hand or by the use of fruit PGRs, such as naphthalene acetic acid (NAA).

Fruit-thinning sprays are most effective in reducing crop load when applied during the June drop period. This application appears to accentuate the natural drop occurring at that time. Water stress during this period may also cause more June drop and reduce crop load. Hedging, topping, and selective pruning are also used to reduce crop load. In fact, hedging and topping are often carried out on some fresh fruit cultivars with the primary purpose of reducing crop load and improving fruit size. When crop load is adjusted, nutritional applications also should be adjusted to match the anticipated crop load.

Any practice that increases flowering or fruit set during the off year also reduces alternate bearing. However, attempts to increase flowering and fruit set during an off year are generally unsuccessful. It is especially important to avoid hedging or topping that reduces crop load during a light crop year. A few cultivars with poor fruit set respond favorably to the application of PGRs during full bloom.

Climatic events can trigger alternate bearing that continues for a number of years before production stabilizes again. However, once regular cropping resumes, special management often is not required until some unusual event triggers another light or heavy crop.

Alternate bearing varies in severity by cultivars and location, but happens all over the world.

Alternate bearing is usually a minor problem in most oranges, but can be a big concern in Pineapple orange.

On a regional basis, alternate bearing can decrease marketable fruit and total fruit production.

Causes are not fully known, but tree carbohydrate status, extreme climactic conditions and management factors play a role.

GROWTH, CROPPING AND FLOWERING RELATIONSHIP

While adequate new shoot growth is necessary each year to achieve commercial cropping, excessive vegetative growth frequently compromises cropping. This is observed in grove blocks where spacing is too tight for the rootstock/scion/site/cultural practice complex. Excessive hedging and topping usually results in rank vegetative growth that inhibits cropping. Similarly, excessive nitrogen (N) fertilization and/or water availability can lead to imbalanced vegetative growth at the expense of acceptable fruit production. Clearly, many factors may contribute to poor production in an otherwise healthy block. To identify corrective measures, ICM must evaluate flowering, vegetative growth, fruit set, fruit drop, and yield.

PRODUCTIVITY

One way to understand productivity in citrus is in terms of four yield levels, each determined by a particular set of constraints. These levels of yield can be labeled as the Absolute, the Attainable, the Affordable, and the Actual (Figure 13.1). The first three of these are theoretical yields, useful in discussion of the factors that influence actual yield.

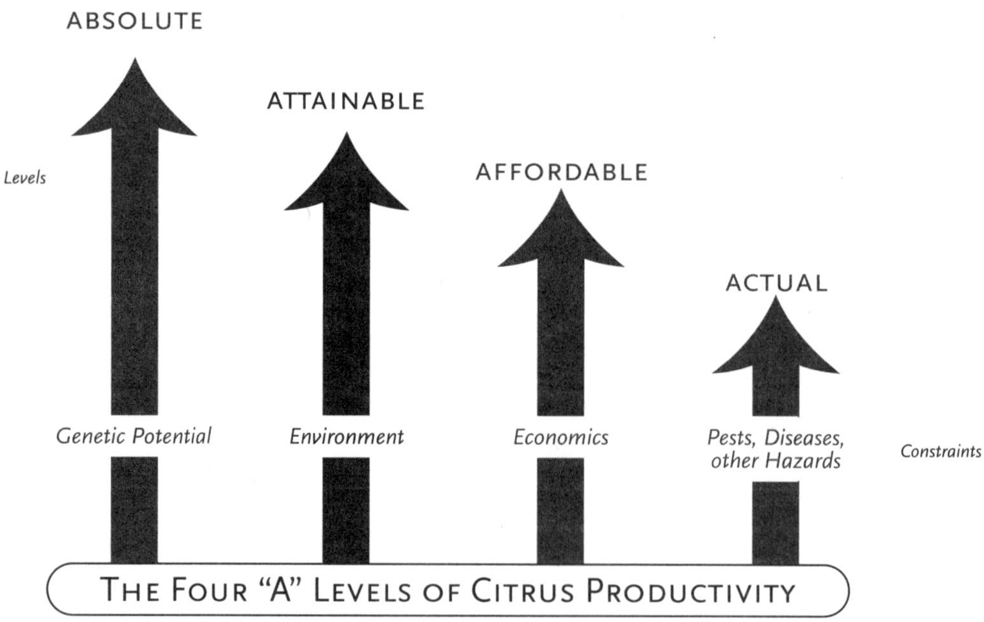

Figure 13.1 Maximum potential yields are always constrained by environmental, biotic and economic factors.

Absolute yield

Citrus cultivars in commercial production have been selected for their potential to produce economic yields of acceptable quality fruit. Every cultivar has an absolute yield determined by genetic potential and achievable only in an idealized environment.

Attainable yield

The attainable yield is the maximum yield that can actually be obtained in any particular location or year. The absolute yield is based on an idealized environment. However, in reality, at least one environmental factor can be expected to be suboptimal. Good management practices can minimize the impact of physical and biological limitations in the environment, but can never completely eliminate them; hence the attainable yield is always lower than the absolute yield.

Affordable yield

The affordable yield is the yield that balances profits and costs, and can be achieved using sustainable, environmentally sound growing practices. Affordable yield is almost always less than attainable yield. The economic value of the attainable yield must be balanced against production costs, including environmental costs. Maximum attainable yields may not be supported by world market prices, or may involve unacceptable environmental degradation.

Actual yield

The actual yield is that which is available for marketing in any given crop season. This depends on a complex set of factors, including the genetic potential of the particular variety, the physical and biological environment of the trees, growing conditions during the production season, management practices, economic factors that influence ICM decisions, and a broad spectrum of environmental protection concerns. We strive to achieve an attainable yield that approaches the absolute yield using ICM and best management practices (BMPs).

REFERENCES

Whitney, J. D., T. A. Wheaton, W. S. Castle, and D. P. H. Tucker. 2003. *Canopy hedging, topping and skirting effects on yield and fruit quality of valencia oranges.* Proceedings of the Florida State Horticultural Society. 116: 215-218.

14

Influence of Environmental Stresses on Tree Growth and Productivity

J.P. SYVERTSEN

INTRODUCTION

In spite of the best efforts being made to maintain citrus tree health and productivity, there will always be events that compromise the ability of a citrus manager to ensure a profitable crop. These events include environmental stresses on a tree brought about by the physical environment. There are cultural activities that can help during times of environmental stress, and citrus managers work to minimize environmental limitations on tree growth and yields. For example, healthy, well-watered trees are usually more tolerant of other physical and biological stresses than are drought-stressed trees. Yet all these factors are interrelated. A change in any component of the soil environment, whether by irrigation, fertilization, cultivation, or the application of soil pesticides and organic matter, has an effect on other environmental factors. The citrus manager must be aware of these interactions and act accordingly. This chapter reviews these factors as they relate to the management of environmental stresses in citrus trees.

It is impossible to isolate and change one factor in the environment without affecting other factors. The environment in which a citrus tree grows consists of many distinct, closely related components that interact with the tree and each other (Plate 14.1).

In order to be effective, managers need to understand how these components are integrated and how they affect tree health and productivity. ICM strives to minimize environmental limitations on tree growth and yield through best management practices. However, citrus trees are always enduring some sort of limiting stress. We can artificially divide the environment into its physical components (light, heat, soil, etc.) and its biological or biotic components (pests and diseases). Physical and biological components can combine to produce visible signs of stress and twig dieback (Plate 14.2). Another example is young seedlings in a greenhouse, which are especially susceptible to heat stress (Plate 14.3).

OVERVIEW OF PHYSICAL FACTORS

Citrus leaves utilize sunlight energy to combine water and mineral nutrients from the soil with carbon dioxide (CO_2) and oxygen (O_2) from the air in order to produce carbohydrates, the building blocks for roots, wood, leaves, and fruit. The majority of the dry weight of a citrus tree is composed of carbon (C). This carbon originally comes from atmospheric CO_2 that is transformed first into sugars by chlorophyll in the leaves through photosynthesis. Stomatal opening is sensitive to environmental conditions so the entry of CO_2 into the tree and the exit of water vapor from leaves are affected by light intensity, tree water status, wind, air temperature, and humidity. Since CO_2 concentration in the air almost always limits photosynthesis, these environmental factors can be important to productivity.

Solar energy, essential to the citrus tree system, is composed of visible light, ultraviolet, and thermal (heat) radiation. Visible light, the effective energy used in photosynthesis, varies daily with cloud cover and seasonally with geographic location. The structure of the tree canopy influences the amount of light that reaches the leaves and the temperature of the leaves. Worldwide, thermal energy evaporates water from the surface of the earth, from vegetation surfaces and through stomata of vegetation providing the water vapor that eventually returns to earth as rain.

The soil acts as a reservoir for water and mineral nutrients. The capacity of the soil to hold water depends on its particle size (texture) and structure. The nutrient-holding capacity of the soil is related to the amount of organic matter present, its ion exchange capacity and pH. Microbiological activity and daily and seasonal changes in soil temperature and moisture facilitate the release of nutrients through mineralization.

Judicious irrigation and fertilization can prevent or relieve drought and nutrient deficiency stresses. In fact, the management of drought stress is one of the most important tasks of the integrated crop management (ICM). Consider that plants transpire 300 times more water than the amount of CO_2 fixed into sugars during photosynthesis. Healthy, well-watered trees not only are more tolerant of other physical stresses, but also are more tolerant of biological stresses than drought-stressed trees. ICM balances the tree with its environment. For example, too much water or poor drainage can lead to flooding stress, which can be as detrimental as drought stress. Appropriate fertilizer programs prevent mineral nutrient deficiency, but too much fertilizer can result in salt stress, nutrient imbalances, and excessive nitrate (NO_3) leaching losses to groundwater.

The soil harbors both beneficial organisms, such as mycorrhizal soil fungi that grow intimately with roots and enhance mineral nutrient uptake, and also competitors, such as weeds, various pests and pathogens. Cultural programs to control harmful soilborne organisms should consider the impact on soil organisms that are beneficial.

It is important to understand that a change in any component of the soil environment, whether by irrigation, fertilization, cultivation, or the application of soil pesticides and organic matter, will have either a direct or an indirect effect on the soil environment and possibly other environmental factors.

The goal of the ICM is not to eliminate tree stresses, but to manage them so they have minimal negative impact. Indeed, physical environmental stresses are not always entirely negative. Moderate stress can predispose the tree to be more tolerant of repeated or more severe stresses. For example, freezes can be devastating, but cool winter temperatures increase cold hardiness, enhance flower induction, and improve fruit color development. Moderate winter drought stress can also increase cold hardiness and increase flower induction.

Management of drought stress is one of the most important tasks of the grower.

Well-watered trees are more tolerant of physical and biological stresses than severely drought-stressed trees.

Too much water or poor drainage can lead to flooding stress.

A change in any component of the soil environment, whether by irrigation, fertilization, cultivation, or soil pesticide application, will have an effect upon the soil's other environmental factors.

The goal of the grower is not to eliminate tree stress, but to manage it.

OVERVIEW OF BIOTIC STRESSES

The citrus tree interacts with a host of other biological elements in its environment, including insects, mites, foliar pests, soilborne fungal pathogens, virus complexes, other decline-inducing disorders, as well as neighboring citrus trees, weed competitors, and wildlife.

Foliage accounts for up to 25 percent of tree fresh weight. Since leaves are present year round and have a high nutritional content, they are an attractive food resource for pest species. Similarly, sugars in fruit are a resource for pests and diseases. Some biological pests, such as insects, mites, and nematodes, can cause an obvious deterioration in tree vigor and fruit quality. Fungal, bacterial, and viral diseases, although not always easily visible, can also lead to significant fruit and tree loss. Some of the most troublesome citrus production problems belong to the group of diseases known as tree declines. The two most important tree declines are citrus blight, which is still not linked to a causal agent, and citrus tristeza virus (CTV). Again, not all disease stress is entirely negative. The intentional infection of citrus trees with mild strains of CTV can limit tree size and provide limited protection against more severe strains.

Cultural practices can alleviate stresses from biological factors. Healthy trees that are correctly irrigated and appropriately nourished are the best prepared to withstand pest and disease pressures. Chemical control of citrus insects and disease has long been an important aspect of citriculture. Alternatively, biological control of pests by natural enemies or through cultural practices may control pests and reduce the use of spray materials that can have unwanted environmental effects. Some progress has also been made in developing rootstocks of citrus cultivars that are tolerant or resistant to specific pests or diseases. Our current understanding of how cultural, chemical, and biological manipulations of environmental factors interact, and modify the expression of damage caused by particular pest and pathogen agents is incomplete. Our current understanding, however, provides useful insights into beneficial methods of efficiently achieving desired production goals. These benefits provide the rationale for addressing the plant/environment/pest-disease complex as an overall system, which is the basis of ICM.

Weeds are also biological pests that compete with citrus trees for available soil moisture, nutrients, and light. In addition, they can act as alternate hosts or reservoirs for plant pests and pathogens and some beneficials. However, some nitrogen-fixing weeds may be beneficial in the orchard. Weed competition for water and nutrients can be minimized using good water and nutrient management practices.

Within the tree itself, there is competition for carbohydrates, which manifests itself in the typical alternation of growth flushes between shoots and roots, and even in alternate bearing from year to year. Again, management practices aim at striking the right balance with environmental factors to maximize consistent good-quality fruit yields while maintaining high efficiency.

> The citrus tree interacts with a host of other beneficial organisms, pests, diseases and other citrus trees.
>
> Cultural practices can alleviate stresses from biological factors
>
> The plant/environment/pest-disease system is the basis of integrated crop management.

WIND AND MECHANICAL INJURY

In 2004 and 2005, Florida experienced four powerful hurricane events within a period of about two months. Three of these storms crossed the state, intersecting in southern Polk County (Plate 14.4) and severely impacted the citrus industry and Florida agriculture as a whole. Separate hurricanes struck the southwest and southeast coasts and two storms traversed the state from east to west along remarkably similar paths leading to cumulative destruction. Not since hurricane Donna in 1960 has a storm of such magnitude traversed Florida, let alone three such storms. The economic impact was considerable, the full effect of which will not be accurately known for some time. Immediate observations focused on crop loss, fruit damage, defoliation, tree loss, and potential flooding damage.

There will be many questions asked and answers to be provided following such hurricane-generated winds and rainfall. Consequences to be considered should include:

1. the disruption of vegetative and fruiting cycles;

2. the potential for subsequent freeze damage with newly emerging tree canopies;
3. the disruption of disease and insect pest cycles along with the potential for the spread of diseases such as citrus canker from infected areas;
4. the potential for root system damage following extensive and prolonged flooding in some areas;
5. modification of cultural practices including irrigation and nutrition; and
6. immediate and long-term marketing strategies for market preservation.

Much of the information on which we based our estimates of posthurricane tree damage and recovery comes from previous work on defoliation and minor root damage after mechanical harvesting. Many people are alarmed the first time they witness the apparently violent manner in which trees are treated during mechanical harvesting with trunk and canopy shakers. Visible injuries to tree can include leaf and twig loss, exposure of shallow fibrous roots, scuffing of the trunk bark, limb breakage, and the removal of flowers and young green fruit. Previous research, however, has provided no evidence that mechanical harvesting reduces yield or increases tree mortality. Our work has shown that are no major negative effects of mechanical harvesting on physiological or growth responses in "Hamlin" and early season "Valencia." Healthy citrus trees in a well-managed grove appear to be able to tolerate any visible injury from trunk or canopy shakers and recover from any leaf loss and root injury.

INTEGRATING STRESS MANAGEMENT AND PRODUCTIVITY

Before the adoption of new production practices, their possible impacts on development and productivity, pests, disease incidence, and severity, must be carefully evaluated. No individual factor can be modified without influencing others within the system. Grove floor management, irrigation, and nutrition practices are known to influence grove microclimate, pest and disease populations, fruit quality, postharvest fruit disorders, soil properties, and root behavior. To be effective, the crop management system must consider geographic and topographic location, climate and weather, citrus cultivars, and the physical and biological environment.

REFERENCES

Bower, J. P. 2003. *Water stress in citrus and its alleviation. International Society of Citriculture Congress.* (9th 2000: Orlando, Florida), Orlando, Florida USA 2: 630-633.

Crane, J. H., R .J. Campbell, and C. F. Balerdi. 1993. *Effect of hurricane Andrew on tropical fruit trees.* Proceedings of the Florida State Horticultural Society. 106: 139-144.

Graham, J. H. 1986. *Citrus mycorrhizae: Potential benefits and interactions with pathogens.* HortScience. 21: 1302-1306.

Jifon, J. L., and J. P. Syvertson. 2001. *Effects of moderate shade on citrus leaf gas exchange, fruit yield and quality.* Proceedings of the Florida State Horticultural Society. 114: 177-181.

Reuther, W., E. M. Nauer, and C. N. Roistacher. 1979. *Some high temperature effects on citrus growth.* Journal of the American Society for Horticultural Science 104: 353-356.

Rouse, R. E. 2001. *Hurricane management planning for Florida citrus groves.* Proceedings of the Florida State Horticultural Society 114: 135-137.

Smith, P. F. 1969. *Nitrogen stress and premature leaf abscission in citrus.* Hortscience 4: 326-327.

Syvertsen, J. P. 1997. *Water stress and carbon budgets.* Eighth International Citrus Congress, 1996, Sun City, South Africa 1: 46-50.

Yelenosky, G. 1991. *Responses and adaptations of citrus trees to environmental stresses.* Israel Journal of Botany 40: 239-250.

15

Cold Protection

L.K. Jackson

INTRODUCTION

One of the most influential environmental stresses is freezing weather. Many different methods of freeze protection have been used over the years, ranging from fuel-based heaters to tree wraps and soil banking. However, one of the most effective freeze protection methods is to apply water to the trees at as high a volume as economically feasible. In the early 1980s, for example, growers experienced considerable protection from a relatively small volume of water (1,400 to 2,100 gallons per acre per hour) applied under citrus trees. Increased numbers of growers used microsprinklers for cold protection during the freezes of 1985 and 1989, and the University of Florida conducted research to examine the effectiveness of small or moderate volumes of water applied under and within the tree canopy. As a result of this research and grower experience, the most common method of cold protection used by Florida citrus growers since the mid 1980s has been the application of moderate amounts of water delivered under or within the tree canopy. This chapter reviews the current methods of freeze protection and explains the advantages and disadvantages of each method.

Cold weather is one of the most damaging environmental events that can affect the citrus industry. A single night of severe temperatures can kill millions of trees. Freezes have had a tremendous economic impact on the Florida citrus industry, often forcing growers to plant in sites with marginal soils because of their warmer locations. Few areas in Florida are free from the risk of cold damage, and most growers need to be prepared to deal with cold weather. The loss of revenue caused by even a moderate freeze and the very expensive prospect of tree attrition warrant the implementation of a sound cold-protection program customized to each particular grove.

TYPES OF FREEZES

Two major types of cold weather events occur in Florida. Calm winds and a rapid drop in temperature characterize a "radiation" freeze after sunset, followed by a more gradual temperature drop during the night. The earth's surface cools, causing temperatures close to the ground to be colder than those above it. The majority of freeze situations in Florida are radiation events.

An "advection" freeze brings strong winds as a cold air mass moves into the area. In an advection freeze, northern areas are colder than southern areas, but there is little temperature difference between high and low elevations. Many cold periods start with an advection freeze the first night, followed by one or two nights of radiation freeze. (The freezes of 1962, 1983, 1985, and 1989 were all advection freezes that killed millions of trees, especially north of Interstate 4).

PREDICTION AND FORECASTING

It is difficult to predict how often a major freeze will occur in Florida. It appears that the 100-year period prior to 1835 was relatively warm, with no record of severe freeze. However, severe freezes occurred in 1886, 1894 through 1895, 1899, 1934, 1940, 1962, and 1977; and five severely cold weather events occurred in the 1980s (1981, 1982, 1983, 1985, and 1989).

Prior to its cessation of service to agriculture on April 1, 1966, the National Weather Service (NWS) issued annual reports and information to Florida growers in cooperation with the University of Florida (UF). Budget constraints forced both the NWS and UF to discontinue the annual reports, although weather information continues to be collected by both institutions. In 1998, UF established the Florida Automated Weather Network (FAWN) to collect real-time weather information and provide it to the Florida agricultural industry. This network supplies data from agricultural locations throughout Florida citrus-growing regions. Data is collected electronically and updated every 15 minutes. Such data is invaluable in planning for frost protection, and has helped citrus growers save hundreds of thousands of dollars. The detailed weather data generated by FAWN and satellite imaging (Plate 15.1) are also useful in expert systems for disease control and flowering management.

TOLERANCE TO COLD WEATHER

Citrus cultivars vary tremendously with respect to cold tolerance. Generally speaking, limes and lemons are the most sensitive and temperatures in the mid to upper 20s can cause extensive damage to these trees. Mandarins are the cold-hardiest, and can tolerate temperatures into the upper teens. Oranges and grapefruit fall between these extremes in cold hardiness. Within each variety group there are differences as well. The Satsuma mandarin is extremely cold hardy (it has been grown commercially in southern Georgia), while the Fallglo tangerine appears to be moderately cold sensitive. Rootstocks also have a major influence on cold hardiness. The more vigorous and vegetative the stock, the less cold tolerant the scion will be. Trees budded on rough lemon, Volkamer lemon, or macrophylla are more cold sensitive than those on Swingle, sour, or trifoliate. Rootstocks such as Cleopatra, Carrizo, and sweet orange are somewhat less cold tolerant than Swingle, but much better than rough lemon.

Even though citrus is a subtropical plant, it has the ability to acquire considerable cold tolerance; this is a major factor in determining the extent of damage suffered during a cold weather event. However, citrus does not enter a true dormancy period, but goes through a period of no apparent growth called quiescence, which can be interrupted at any time during the winter by warm weather. Exposure to cool night temperatures (it is generally thought that 50°F is adequate) allows the tree to gain some degree of cold tolerance. The amount of cold tolerance developed may be only a few degrees, but this can be very significant. A tree with abundant new flushing often sustains considerably more damage than one that has become quiescent and somewhat cold tolerant. Moderate moisture stress induces some degree of cold hardiness as well. However, it has been demonstrated many times that a healthy tree (one receiving adequate levels of nutrition, moisture, pest control, etc.) sustains less cold damage than a tree that is stressed. The grower strives to keep trees healthy and hopes for a cool, but not cold, winter.

> Limes and lemons are the most sensitive to cold (temperatures in the mid 20s cause damage).

Mandarins are generally the most cold-tolerant citrus selections.

The more vegetative and vigorous the rootstock, the less hardy the scion will be.

Although citrus is a subtropical plant, it can acquire some cold tolerance.

Exposure to cool night temperatures (50°F) can help trees gain hardiness.

A healthy tree sustains less cold damage than a tree that is stressed.

PASSIVE COLD PROTECTION

There are several cultural practices that can provide protection from cold weather. As mentioned previously, a healthy tree handles cold weather better than one that is stressed. Therefore, maintaining a sound cultural program by providing adequate nutrition, maintaining soil moisture to avoid prolonged drought, and controlling pests is one of the most important components of a sound cold-protection program. Most research indicates that freeze damage is not related to any specific mineral element; however, following a sound nutrition program that produces a dense canopy is recommended. Once thought to be related to cold hardiness, severe leaf loss can be caused by magnesium deficiency. However, it appears now that the lack of leaves, rather than the magnesium deficiency, influences cold damage, primarily through thermal effects. Likewise, it is well documented that high levels of potassium do not increase hardiness.

Soil is a major reservoir for heat. Practices that increase the amount of heat that can be stored during the day provide better cold protection during the night. A bare or closely mown grove, a compact soil, and a moist soil all can store more heat than a dry, loose soil with a high weed cover. Therefore, ensure adequate soil moisture and maintain a closely mown grove or compact bare ground. Little data exists on the impact of soil management on cold protection, but numerous observations support the above recommendations. There is very little difference in the amount of heat stored by bare ground and that stored by a closely mown grove floor.

Pruning practices can also significantly affect the impact of cold weather. Certainly, all hedging and topping activity should be complete well before winter arrives. Since pruning encourages growth and new growth is far more susceptible to cold damage, it is best to prune in the spring and early summer. Young trees are more susceptible to cold damage then mature trees (Plate 15.2). Another aspect of pruning to consider is air drainage. Cold air is heavier than warm air and flows downhill, so groves should be hedged to allow the cold air to flow off the site. This is especially applicable on the Central Florida Ridge, where "cold pockets" can be found. These areas have no outlet, and cold air builds up during a radiation freeze. For hedgerow groves, it is best to have the rows run with the slope to maximize cold air drainage.

A common practice during the development of groves along the Ridge was to select sites located on the south side of lakes. Since damaging, advecting freezes generally bring cold air in from the northwest, the southeast side of a lake is generally warmer. The larger and deeper the lake, the less the freeze impact on groves in the downwind locations. There are regions within the citrus industry that are historically cold, especially during a radiation freeze. Thermal maps clearly show these areas. However, within a cold region there are warm sites, just as there are cold sites within a warm region. It is important to look both at the region as well as the individual site to determine the natural or passive cold protection present.

Wind is a major concern during severe advection freezes. Many years ago, windbreaks were commonly used to reduce the rate of heat exchange within a grove. Windbreaks are also found in many citrus growing areas outside Florida. It is extremely important to select the proper plant material for windbreaks. It is best to have a hardy plant that is not highly competitive with citrus. Close-set groves provide an internal windbreak and, due to the increased canopy area, are able to intercept more of the radiant heat given off from the soil. Therefore, during an advection freeze, a closely spaced grove may generally remain somewhat warmer than a widely spaced one.

Any or all of the above passive cold protection measures can play a significant role in mitigating the impact cold weather will have on a grove. In warmer locations, they may be the only methods of cold protection used.

Soil is a major heat reservoir.

A bare or closely mown grove, compact soil, or moist soil store heat better than a loose soil with high growth.

It is best to prune a grove in the spring or early summer.

Groves should be hedged in such a way as to allow cold air to flow out of the grove.

Placing a grove on the south side of a lake is a common practice; since advecting freeze winds flow from the northwest, the southeast side of the lake is warmer.

If windbreaks are to be used, choose a hardy plant that does not compete vigorously with the citrus.

Closely spaced groves are warmer than widely spaced ones.

ACTIVE COLD PROTECTION

Methods used to protect citrus trees from freezing temperatures have changed over the years. Economics and regulations have played a major role in this shift. In addition, citrus plantings have moved farther and farther south to avoid cold weather. An active cold protection program is not as necessary in Immokalee or Ft. Pierce as it is in Umatilla or Wiersdale. Detailed information is available for the primary methods of cold protection listed below.

IRRIGATION

Water has been used to protect crops from freezing temperatures for many years. In the early 1960s the University of Florida evaluated overhead application of water in a grove environment. Results were very poor; as were field experiences throughout the industry during the 1962 freeze. Application rates were too low and sprinkler rotation was slow, resulting in excessive ice loading, which produced extensive limb breakage. The 1962 experience caused many citrus growers to abandon the use of water for cold protection. Citrus nurseries, however, successfully used large volume applications to protect young stock, with application rates in the range of 13,000 to 20,000 gallons per acre per hour.

The introduction of microsprinklers to the Florida citrus industry caused some growers to once again consider the use of water for cold protection. In the early 1980s, growers experienced considerable protection from a relatively small volume of water (1400 to 2100 gallons per acre per hour) applied under citrus trees. Increasing numbers of growers used microsprinklers for cold protection during the freezes of 1985 and 1989, and the University of Florida conducted research to examine the effectiveness of small or moderate volumes of water applied under and within the tree canopy. As a result of this research and grower experience, the most common method of cold protection used by Florida citrus growers since the late 1980s has been the application of moderate amounts of water delivered under or within the tree canopy (Plates 15.3, 15.4). The standard practice used by most growers in the central and northern portions of the citrus industry today is to elevate microsprinklers 24 inches above the ground, placing the nozzle as close to the tree trunk as possible. The more water that can be applied, the better the level of protection. Rates below 1,400 gallons per acre per hour are not recommended. A Navel orange block containing 140 trees per acre received 23 gallons per hour per tree (3,220 gallons per acre per hour) during the 1989 freeze. Air temperature went to 16°F and the 12-foot trees only suffered defoliation and minor twig damage. No pruning was necessary and the trees were able to set a crop the following season. Lower application rates applied in the same manner resulted in more extensive limb damage, but all trees recovered, even with as little as 8 gallons per hour per tree (1,120 gallons per acre per hour).

A major concern with the use of water for cold protection has been evaporative cooling, which results from the evaporation of water from the leaf or wood. This situation can result in actual leaf or wood temperature being several degrees colder than ambient air temperature. Factors that increase the possibility of evaporative cooling are low dew point (dry air) and high winds. It is generally accepted that if temperatures will drop below 23°F for an extended period, then an irrigation system should be operated, no matter what the dew point and wind speed. A minimum temperature of 23° to 28°F suggests possible operation, based on tree condition. If the minimum temperature does not reach 28°F then it is not necessary to apply water for cold protection (the critical value for lime groves is 32°F).

A number of issues arise in the management of an irrigation system for cold protection. First, a reliable power source must be obtained. During the 1989 freeze, rolling blackouts were common and many growers with electric motors were without power long enough for their system to freeze. It is recommended that a diesel, liquid propane, or gasoline motor be used as a main or back-up power source. Obviously, in order for an irrigation system to be effective, it must be in excellent working order. The emitters need to be checked periodically; a plugged microsprinkler can mean a dead tree. Grower experience indicates that it is advisable to have microsprinklers elevated no more than 30 inches above the ground. Fortunately, there have been no severe freezes since 1989, and it has not been possible to further evaluate microsprinklers elevated above 30 inches. Grower experience also suggests that it is advisable to select an emitter that has a continuous pattern rather than a spoke (numerous tiny streams of water). The spoke pattern generally provides a greater coverage area for irrigation, but may not deliver sufficient water for cold protection to the inside canopy of the tree.

The last component of the management system is operation. In order to be sure that none of the microsprinklers freeze, systems should be started when the air temperature reaches 34°F to 36°F. Since the aboveground portion of the irrigation system is black and located on the soil surface, it may be a few degrees colder than air temperature. The most difficult aspect of managing an irrigation system for cold protection is making the decision to turn on the pumps. Growers rely on a minimum temperature prediction to determine if there is a possibility for cold damage. Actual temperatures in the area must then be measured and recorded to verify if the forecast appears to be correct. The decision to operate the irrigation system is a difficult, but very important one. Many dollars are wasted if the system is operated during periods when critical temperatures are not reached, but even more money may be lost if the system is not operated when needed. Experience and an accurate thermometer are essential.

The most common method of cold protection used in Florida is applying moderate amounts of water under or within the tree canopy.

If temperatures will drop into the low 20s for an extended period of time, an irrigation system should be used no matter the dew point or wind speed.

If the minimum temperature will not reach 30°F, water is not necessary for cold protection.

To compensate for power outages during a freeze, the use of a diesel-, liquid propane-, or gasoline-powered motor is recommended.

Grower experience suggests that microsprinklers should be elevated no higher than 30 inches above the ground, placing the nozzle as close to the tree trunk as possible, and that the emitters should spray with a continuous pattern.

Irrigation systems should be started when temperatures reach 34° to 36°F.

Money can be lost if irrigation is used when critical temperatures are not reached, but even more money can be lost if no water is applied during a major freeze.

SOIL BANKS AND TREE WRAPS

Soil banks and tree wraps can be used with or without water to protect young citrus trees. Like all methods of cold protection, they have their advantages and disadvantages. Protecting the bud union with an earthen mound of soil almost guarantees the survival of the scion, since a well-constructed bank provides the maximum degree of cold protection. However, soil banks have several major drawbacks. They are difficult and labor-intensive to construct, especially when a sod-middle is present, and they must generally be in place from December 1 until March 15. More significantly, they may create problems such as insect damage and fungal infection to the trunk. Tree wraps were very popular during the 1970s and 1980s because they provided sprout control and protected the trunk from contact with herbicides. In addition, the wraps provided some degree of cold protection. The exact mechanism for the protection was never identified, but it was thought that insulation from the wrap kept the tree trunk cool and less vigorous, therefore, more hardy. Wraps were effective during winters when the trees had time to develop hardiness, but were less effective when a freeze arrived early in the season, as in 1983 and 1989. Since wraps may cause problems similar to those associated with soil banks, namely insect and disease damage, they are declining in popularity. Some wraps incorporate insecticides.

Soil banks and tree wraps:

reduce cold injury from late freezes (thought to keep tree quiescent by preventing warming),

reduce trunk sprouting,

protect against herbicide damage, and

may enhance insect and disease problems.

WIND MACHINES

Wind machines have been used in the Florida citrus industry with varying degrees of success. They cannot be used during windy conditions (generally, wind speeds in excess of 5 mph) but can be effective in protecting trees from cold pockets during a radiation freeze. The high initial cost of a machine, problems with theft and vandalism, and the machine's inability to protect trees during windy conditions have caused most growers to select another method of cold protection. A wind machine provides protection for approximately 10 acres and is best suited to areas with marginal or poor air drainage.

HEATERS

Heaters are another method of active cold protection that was used for many years to protect Florida citrus. However, they are rarely used today due to the high cost of operation and potential for air pollution. Recommendations call for approximately 35 heaters per acre, with additional heaters on the north and west sides of a grove. Since most grove heaters have a 10-gallon capacity (once a heater is lit, it is rarely extinguished; therefore the entire fuel supply is burned), and 35 to 40 are used per acre, the volume of fuel required per acre per night is 350 to 400 gallons. With the additional cost of the labor needed to light, refuel and move the heaters, the equipment to refuel the heaters, and summertime storage, it is evident that heating is an expensive operation. However, it can be very effective in protecting citrus from cold weather. Today, the use of heaters is limited to unusual situations, such as the protection of rare or extremely valuable trees, research plots, germplasm, etc.

FOG MACHINES

Artificial fogs have been experimented with in Florida citrus, but never used commercially. A water barrier is generated that traps the radiant heat close to the surface. Several major problems exist with artificial fog. It is difficult to control the cloud, meaning that the fog may drift off-site and provides protection for a neighbor rather than for the intended grove, or may create a driving hazard on nearby roads. In addition, the cost per acre for a fog system is high. Commercial fog systems are used in greenhouses and were used in a field nursery, but not in the grove.

Microsprinklers are the standard in Florida citrus.

Wind machines are expensive but effective in limited circumstances.

Heaters are very expensive, but effective; used for extremely valuable trees or crops.

Fog machines are expensive and the fog is difficult to control; they are better for greenhouse use.

EVALUATION OF COLD DAMAGE

The determination of fruit damage generally starts early in the morning during or following a freeze event. Fruit is cut and the presence of ice noted. If the freeze is not severe, little to no ice will be present; in a severe freeze, the fruit may be frozen solid. The rate of fruit breakdown is related to the extent of damage and the postfreeze temperature pattern. Warm weather after a freeze hastens fruit loss, while cool weather allows more of the damaged fruit to be salvaged for juice.

Most growers tend to evaluate the extent of cold damage to the trees much too soon after a freeze. It may take 3 to 6 months for large limbs to start to die back. There may even be new foliage development in the spring on limbs that appear dead, but are able to support growth with the reservoir of nutrients in the limb. Growers should wait until early summer before checking groves to determine the extent of serious cold damage. Twig and small limb damage can be determined in the spring by simply making a shallow cut in the bark and checking for a dark green color. If the bark is brown, the twig or limb is dead, even if new growth is present. Once the extent of the damage has been determined (Plate 15.5), appropriate action can be taken.

MANAGEMENT OF COLD-DAMAGED TREES

There are no established guidelines as to the size of damaged wood that requires removal. Defoliation and small twig damage require no pruning at all. Some growers prefer to remove all dead wood larger than ½ inch in diameter, while others prune only those limbs 1 inch in diameter and larger. If the damage to the trees is extensive, severe pruning is required. It is suggested that growers "hatrack," or cut the tree back to a uniform height. Experience has shown that trying to leave every live limb produces less new growth, and that the resulting tree will be more open and thus, susceptible to cold damage the following year. It is best to develop as dense a canopy as possible. Note that pruning cuts do not have to be painted.

Citrus trees are vigorous and recover quickly from cold damage if given proper care. The extent of damage dictates the rate of recovery and the appropriate cultural practice to follow. In the case of trees that have suffered partial defoliation, few major changes will need to be made to the cultural program. It is very important to avoid stress during spring when new growth is developing. By summer, the trees will be fully recovered, though special care should be given to ensure that mites, greasy spot, etc., do not attack the foliage. It should be noted that melanose could cause serious damage to new growth. It is extremely difficult to control this disease if a substantial amount of dead wood is present. Trees with severe defoliation and twig damage need the same type of care. Spring fertilizer can be reduced according to the extent of damage (100 percent defoliation would justify a reduction of nitrogen and potassium by 50 percent). If regrowth is substantial, the prefreeze nutrition program can be resumed in the fall.

If the tree suffers extensive limb damage, the cultural program should be adjusted more dramatically. Irrigation requirements are reduced, since the transpiring surface area is smaller. Nutrition levels should be reduced for the year based on the extent of damage and rate of recovery. Every leaf produced needs protection so that maximum canopy regrowth can take place. Sunlight reaches areas under the tree that were previously in the shade, and weed pressure increases. On rare occasions trees do experience sunburn, so whitewash may have to be applied to trunk and limbs. Regrowth takes place quickly and protects the exposed limbs. Trees with extensive regrowth are particularly vulnerable to freezes the following winter. Therefore, management should be prepared to take action in anticipation of freezing temperatures to protect tender growth on trees with a reduced canopy.

The severity of cold damage should dictate pruning severity.

Reduce nitrogen (N) and potassium (K) 50 percent for complete defoliation.

Take special care to avoid stress to cold-damaged trees.

Cold-damaged trees are likely to be more sensitive to cold the following year.

EXPECTATIONS FOR DAMAGED TREES

The extent of damage suffered and the level of care provided during the subsequent year dictate the recovery rate for trees following a freeze. Those suffering defoliation and minor twig damage may have a reduced crop the season following the freeze. With proper care, the trees can be expected to return to prefreeze condition by the second season following a freeze. Trees that suffer wood damage on limbs up to one inch in diameter but have no damage to the trunk and scaffold limbs can recover remarkably well in one season. By the second year following a freeze, they should be restored to prefreeze condition as well. Even trees with extensive damage to larger limbs but no damage to the trunk should be able to support 50 percent of their prefreeze crop the second year after a freeze, and by the third year they should be in prefreeze production. These recovery estimates assume no additional cold damage to otherwise healthy trees.

TREE MORTALITY

Growers are faced with difficult decisions when trees are killed by cold weather. Economics determines the best course of action. For many the only option is to abandon the grove, especially if the location is desirable for another land use.

If the decision is made to replant, several distinct options are available to the grower. It may be best to push out and burn the dead trees before replanting, or the dead trees may be clipped to the ground and new trees planted between the stumps. Several factors should be considered when making choices. Is there an irrigation system in place, and if so, what will be the impact on the system? Would tree removal destroy beds, drainage, etc., and result in additional expenses? Is the planting density and rootstock suitable for continued use? Can the grower bear the expense of tree removal and replanting? Each situation is highly individual in both the economic and physical considerations.

REFERENCES

Grosser, J. W., and J. L. Chandler. 2003. *Development of improved citrus rootstocks via somatic hybridization.* International Society of Citriculture Congress. (9th 2000: Orlando, Florida), Orlando, Florida. 1: 47-50.

Martsolf, J. D. 2000. *Diversity of frost protection methodology.* Proceedings of the Florida State Horticultural Society 113: 133-137.

Martsolf, J. D. 2001. *Relationship between florida freeze incidence and sea surface temperature in the tropical pacific.* Proceedings of the Florida State Horticultural Society 114: 19-21.

Martsolf, J. D., and R. Peart. 2003. *Frost control rules.* Proceedings of the Florida State Horticultural Society. 116: 282-285.

Oswalt, W. C., W. J. Wiltbank, and F. S. Davies. 1988. *Tree wraps, trunk temperatures and growth of young citrus trees.* Proceedings of the Florida State Horticultural Society 101: 48-53.

Parsons, L. R. 2003. *Microsprinkler irrigation for citrus freeze protection.* International Society of Citriculture Congress. (9th 2000: Orlando, Florida), Orlando, Florida 1: 390-392.

Wright, G. C. 2001. *Protecting a citrus tree from cold.* AZ1222:4.

Zekri, M. 2003. *The basics of cold hardiness and cold protection.* Citrus Industry. 84: 18-20.

16

Water Management

B.J. Boman, K.T. Morgan and D.P.H. Tucker

INTRODUCTION

Florida has such abundant rainfall that a topic such as "Water Management" might seem unnecessary to nongrowers. Nevertheless, any particular grove can suffer from too much or too little rainfall. Rainfall in Florida is usually inadequate for citrus in the spring and fall but excessive in summer. Ideal water management keeps root zone water content in an optimal range for fruit production, minimizes water use, and avoids leaching of agricultural chemicals into surface water or groundwater. A water management program provides irrigation to supplement rainfall as needed and may include an engineered drainage system to remove excess water from the root zone. Adequate water is essential for efficient uptake and utilization of applied fertilizers, but excessive irrigation may result in nutrient leaching, meaning that nutrients—especially nitrates—move out of the root zone and into the groundwater.

Clearly, there are many considerations to be made when ensuring that citrus trees will obtain the water they need to develop and produce a high-quality crop. This chapter covers the various aspects of managing a water system in a citrus grove. It is important to note, however, that irrigation is only one component of the complete water management system. Other factors to consider include evapotranspiration rates, soil moisture retention, rainfall, and drainage. Proper soil water management affects horticultural plant protection practices and postharvest behavior of fruit as well.

Water management (Figure 16.1) and nutrient management are closely related, especially when it comes to water quality protection. Citrus trees are evergreens and, therefore, require water for transpiration throughout the year. Water requirements vary with soil type, climate, ground cover, cultivation practices, weed control, tree size, age, scion, and rootstock. A water management program provides irrigation to supplement rainfall as needed, and may include an engineered drainage system to remove excess water from the root zone. The need for drainage should be determined and addressed at the time of site development, before trees are planted. The irrigation system (Plates 16.1a-c) should also be installed at this time. Both the irrigation and drainage systems will require continuous monitoring and maintenance. The water requirements of the trees must be continually assessed, and irrigation should begin prior to the onset significant of water stress. A water management BMP maintains the root zone water content in an optimal range for fruit production, minimizes water use, and avoids leaching of agricultural chemicals (Plate 16.2).

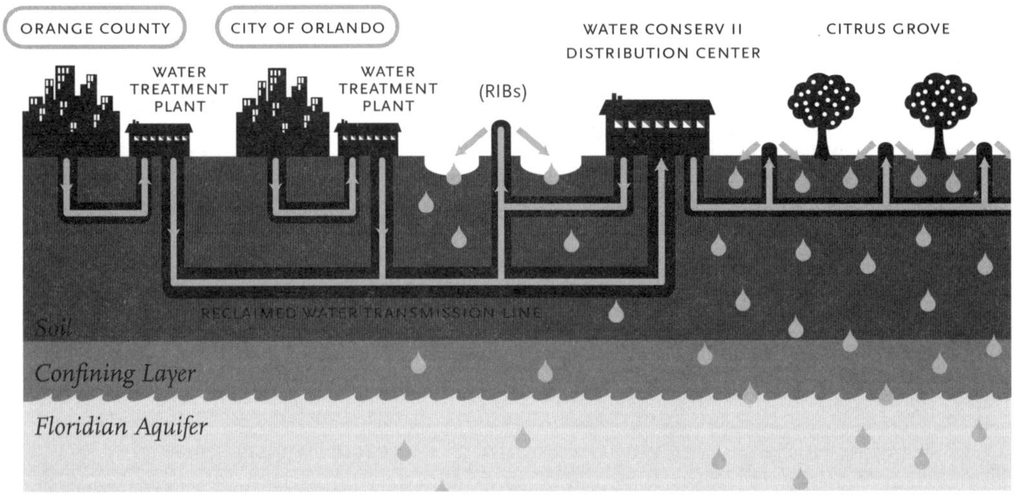

Figure 16.1 Schematic showing the Orange County/Orlando Conserv II recycled water distribution system. Conserv II is an outstanding example of a cooperative venture benefiting both the urban sector and commercial agriculture.

IRRIGATION

Water Stress and the Benefits of Irrigation

Water deficits in citrus trees can retard the development of vegetation and fruit, reduce leaf development, and, in severe cases, wilt leaves, which may then abscise and drop. Fruit from severely stressed trees generally has low juice content and may be of inferior quality. Moderate to severe drought stress may also reduce fruit set and vegetative growth. When soil moisture deficits are severe, tree growth terminates and fruit drops depending on the stage of development and cultivar. Leaves become desiccated, and twig and branch dieback follows. Under conditions of high evaporative demand, citrus stomata (leaf openings that permit CO_2 entry) close in order to reduce water loss, so wilting of mature leaves is seen only during severe drought stress.

Even modest water deficits may reduce tree growth and yield. Vegetative growth determines the canopy size and future yield potential of young trees; healthy vegetative growth to maintain and replace fruit-bearing branches is essential to continued productivity. During periods of rapid vegetative growth, water stress that may affect root and leaf development should be particularly avoided.

Water shortages in citrus trees during bloom, fruit set, and early development can dramatically affect productivity and quality. Water stress during flowering directly reduces fruit set. Continued water stress during the fruit set period is likely to exacerbate droppage of small fruit until June and reduce the rate of fruit development. The increase in fruit size from June through December is also highly related to water uptake, but summer rain is usually adequate for fruit development.

It should be noted that water stress and irrigation influence fruit quality in several ways. For example, although irrigation can increase fruit size, it can also increase the number of fruit per tree. Although this may decrease fruit size, it may also increase the overall crop yield. In general, drought increases citrus juice Brix and, usually acidity, potentially resulting in little change in Brix/acid ratio. Drought increases peel thickness and fruit softening and decreases juice content. However, an excess of water can diminish fruit color and increase peel breakdown along with reducing soluble solids and acid, which can lower the quality of the crop.

Proper water management also influences the effectiveness of other crop management practices. Adequate water is essential for efficient uptake and utilization of applied fertilizers, but excessive irrigation often results in nutrient leaching or run-off, meaning that nutrients—especially nitrates—move out of the root zone and into the groundwater. Fertigation (the practice of injecting water-soluble fertilizers into the irrigation water) can be an economical and effective method of fertilizer applications. Some nematicides and insecticides can also be applied efficiently and economically through the irrigation system. This practice is referred to as chemigation. Water stress increases susceptibility to some pests and diseases; however, sustained high soil moisture levels can encourage soil-borne pathogens. Water is also important in the activation of some herbicides, since it is responsible for moving the applied materials into the zone of germinating weed seeds or the root zone of actively growing weeds. Again, excessive irrigation may leach the herbicide beyond the weed root zone, resulting in ineffective weed control and possible contamination of groundwater. Irrigation can also be used effectively for citrus frost and freeze protection if it is properly managed (see Chapter 15).

CITRUS WATER USE

The Florida citrus belt is located in a humid subtropical climate zone that receives an average of 51 to 59 inches (1,300 to 1,500 mm) of rainfall annually. Overall, this amount of rainfall should be adequate to meet the requirements of citrus production in Florida's sandy Ridge soils and poorly drained Flatwoods soils. However, its annual distribution does not usually satisfy peak seasonal demands during the critical fruit-set and early fruit-development phases of March through early June. During this period, good irrigation management is particularly critical to reducing stress and associated yield reduction.

Water use rates and irrigation requirements are usually indicated in terms of inches of rainfall. Citrus trees have lower evapotranspiration (ET) rates than most other crops, indicating efficient water use. ET requirements for healthy, mature citrus trees generally range from 40 to 48 inches per year, no matter where they are grown. To manage microirrigation systems that water only a portion of the grove floor, this must be converted into gallons per tree. These rules of thumb can be used to convert an ET rate expressed as depth, to an approximate volume of water per tree:

Gallons of water per tree per day
For < 125 trees/acre =ET x s50
For 125-160 trees/acre = ET x 200
For 160-200 trees/acre = ET x 150

(Examples: 0.10 inches ET = 25 gal/tree; 0.20 inches ET = 50 gal/tree).

This conversion is only valid for healthy, mature trees. The actual water requirement depends on the age and size of the trees, planting density, soil type, rootstock and scion, and the general health and condition of the trees. For mature trees, the recommended irrigation requirement is 0.18 to 0.20 inches per day (45 to 50 gal/tree/day) during the spring dry season. Daily citrus water requirements for healthy mature Florida citrus in the winter months normally range from 10 to 25 gallons per day.

Newly planted trees require only a fraction of the water needed by mature trees, and the amount of water applied to a young grove should be adjusted for the canopy size. With modern microirrigation methods, it is possible to irrigate individual trees and increase the size of the wetted area or the numbers of emitters as the trees develop within the limit of the pump capacity. Average water use during the first few years is normally only about 1 to 2 gallons per day. Studies have shown that groundcover can greatly affect water use by young trees, with two-year-old citrus trees with grass cover using nearly twice as much water as trees with bare soil. As young trees develop, water usage has been reported to increase about 20 percent per year for years four through six. Recommended spring dry season application rates are 2, 4, 6, 10, and 14 gallons per day, from the second to the sixth year, respectively. Depending on planting density, it may take four to 10 years to achieve maximum ground cover and maximum ET for a newly planted orchard. During the first several years, trees budded on fast-growing rootstocks, such as rough lemon, require more water per tree than trees on slower-growing rootstocks.

Groves planted at high densities have a higher per-acre demand for soil moisture during the first several years. An irrigator may need to modify irrigation practices in the early years in high-density plantings in order to compensate for the higher root concentrations as compared to wider tree spacing. Once citrus trees form hedgerows and hedging and topping begins, the water use per acre will be similar for various planting densities with similar production.

Large trees require more water than smaller trees. Due to their larger size, grapefruit trees typically require more water than sweet orange cultivars, and trees that are not topped or hedged to control tree size have higher ET rates, thus requiring more water.

Some rootstocks (such as rough lemon) have an efficient and abundant fibrous root system that is also deeper and allows extraction of water more readily from the soil than less efficient root systems (such as those on sweet orange). Trees on rootstocks with efficient root systems are more tolerant of dry conditions.

Early citrus production in Florida was accomplished with essentially no irrigation in established plantings. Those early groves were generally planted at low densities along the Ridge on rough lemon rootstock. They developed extensive root systems that foraged widely through the soil profile. Modern groves are typically much more closely planted on poorer soils. Acceptable levels of production are simply not attainable in modern groves without some supplemental watering. In addition, modern nutrient and pest management practices are most effective when coupled with active water management.

> WATER USE IN MATURE FLORIDA CITRUS
>
> Healthy trees need about 7,500 gallons of water per tree each year.
>
> While rainfall in Florida exceeds this amount, its distribution doesn't meet tree needs.
>
> Trees require 45 to 50 gallons per day during the spring.
>
> Trees require 10 to 25 gallons per day during the winter.

IRRIGATION SCHEDULING

The objective of irrigation is to provide trees with sufficient water to minimize stress that may cause reduced yield or poor quality fruit. The volume and timing of water applications depend on climatic conditions, the moisture-holding capacity of the soil, and the stage of tree growth, root development, and fruit development. Studies have shown that irrigating when 25 percent to 50 percent of the total moisture available in the soil is depleted during the period from January through June, and at 50 percent or more depletion the remainder of the year, can provide efficient use of water while maintaining high yields.

The optimum interval between irrigation events depends on the design of the irrigation system, time of year, and soil characteristics, as well as tree size and condition. Citrus ET rates are highly dependent on climatic conditions. Higher evaporation and transpiration rates occur on clear, sunny spring days with low humidity and high winds than on cool, damp, overcast winter days. Florida summers are usually quite wet, but the timing of summer rains may not satisfy citrus water needs even in the wet season, and supplemental irrigation may be needed. Clearly, irrigation scheduling must be adjusted through the year (Table 16.1).

Table 16.1 Suggested citrus microsprinkler irrigation schedule for average conditions.

Young (age = 1 to 3 years)

Month	*Allowable soil water depletion (%)*	*Duration (hours)*	*Interval (days)*
January	50	3-4	7-8
Februrary	25	2-3	3-4
March	25	3-4	3-4
April	25	2-3	2-3
May	25	2-3	2-3
June	25	2-3	2-3
July	25	2-3	2-3
August	25	2-3	2-3
September	25	2-3	2-3
October	25	2-3	3-4
November	50	3-4	2-6
December	50	3-4	7-8

Mature Trees

		Ridge*		**Flatwoods***	
Month	*Allowable soil water depletion (%)*	*Duration (hours)*	*Interval (days)*	*Duration (hours)*	*Interval (days)*
January	50	4-6	7-8	4-6	7-8
Februrary	25	3-4	3-4	3-4	3-4
March	25	3-4	3-4	3-4	3-4
April	25	3-5	2-3	3-5	2-3
May	25	3-4	2-3	4-5	2-3
June	25	3-4	2-3	4-5	2-3
July	25	4-6	3-4	4-6	2-3
August	25	4-6	3-4	4-6	2-3
September	25	4-6	3-4	4-6	2-3
October	25	4-6	4-5	4-6	3-4
November	50	4-6	6-7	4-6	2-6
December	50	4-6	7-8	4-6	7-8

*Astatula and Candler soils were used for Ridge soils; Immokalee soil was used for Flatwoods root depth-ground cover-tree area. Irrigation efficiency was set at 75 percent. Irrigation frequency and duration depend on tree size, soil type, rooting depth, climatic factors, microsprinkler volume, and spray diameter. Water usage by ground or weed cover in the tree rooting zone will be a factor in the soil moisture equation. See EDIS Fact Sheet: http://edis.ifas.ufl.edu/HS204.

A tensiometer can be a valuable tool for irrigation scheduling, especially in sandy soils. Tensiometers should be installed in the root area wetted by the microsprinkler or dripper, near the drip line of the tree. They should be calibrated annually to ensure that they are working properly, and they require periodic maintenance and cleaning. In most of the sandy soils on which citrus is grown in Florida, half of the available soil moisture will be depleted by the time tensiometers read 20 to 30 centibars. Therefore, during the period from prebloom through June drop, irrigation should commence before soil tensions reach 10 to 15 centibars. During the rest of the year, it should commence when readings are no more than 30 centibars. Other soil moisture measuring devices provide automatic readings than can be coupled to the irrigation system by computer for automatic irrigation scheduling. Care should be exercised regarding the depth and placement of all devices (Plate 16.3), and enough of them should be used to represent the whole grove.

On sandy Ridge soils, irrigation applications should normally provide sufficient water to wet the upper 2 to 3 feet of the root zone. For shallow-rooted trees in the Flatwoods, the entire 12- to 24-inch root zone (often 14 to 16 weeks) should receive adequate moisture. On sandy soils, irrigation is typically required every one to three days during peak water use times.

Application rates of microirrigation emitters are governed by flow rate, pressure, diameter of coverage, and run time. It is a good practice to measure application rates of the emitters as installed in the field. The efficiency of the application is affected by various environmental factors, including wind and relative humidity. Longer duration or nighttime irrigation may be needed to compensate for low application efficiency. One of the most important elements of irrigation management is a frequent field check of soil moisture. Use a shovel or auger to dig down into the root zone to assure that the irrigation program is supplying the required amount of water to the trees without overwatering them (Table 16.1).

Efficient Water Use and High Yields in Sandy Soils

Irrigate at 25-percent or more soil moisture depletion (10 to 15 centibars in sandy soils) from January through June.

Irrigate at 50-percent soil moisture (30 centibars in sandy soils) for the remainder of the year.

Soil moisture measuring devices are especially useful for scheduling irrigation in sandy soil.

ROOT SYSTEMS AND THE WATER TABLE

The root systems of citrus trees are usually relatively shallow compared to those of deciduous fruit trees. Even in the well-drained soils of the Central Florida Ridge, the main root system of citrus trees is concentrated in the upper 2 to 3 feet of soil. Citrus trees grown in the coastal Flatwoods areas of Florida typically have rooting depths of less than 18 inches. Shallow water tables can provide significant amounts of water for root uptake. High water tables permit upflux of water from the water table into the root zone, and thus decrease the need for supplemental irrigation.

Capillary rise can be up to 12 inches above the free water surface in loamy fine sand, but less than 6 inches in fine sand. Generally, water table depths within 20 to 30 inches of the surface are adequate to maintain sufficient moisture in the root zone so that ET is not limited. This is often the case in the Flatwoods. When the water table in most Flatwoods soils drops 20 inches or more below the midroot zone (normally 30 to 36 inches below the surface), irrigation should commence, since the rate of water movement into the root zone by capillary action is insufficient to meet the water demands of the trees.

Root hydraulic (water) conductance is continuously reduced as soil temperature decreases from 95° to 41°F, even when the soil water supply is adequate. Consequently, citrus trees may suffer from drought stress when the soil is cold, and there is a high evaporative demand due to low humidity or high wind. This rarely happens in most Florida Citrus areas.

SOURCES AND QUALITY OF IRRIGATION WATER

Irrigation, made necessary by frequent drought periods throughout the winter and spring months, has made agriculture the single largest user of water in Florida. Water shortages, along with water quality issues, have caused the general public, environmental groups, and regulatory agencies to focus on agricultural practices, affecting the volume and quality of water available for use by citrus growers.

Water sources available for irrigation of citrus include surface water (canals, ponds, and lakes), groundwater (through digging wells into various aquifers), reuse of processing plant water, and recycled wastewater from industry and domestic sources. Water quality in the wells and lakes in the Central Florida Ridge area tends to be excellent, with low salinity levels. But these sources are not abundant, and large plantings may require several wells to supply enough water to meet irrigation demands.

The water from wells in the Floridian and Hawthorne aquifers (in the Indian River region and southwest Florida) can have high salinity levels. Typically, the salinity of the Floridian aquifer increases with depth. Therefore, sacrificing water yield by tapping only the top of the Floridian aquifer is often the best long-term choice for groves where ground water shows significant salinity. Frequently, the best choice for flatwoods growers is a blend of plentiful, poor quality water from the Floridian or Hawthorne aquifers with surface water or shallow groundwater. This yields sufficient water of acceptable quality.

Reuse of processing plant or industrial process water is available in some areas. The quality of this water varies considerably from source to source, and also with time. Caution must be exercised if industrial cleaning agents (frequently sodium- or chlorine-based materials) are contained in the reuse water. At high rates, these chemicals cause defoliation and dieback in citrus trees.

The use of reclaimed sewage wastewater is an option in some areas. Reclaimed water can provide growers with an economical supply of fairly good quality water. Water Conserv II, a combination of agricultural irrigation and rapid infiltration basins (RIBs), is the first project in Florida in which the Florida Department of Environmental Protection (FDEP) permits irrigation with reclaimed water for crops grown for human consumption. The project is jointly owned by the City of Orlando and Orange County. Research on the Conserv II project has shown few detrimental effects to citrus, even when contracts call for the application of 50-plus inches of irrigation with reclaimed water per year (Figure 16.2). Any reclaimed water supplier should guarantee long-term maximum total salinity levels and pledge that Cl, Na, B, and heavy metal concentrations will not rise to levels that adversely affect tree health, growth, and production.

SALT STRESS

Water quality is an important factor in citrus production because citrus trees are more sensitive to soil and water salinity than many other crops. The quantity of salt in the poorer irrigation sources can be very significant. For example, there are about 1.7 pounds of salt in each 100 gallons of water when the salinity level is 2,000 parts per million (meaning that each acre-inch of water applied contains over 400 pounds of salt). In Florida, there is usually adequate summer rainfall to leach out accumulated salts if the soil is sufficiently permeable.

During extended dry periods, however, salts applied in high-salinity irrigation water can lead to leaf drop, twig dieback, diminished fruit set, and reduced fruit size. Salt-injury symptoms on leaves can occur on hot, dry windy days if irrigation water containing even 800 to 1,000 ppm of salts is allowed to intermittently wet leaves or young shoots.

Excessive salts reduce the growth and yield of trees on all rootstocks. However, rootstocks differ in their salinity tolerance. The following rootstocks are arranged from most tolerant to least tolerant: Cleopatra, rough lemon, sour orange, Swingle citrumelo, and Carrizo citrange. Consider the quality of irrigation water available when selecting rootstocks. Rootstocks do not affect the sensitivity to injury of scions through direct foliar contact with saltwater. Young, tender shoots are especially vulnerable to salt burn, and young trees on Swingle citrumelo are especially sensitive to salt spray on their trunks.

It should be noted that salts in solution reduce the availability of water to the plant even when soils are wetted to field capacity. Roots are not able to extract as much water from a solution that is high in salts as from a solution that is low in salts. Consequently, water stress is dramatically increased when the available water is high in salt.

The critical salinity level for citrus varies with the buffering capacity of the soil, climatic conditions, and the soil moisture status. Irrigation scheduling becomes very important when saline water is used. Once salts are in the root zone, they can only be removed through leaching. As the soil dries, the salts are concentrated in the remaining soil water. Therefore, maintaining a high soil water content and applying irrigation quantities great enough to provide a downward movement of salts through the root zone helps to minimize salt damage.

Nutrient management practices have a direct effect on the concentration of salts in the soil. A fertilization program consisting of frequent applications with relatively low concentrations of salts reduces salinity stress, as does the use of controlled-release formulations. In general, natural organic fertilizers, and high-analysis fertilizers that use mostly essential element ingredients, may have a lower salt index than lower-analysis fertilizers and are less likely to produce salt injury. It should also be noted that some nutrient imbalances, especially inadequate levels of K and Ca, might compound the effects of salinity stress.

Integrated Water Management

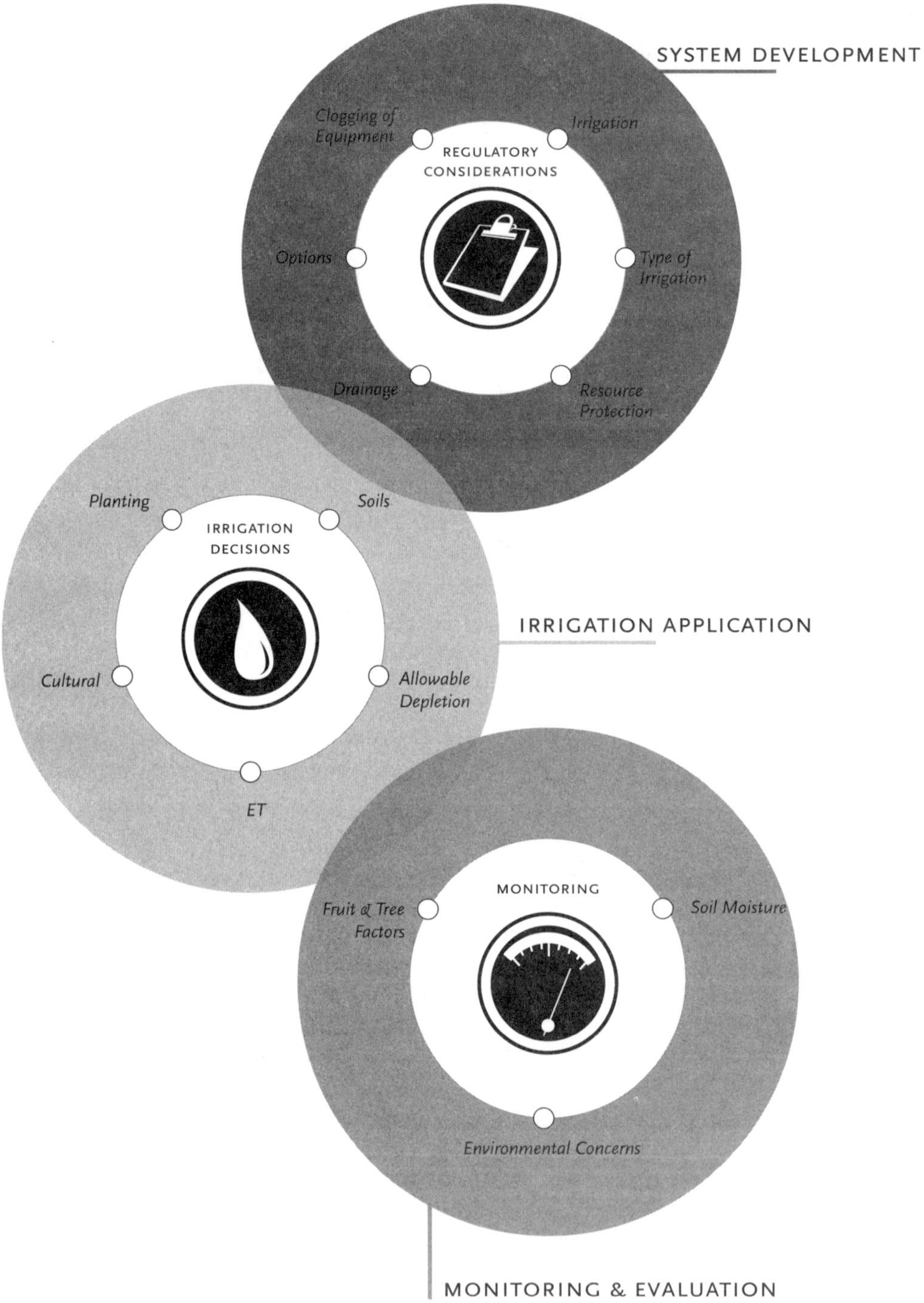

SYSTEM DEVELOPMENT

REGULATORY CONSIDERATIONS
Consumptive use permits
Shortage restrictions
Off-site discharges
Well permits

IRRIGATION
Wetted area coverage
GPH per tree
Efficiency
Uniformity
Water quantity
Water quality
Labor availability

TYPE OF IRRIGATION
Drip
Microsprinkler
Overhead
Flood
Under-tree
Volume gun
Seepage

MAINTENANCE
Filtration
Water treatment
Water source

CLOGGING OF IRRIGATION EQUIPMENT
Insects
Bacteria
Algae
Precipitates

DRAINAGE
Beds and water furrows
Lateral ditches
Perimeter ditches
Tile drainage
Retention/discharge

OPTIONS
Freeze protection
Fertigation
Chemigation
Automation

RESOURCE PROTECTION
Endangered species
Impacts to wetlands
Water quality

IIRIGATION APPLICATION

IRRIGATION DECISIONS
When to irrigate
How much to apply

SOILS
Water holding capacity
Hardpan
Water table

ALLOWABLE DEPLETION
Time of year
Soil characteristics
Irrigation system

CULTURAL
Herbicide
Fertilization
Pests
Chemigation

PLANTING
Rootstock
Scion
Age and size
Root zone

ET
Time of year
Size/condition
Temperature
Humidity
Wind
Radiation

MONITORING & EVALUATION

MONITORING
Soils-trees-fruit
Off-site discharges
Feedback
Adjustments

SOIL MOISTURE
Tensiometers
Sensors
Water budget
Feel method

ENVIRONMENTAL CONCERNS
Off-site water discharges
Nutrient and pesticide leaching
Off-site sediment movement
Off-site nutrients and pesticides discharge

FRUIT & TREE FACTORS
Fruit yield and quality
Tree size, growth, canopy

Deficient
fruit drop, lower yield, small fruit, reduced post-harvest quality

Excess
leaching, lower TSS, increased disease, increased weeds, reduced post-harvest quality

Figure 16.2 Diagram of various factors involved in and influenced by integrated water management programs.

Quality of Irrigation Water

Surface and well water quality are usually very high on the Ridge.

Well water near the coast and ditch water during droughts also may have high salinity.

If a grower is forced to use saline water, irrigation must be more frequent to drive salts out of the root zone, and soil moisture must be higher to avoid water stress.

When using reclaimed or wastewater, ensure that salts, metals, or industrial cleaners are below levels that hamper production.

MAINTENANCE OF MICROIRRIGATION SYSTEMS

Irrigation systems must be maintained in good working order. One of the biggest problems with microirrigation systems is the clogging of emitters, which results in lowered system performance and water stress to the nonirrigated plants. As surface and groundwater are drawn into an irrigation system, changes in chemical balances and bioactivity can result in reactions that cause the formation of precipitated scales and slimes, impairing the irrigation system.

Prevention of clogging problems is best accomplished by frequent water treatment. The chemicals involved are generally composed of a variety of organic and inorganic compounds that have been found to prevent and dissolve scales. Product stability, injection rates, biological control capabilities, and required dosages vary widely. "Shock treatments" are directed towards existing problems. In order for shock treatments to be effective, the system must still allow some flow and be completely flushed after each shock treatment. The two major shock treatments used are chlorination (used to kill bacteria and algae) and acidification (used to dissolve or break up existing scale from the water lines). Acid injection is also used to make chlorination more effective. Preventative treatments usually require less manual labor, since little or no line flushing is required.

An effective weed control program is essential for groves watered with microirrigation systems. Weeds and high grasses disrupt the wetting pattern of microsprinklers and compete for available water. Lack of weed and cover crop control increases irrigation system maintenance costs and reduces the effectiveness of fertigation and chemigation applications.

Maintenance of Microirrigation

Microirrigation is efficient, effective, and inexpensive to install and run, but requires continual maintenance.

Clogging of emitters by scales and slime is a primary problem.

Schedule frequent visual inspection of emitters during operation.

Maintain a program of chlorine and acid flushing to remove materials that accumulate, and flush cleaners from lines.

DRAINAGE

In general, citrus does not tolerate prolonged flooding of the root zone, although the lemon rootstocks (Milam, Volkamer, and rough lemon) and sour orange rootstock are more tolerant than others. Cleopatra mandarin, Carrizo citrange, and citrus macrophylla are the least tolerant. Most flatwoods sites require an active drainage system to rapidly remove excess water from the citrus root zone. The system should be installed after the land is cleared but before the trees are planted. The system then requires regular maintenance and monitoring.

THE NEED FOR DRAINAGE

In flatwoods groves, adequate drainage is essential to ensure that summer sprays for mites, insects, and fungal diseases can be applied at appropriate times. Without drainage, the application of insecticides and fungicides on the water furrow sides of the trees can be difficult or impossible during rainy periods. Applications made when water furrows are even marginally wet may result in equipment damage and disruption of the flow patterns in water furrows, requiring reshaping of the furrows.

MAINTENANCE OF DRAINAGE SYSTEMS

Excessive chemical control of vegetation (via herbicides) in bedded groves results in erosion of soil from bed tops, and reduces the effectiveness of water removal through water furrows and swales. Ditch banks and aquatic weed management are also essential to allow unimpeded water flow.

WATER MANAGEMENT

Integrated Crop Management (ICM) promotes a holistic approach to citrus management. Each practice (for instance, water management) is evaluated not only for its direct effects, but also for its influence on other practices within the overall citrus production system. In today's economic, political, and environmental climate, best management practices cannot be considered independently from water and nutrient issues. Their use, efficiency, and impacts are closely interrelated.

Irrigation is only one component of the complete water management system, which also includes ET rates, soil moisture retention, rainfall, and drainage. These five components affect horticultural plant protection practices and postharvest behavior of fruit by maintaining tree root zone moisture content in the adequate range, while optimizing nutrient and soil pesticide use efficiency. Significant soil moisture deficiencies and excesses result in: tree stress, reduced production and poor fruit quality, increased susceptibility to pests and diseases, and an overall negative effect on tree health and grove profitability. Figure 16.1 shows an overview of Integrated Water Management.

Good water management includes the following:

1. Balanced water inputs and removal, keeping soil type, rainfall, temperature, existing water table levels, tree demands, and other factors in mind.
2. Ensuring a positive effect on tree health, growth, production, fruit size, and quality. For instance, irrigating with the correct amount of water can reduce fruit drop during the dry period from spring to early summer. Water deficiency or excess can reverse such effects.
3. Keeping varietal demands (some need additional water to sustain tree vigor and fruit production) and rootstock preference (rootstocks have different rooting densities and depths, creating different demand) in mind.
4. An awareness of tree planting densities, since greater canopy area and rooting density per acre exert higher moisture demand.
5. Efficient nutrient and soil-applied agrichemical utilization particularly as it relates to reduced root zone leaching and surface runoff.
6. Control of soil salinity, achieved by lowering the concentration of salts in the root zone.
7. An awareness of the impact of irrigation delivery systems on the efficiency and uniformity of application of nutrients and other agrichemicals in the tree-rooting zone.

8. Grove floor vegetation management to (1) minimize competition between trees and groundcover for water and applied nutrients; (2) prevent groundcover interference with emitter operation; and (3) achieve these goals with minimal cost while maintaining sufficient groundcover for erosion control and grove operations.
9. Ditch bank and aquatic weed management, so that water flows unimpeded into system intake areas while minimizing losses due to aquatic weed transpiration.
10. Allowances for seasonal water requirements for cold-protection purposes during freeze-prone periods.
11. Contributions to pest and disease management by (1) enhancing the ability of trees to tolerate pests, such as nematodes; (2) sustaining a soil environment at moisture levels less favorable for soil-borne diseases; (3) minimizing the use of overhead sprinkler systems on cultivars susceptible to foliar and bloom pathogens; and (4) influencing growth of grove floor vegetation.
12. Irrigation scheduling using ET-based water balance models, such as the accounting method, or on some recognized moisture monitoring system, such as tensiometers or capacitance-sensors and where appropriate, the use of water table observation well readings.

Crop water use models estimate ET demand using environmental inputs from weather stations. These parameters are relative humidity, wind speed, air temperature, and solar radiation. Once ET has been estimated, specific water use can be estimated, based on tree size and spacing. Soil water balance models can then be used to estimate soil water status by subtracting daily water use from the sum of the water content the previous day plus rainfall.

Under Florida conditions, ET varies little within a geographical area of 50 to 100 miles. Therefore, ET for a given planting can be estimated using a regional weather station, such as one in the FAWN (Florida Agricultural Weather Network) system. Nevertheless, local rainfall information should still be maintained. Irrigation scheduling using soil water status has been used successfully throughout the world. This requires frequent monitoring of soil water tension (tensiometer) or soil water content (capacitance probes) during periods of high irrigation demand. Soil water status set points for optimum irrigation are maintained by utilizing soil water depletion levels based on crop development. The citrus industry will need to continue to challenge the imposition of future, more restrictive water regulations without information on crop water needs and appropriate irrigation scheduling parameters to support its water demands. Water use based solely on observations of soil and tree conditions cannot be quantified, and is not acceptable to regulators seeking specific information on seasonal tree water requirements and nutrient leaching.

WATER QUALITY MANAGEMENT STRATEGIES

The threat of groundwater pollution has become one of many environmental, economic, and social problems requiring citrus growers to reexamine their traditional growing practices. Many irrigation and drainage practices were developed in an era when the environment's capacity for recovery appeared to be infinite, and man's effect on the environment was considered negligible. Growing practices were designed solely to maximize profit, with little consideration of long-term consequences. Our more modern understanding of agriculture's role in the environment has prompted the citrus industry to develop and implement low-impact growing practices. Many crop management practices have been reevaluated with the specific goal of reducing off-site impacts and their associated threat to groundwater.

There is an inherent risk of groundwater contamination whenever nitrogen fertilizers are applied to the soil. Elimination of nitrate movement through the soil to groundwater is probably impossible without significant yield losses; nevertheless, recommendations for reducing nitrate movement have been developed, and many growers have adopted these recommendations. Growers are advised to avoid unnecessary fertilization by basing their fertilization program on analysis of leaf nitrogen content. Other tactics include foliar applications of nitrogen rather than soil applications, and applying fertilizer in split doses, both to improve plant uptake and reduce excess nitrates in the soil at any one time to minimize leaching. Finally, most growers now avoid applications of nitrogen fertilizers during the rainy summer months and manage irrigation to avoid nutrient leaching.

The application of herbicides and various other soil-borne agrichemicals also presents a threat to groundwater. Methods for reducing the risk of groundwater contamination associated with these products are less well-developed, but the same guiding principles apply. These chemicals move through the soil with percolating water, so applications should be avoided during rainy periods, and irrigation schedules should be coordinated with chemical applications. Many newer postemergence materials are less likely to leach into groundwater and may be good substitutes for established chemicals. Their higher direct costs may be justified by reduced losses to groundwater. Revising the scheduling and placement of herbicide applications and minimizing the quantity applied may reduce the threat to groundwater and improve the effectiveness of the applications.

With the reexamination of traditional fertilizer and herbicide applications, the fundamental philosophy of citrus grove vegetation management has come under scrutiny. Traditionally, a "bare floor" has been maintained in groves. However, vegetation cover plays an important role in cycling nutrients and enhancing soil biological activity. The use of appropriate cover crops can help protect groundwater by removing excess soil nitrate and suppressing aggressive native weed growth, hence reducing the need for herbicides. Unfortunately, cover cropping also has its disadvantages. The greatest drawback may be increased water demand, with its associated economic and environmental costs. It is possible, however, to manage cover crops so that total water requirements during critical shortage periods are only slightly greater than those of a bare grove floor. Other disadvantages include increasing frost hazards due to reduced bare ground as a heat sink, difficulties associated with tillage, difficulties with shading of the groundcover, and the initial cost of cover crop establishment. These problems are challenging, but not insurmountable. Many growers have developed solutions by experimenting with small plots.

Techniques for reducing groundwater contamination must be considered within the entire system of crop management. For example, growers using irrigation for frost protection may need to delay residual herbicide and fertilizer application until after the high freeze probability period. Growers who choose to use only contact herbicides or foliar nitrogen applications must consider the increased travel over the grove floor and increased fuel use, as well as the possible establishment of tolerant weed species. The key to effective modern crop management is the integration of all growing practices.

Since water management is such an important part of production and water uptake is a chief regulator of tree growth, it is useful to view integrated water management from system development, application, monitoring, thorough evaluation of final results. Protection of natural resources has emerged as an important public policy issue, and survival of the citrus industry depends on a prompt response to environmental concerns. Through ICM, growers are working to maintain the profitability of the crop with minimal effect on the environment. It is essential that growers recognize the inevitability of more stringent environmental regulations and develop viable crop management alternatives. Skillful crop management will be required as the availability of many traditional agrichemicals becomes limited and their application becomes more restricted.

REFERENCES

Boman, B. J. 2003. *Lysimeter evapotranspiration measurements for young 'Valencia' orange trees.* International Society of Citriculture Congress. (9th 2000: Orlando, Florida), Orlando, Florida USA 1: 544-548.

Parsons, L. R., T. A. Wheaton, and W. S. Castle. 2001. *High application rates of reclaimed water benefit citrus tree growth and fruit production.* HortScience 36: 1273-1277.

17

Nutrient Management

T.A. Obreza and D.P.H. Tucker

INTRODUCTION

Most Florida citrus trees are planted on sandy soils that contain little clay or organic matter. These soils are low in inherent fertility and retain only small quantities of water and water-soluble plant nutrients against leaching by rainfall or excessive irrigation. Thus, a good nutrition program is required to keep trees at optimum productivity. The nutrient requirements of citrus trees are influenced by many factors including soil type, scion and rootstock selection, pruning practices, weed control, row-middle management, water management, and intended market (fresh or processed). Selection of nutrient sources and rates, timing and frequency of application, and placement and method of application are based on soil type and vulnerability to leaching, scion and rootstock variety, tree age, rooting depth and area, soil and leaf analysis, fruit production and nutrient removal, and economics. This chapter presents the basic principles of citrus nutrition and explains how they relate to the other important aspects of a citrus ICM program.

Citrus trees need a variety of nutrients for good health, maximum yield, and high-quality fruit production. Most Florida citrus trees are planted on sandy soils that contain little clay or organic matter. These soils are low in inherent fertility and retain only small quantities of water and water-soluble plant nutrients against leaching by rainfall or excessive irrigation. A successful nutrient management program provides the trees with essential elements that are not sufficiently available in the soil.

The goal of fertilizing nonbearing citrus trees is to promote vigorous vegetative growth that rapidly produces a canopy volume with high fruit-bearing capacity. After trees begin to bear substantial fruit, fertilizer is applied to replace nutrients lost with the harvested fruit and to support only that amount of new vegetative growth necessary for the subsequent crop. The nutrient requirements of citrus trees are influenced by many factors, including soil type, scion and rootstock selection, pruning practices, weed control, row-middle management, water management, and intended market (fresh or processed). A sound nutritional program uses no more fertilizer than is necessary to achieve the desired crop response.

Selection of nutrient sources and rates, timing and frequency of application, and placement and method of application are based on soil type and vulnerability to leaching, scion and rootstock variety, tree age, rooting depth and area, soil and leaf analysis, fruit production and nutrient removal, and economics. Potential nutrient sources consist of chemically synthesized products and formulations specifically developed for crops, and include animal waste products, municipal wastes, urban plant debris, and mined minerals.

The economics of citrus nutrient management depend on the rate of application of fertilizer, the cost of the nutrient sources, and the cost of application. Further, the environmental cost of fertilization, especially the effect on water quality, must be considered. These costs must be balanced by the value of the fruit or trees.

SOIL QUALITY

Most Florida soils in which citrus is grown may be classified into one of three soil orders: Entisols, found predominantly on the Central Florida ridge; Alfisols, found predominantly on the central and lower east coast of Florida; and Spodosols, which are numerous in Southwest Florida, but are also common on the east coast (Plate 17.1). Entisols are deep, well-drained sandy soils. Alfisols are lowland soils with a sandy surface layer typically overlying a loamy layer that may be composed of calcareous (high calcium-containing) clay. Spodosols are sandy acidic soils overlying an organic hardpan, with perhaps a layer of calcium carbonate rock in the lower soil profile. Of the three, Alfisols are the most naturally fertile, and Entisols are the least fertile.

Soil is a dynamic, living body that plays a key role in water and nutrient management in agriculture. Soil is comprised of inorganic mineral matter (sand, silt, and clay particles), organic matter, water, gases, and living organisms, such as earthworms, insects, bacteria, fungi, algae, and nematodes. There is a continuous interchange between the solid, liquid, and gaseous phases of soil elements, mediated by a host of physical, chemical, and biological processes. Soil organic matter is a major pool for carbon (C), nitrogen (N), sulfur (S), and phosphorus (P). Essential phases of the carbon, nitrogen, sulfur, and phosphorus cycles occur through microbial interactions within the soil. The availability of nutrients is constantly altered by microbial mineralization and immobilization. Inorganic minerals in the soil play a major role in retaining cations—hydrogen (H), calcium (Ca), magnesium (Mg), and potassium (K) through ion exchange, while organic components and anions like phosphate (PO_4) and sulfate (SO_4) are retained through sorption reactions. Soil also acts as an essential reservoir of water and can also function as a purifying medium.

In Integrated Crop Management (ICM), soil quality refers to the capability of the soil to sustain long-term production of safe and nutritious crops while minimizing negative environmental impacts. Nutrient and water management in Florida's sandy soils focuses on the volume of soil occupied by citrus roots. This zone is also inhabited by competing weed roots and pathogenic microorganisms, and is often characterized by limited nutrient-holding capacity and extremes of soil moisture. Soil management must consider the infiltration rate and water-holding capacity of the soil, the soil depth to the surficial water table, nutrient concentrations, soil pH, and percentage organic matter. Organic matter is almost always low in the sandy soils where most Florida citrus is grown. Amending the soil with organic matter is sometimes recommended to enhance soil quality and increase the availability of water and nutrients.

SOIL PH

Soil pH is a measure of the concentration of hydrogen ions in the soil solution, which is the water that resides in the pores between soil particles. The pH scale ranges from 0 to 14, with values below 7 considered acidic and values above 7 considered alkaline. The pH of Florida soils is typically between about 5.0 and 8.4. Highly acidic soils (pH < 5.5) usually need to be limed for optimum citrus production. Citrus nutrient availability is optimal in soils with a pH of about 6.0.

Unfortunately, it is not economically feasible to lower the pH of alkaline, calcium carbonate-based soils.

Soil pH should be tested every one to two years. Applying either calcitic lime (calcium carbonate only) or dolomitic lime (calcium plus magnesium carbonate) may increase the pH if it is too low. The addition of dolomitic lime is recommended if the soil is also low in Mg. Growers should be aware that deep groundwater usually contains dissolved limestone and can increase soil alkalinity when used for irrigation.

ESSENTIAL NUTRIENTS

On a fresh weight basis, water is by far the largest component of the citrus tree. On a dry weight basis, however, the primary materials are carbon-containing compounds like cellulose, lignin, sugars, starches, and proteins. As the tree grows, a large amount of carbon must be taken up as carbon dioxide from the atmosphere and used to create the carbon-containing compounds that make up tree tissues. This process is accomplished by photosynthesis, using energy from the sun, chlorophyll in green leaves, and water to produce life-sustaining sugars (also known as photosynthates) needed to create other compounds and fuel tree processes.

Sugars from photosynthesis are needed for growth, and the major tree components compete for their share. The entire system depends on leaves as the producers of sugars for growth and maintenance. Consequently, within the system of the tree, leaves have the highest priority for photosynthates; roots are secondary. When leaves or roots are lost or damaged, photosynthates are allocated for their regrowth. Fruit has the lowest priority, and fruit yield declines if the supply of photosynthates is limited.

Vegetative growth, especially additional root growth, may be triggered by moderate nutrient stress as the tree seeks to increase nutrient uptake. When this occurs, photosynthates are less abundant for fruit, reducing yield and diminishing fruit quality. The grower strives to assure that a sufficient quantity of nutrients required by a bearing tree is available so that the supply of photosynthates is not limited for fruit growth and development.

Seventeen chemical elements are essential for satisfactory citrus tree growth. Three of these—carbon (C), hydrogen (H), and oxygen (O)—are obtained from air (CO_2) and water (H_2O). Plants absorb the remaining fourteen mineral nutrients as dissolved ions, originating from inorganic or organic sources. Some of these nutrients are not always sufficiently present in Florida soils and may need to be applied as fertilizer.

Six of the mineral nutrients that are frequently applied as fertilizer are called macronutrients, indicating that trees require them in relatively large quantities. The primary macronutrients are nitrogen (N), phosphorous (P), and K. An inadequate supply of any of these three nutrients has a dramatic influence on fruit quality and yield. It is important to note, however, that excess levels of these three nutrients can affect fruit adversely.

The secondary macronutrients are calcium (Ca), magnesium (Mg), and sulfur (S). Six of the other nutrients that may be needed in fertilizer are part of the nutrient group called micronutrients, indicating that they are required in very small amounts. These nutrients are iron (Fe), zinc (Zn), manganese (Mn), copper (Cu), boron (B), and molybdenum (Mo). The remaining two micronutrients that are of no concern in fertilization are chlorine (Cl) and nickel (Ni). The application of fertilizer represents additional citrus management expense. In each case, the grower must determine the application rate that maximizes economic return. This rate depends on citrus tree response to the nutrient and the economic value of the fruit, and may be lower than a theoretically optimal rate.

Severe shortage of an essential element usually produces a characteristic deficiency symptom in leaves, twigs, or fruit. Observation of these systems and analysis of leaves or soil indicate the need for fertilization to supply the deficient nutrient. Soil analysis is most useful to evaluate the availability of phosphorus, calcium, magnesium, copper toxicity, and, in some cases, the availability of aluminum (Al) and manganese. For the other nutrients, leaf analysis is the only reliable method of evaluation. An annual testing program should be maintained so that trends in nutrient concentration can be determined. Foliar nutrient concentrations change throughout the growing season; therefore, sampling specific types of leaves at a specific time of year and observation of symptoms can make good comparisons. Indeed, Florida recommendations rely on specific sampling timing and protocol. The guidelines for this protocol may be easily obtained from numerous publications. It is generally unrealistic to sample from the entire acreage of a large commercial grove, so indicator blocks can be identified and sampled annually.

Nitrogen

Of the macronutrients applied to citrus, N is required in the greatest amount and exerts the most influence on tree growth, yield, and fruit quality. Yellow leaves, poor shoot growth, and poor fruit set characterize N-deficient trees. To achieve high yields of good-quality fruit, leaf N concentration should be maintained between 2.5 percent and 2.7 percent. Tree growth and fruit yield are maximized when N is maintained in this optimum range. The supply of N is increased through the addition of fertilizer, (manufactured fertilizer salts, manure, organic compost, or biosolids), organic N mineralization resulting from decomposition of leaf and root biomass, nitrate present in the irrigation water, and N present in rain water (about 5 pounds of N per acre per year is deposited in rainfall) (Plate 17.2). The supply is reduced when N is used by citrus for leaf, wood, and root growth or incorporated into fruit (about 12.5 pounds of N per 100 boxes of oranges). Nitrogen is also lost to microbial use, ammonia volatilization (conversion of the ammonium ion to ammonia gas), and denitrification (conversion of nitrate to gaseous nitrogen in the absence of oxygen). Finally, some N is leached below the root zone. The response of fruit yield to the supply of N is not linear (Figure 17.1). Hence, a very large decrease in the supply of N may be required to produce a significant decrease in yield, while N in excess of required amounts will not increase fruit yield and may even adversely affect fruit quality and yield. Environmental damage may also result if nitrate leaching is a problem in the area where the fertilizer is applied.

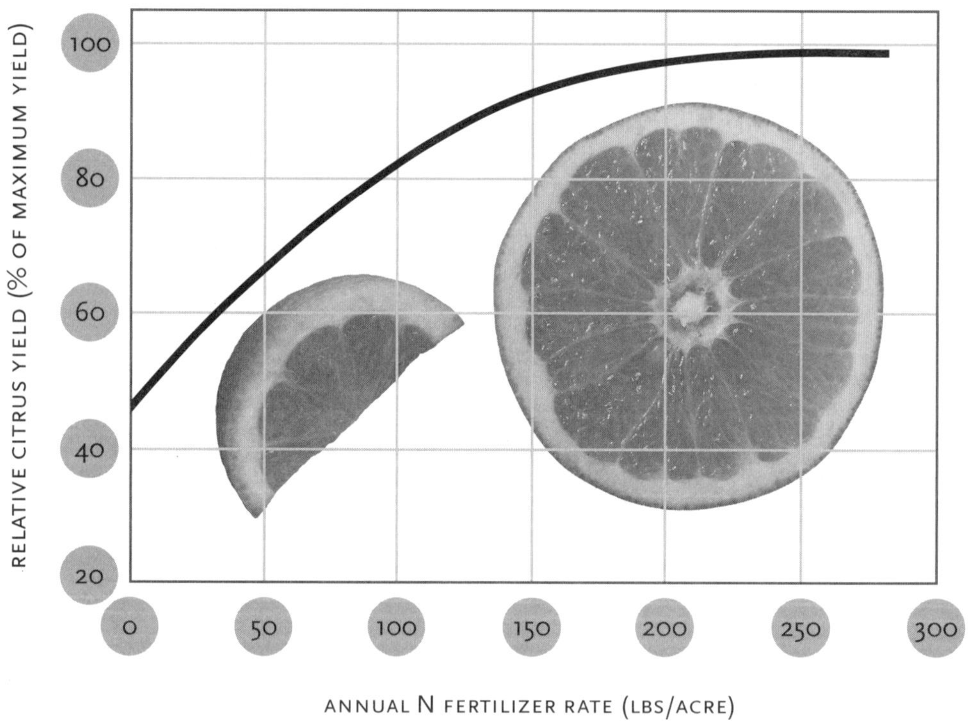

Figure 17.1 Typical response of Florida orange production to N fertilizer rate (Thomas Obreza).

Phosphorus

For best citrus yield and fruit quality, leaf P concentration should be maintained between 0.12 percent and 0.16 percent. Phosphorus does not readily leach where the soil pH is 6.0 or higher, and the fruit crop removes very little P. Therefore, regular P fertilizer applications are usually not necessary. In fact, most sandy soils used for citrus production contain sufficient P that accumulated from fertilizer applications when trees were young. Nevertheless, some soils used for new citrus plantings are low in P and will require P fertilization during the grove's first few years.

Potassium

Leaf K concentration should be maintained between 1.2 percent and 1.7 percent. Potassium plays an important role in determining citrus yield, fruit size, and fruit quality. Potassium can leach from sandy soils where clay and organic matter are low. Citrus has a relatively high K requirement, so it is normally applied annually at a rate approximately equal to or slightly higher than the N fertilizer rate. Potassium deficiency is not common, but may develop with high N rates and high fruit production, especially on calcareous soil.

Calcium

Calcium levels are rarely insufficient. Alkaline soils dominated by lime rock or shell contain an abundance of calcium, and lime ($CaCO_3$) is usually added to acidic soils to raise soil pH. In addition, calcium is present in most irrigation water. Excessive calcium can reduce the uptake of other nutrients and impair the growth of some rootstocks.

Magnesium

Citrus requires Mg in order to produce chlorophyll; a deficiency causes leaf yellowing, known as chlorosis, as a result of low chlorophyll production. Dolomitic lime provides a slowly soluble source of Mg and will also raise soil pH. To treat Mg-deficient citrus trees planted on calcareous soils, foliar Mg sprays are more effective than soil applications.

Sulfur

Citrus trees obtain S from a variety of sources, including the decomposition of organic matter, the atmosphere, and irrigation water. Often, S is applied to the soil with other nutrient fertilizers, such as ammonium sulfate, ordinary super phosphate, potassium-magnesium sulfate, and sulfates of micronutrient metals. Sulfur is not usually applied alone as fertilizer to Florida citrus.

Iron

Iron deficiency occurs in citrus trees planted in calcareous soil, waterlogged soil, soil very low in organic matter ("sand ponds"), or soil high in Cu. The first indication of iron deficiency is the appearance of chlorotic patterns on young leaves; severe deficiency can lead to dieback of twigs. Citrus Fe deficiency is the most difficult micronutrient deficiency to cure. Foliar sprays do not usually provide relief; Fe deficiency should be addressed by soil application of an appropriate Fe chelate fertilizer.

Zinc

Zinc deficiency in citrus trees is expressed as extreme chlorosis; leaves become nearly white, except for green veins. Zinc deficiency in citrus trees is not easily corrected by soil applications, but zinc is readily absorbed from a foliar spray.

Manganese

Manganese deficiency may occur in either very acidic sandy soils or in calcareous soils. A temporary mild manganese deficiency pattern in new shoots is not detrimental to citrus growth or yield, and Mn fertilizer should only be applied if deficiency symptoms persist.

Copper

Copper is usually applied only to newly planted citrus trees in virgin soil. Copper applied in fungicides contributes to Cu accumulation in the soil and readily supplies sufficient Cu for citrus. After numerous repeated applications of Cu, soil accumulation may become excessive. Potential Cu toxicity can be avoided by maintaining soil pH at 6.5 or higher to render the Cu insoluble.

Boron

Boron leaches readily from sandy soils, and a deficiency can develop after several years if a fertilization program does not include it. Boron is easily applied to citrus trees using a water-soluble borate source. It should be applied in alternate years, either to the soil or foliage, but not to both. Boron should be applied very carefully; the difference between beneficial and toxic concentrations in the soil is small.

Molybdenum

Molybdenum (Mo) is more readily available to citrus in alkaline soils than in acidic soils. A foliar deficiency is indicated by interveinal yellow spots. Molybdenum concentration can be increased by the addition of lime or the application of water-soluble molybdenum to foliage.

NUTRITION AND FRUIT QUALITY

As mentioned previously, N, P, and K have the most significant impact on fruit quality and yield. It is important to note, however, that increasing the supply of any of these nutrients may have both desirable and undesirable consequences.

As the supply of N increases, juice volume and color, soluble solids, and acid concentration are increased. However, fruit size, weight, and peel thickness are decreased, and green fruit color at harvest becomes more common. Creasing and scab also increase, but peel blemishes like wind scar, mite russeting, and rind plugging decrease. Adequate concentrations of N also reduce stem-end rot and green mold of fruit in storage. Excessive N can decrease soluble solids per acre and total production.

The solids-to-acid ratio is also increased when the supply of P increases. However, the number of green fruit at harvest also increases, and peel thickness is reduced. Wind scar is increased, but russeting of fruit is decreased.

As the supply of K is increased, fruit size, weight, and peel thickness all increase, as do peel coarseness and the number of green fruit at harvest. Juice content, soluble solids, solids to acid ratio, and juice color are all decreased. Acid concentration is increased.

To maximize soluble solids, citrus trees bearing fruit destined for processing should be fertilized using N and K rates near the high end of the recommended range. If fruit is being grown for the fresh market, however, the N rate should be decreased so that desirable fresh fruit qualities such as large size, natural color, and early maturity are maximized.

METHODS OF NUTRIENT APPLICATION

Fertilizer may be applied to the soil as dry material or as an aqueous suspension, or it may be injected into irrigation water in a process called fertigation. The method of application influences the cost and effectiveness of fertilization and the potential for environmental damage.

Any nutrient may be applied as dry material to the soil surface, but it is recommended that P and some micronutrients be applied in this way under the tree canopies. The fertilizer should be applied over or within the citrus tree root zone, and fertilizer spreaders should be adapted for young trees so that fertilizer is not applied to the open space between the trees.

Fertilizer can also be applied in solution or suspension with an under-tree boom. This method applies nutrients where the most active roots are located. Other agrichemicals, such as herbicides, insecticides, and fungicides, can be applied simultaneously, and fertilizer salts may complement residual herbicides by causing some burn-down of existing weeds. The grower must ensure that the chemicals being mixed in the tank are compatible, and the boom equipment must be constructed of materials that are resistant to fertilizer-induced corrosion.

Fertigation is a popular method used to apply water-soluble fertilizer, especially N and K (Plate 17.3). This method also applies the nutrients where the most active roots are located. The cost of such applications of fertilizer is relatively small, and little travel through the grove is required. However, the cost per unit of fertilizer may be much higher. The grower should note that the uniformity of the fertilization application depends entirely on the uniformity of the water application. To avoid clogging of the irrigation equipment, fertilizer materials that precipitate when combined with calcium in the irrigation water (particularly P) should not be used in fertigation.

Finally, nutritional sprays may be applied to tree foliage in cases where soil applications are not effective. The citrus tree root system is its most efficient nutrient absorption site, so fertilization should primarily be applied to the soil. Foliar applications are usually recommended if adverse soil chemical conditions prevent nutrient availability and tree uptake. This method has traditionally been used only to apply micronutrients, because only a limited amount of fertilizer can penetrate the leaf surface. In modern grove management, this method may be used to apply urea-N and soluble K compounds particularly to enhance flowering and fruit growth. Fertilizers or fertilizer-fungicide combinations applied directly to foliage can burn leaves or fruit, and should be used with extreme caution.

NUTRIENT AND WATER MANAGEMENT

Nutrient management and water management are closely related. The irrigation and drainage management of the root zone directly influences the ability of citrus trees to absorb applied nutrients (Figure 17.2). Further, preservation of water quality is an important concern in nutrient management.

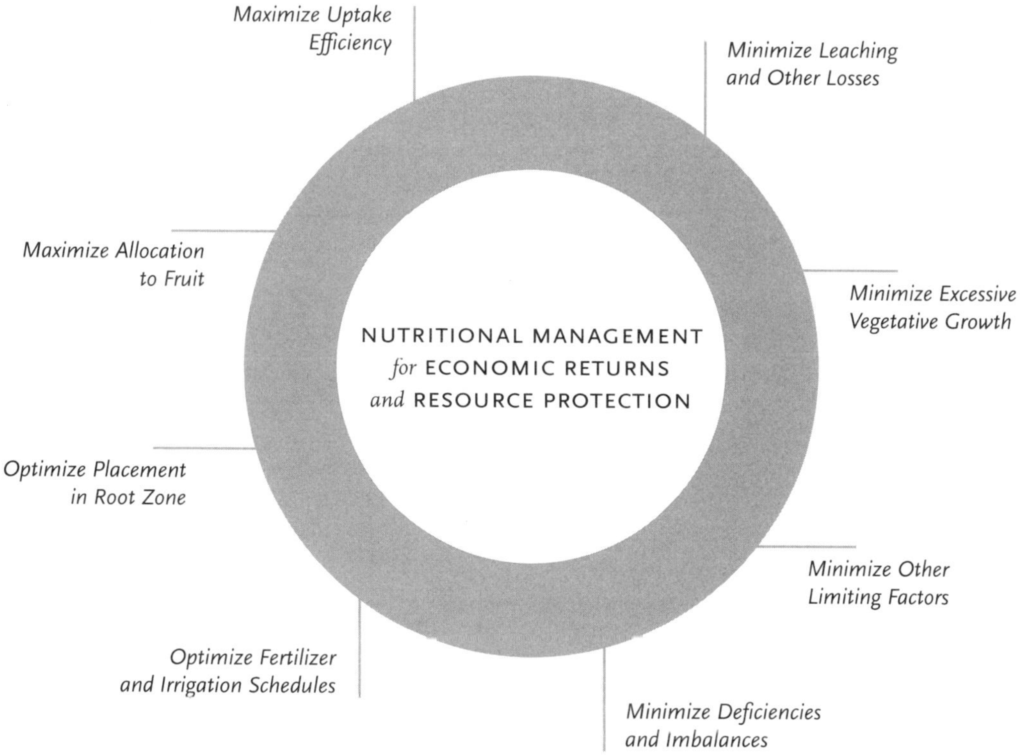

Figure 17.2 Factors affecting nutritional management for increasing nutrient uptake, high economic returns, and resource protection.

Nutrient movement in the soil occurs through water, either by diffusion or mass flow. Mass flow occurs following a rainfall or irrigation that wets the root zone above field capacity. Mass flow can move water-soluble nutrients (for example, N, K, or B) a long distance in a short time and is responsible for leaching nutrients below the root zone. Since the frequency and volume of rainfall are highest in summer, application of water-soluble N should be avoided during that time. In the dry season, the frequency and duration of irrigation should be determined based on evapotranspiration (ET) or soil moisture status to prevent excessive water application and leaching of N.

The presence of nitrate in groundwater has become an issue in some parts of Florida where the soil is vulnerable to downward movement of N directly to a drinking water aquifer. The national drinking water standard for nitrate-N is 10 ppm. More than 90 percent of Floridians rely on groundwater supplies for drinking, so preservation of groundwater quality is essential. Nitrate pollution of groundwater is more likely on deep sandy soils where downward water movement is unrestricted.

Soils commonly used for citrus production on the Central Florida Ridge are particularly vulnerable to nitrate leaching. These soils have a very coarse texture, are excessively well drained, and are low in organic matter. Nitrate-N can move readily past the citrus tree root zone with deep-percolating water from excessive rain or irrigation. Loss of nitrogen by denitrification is minimal because soil aeration is high and denitrifying microorganisms function in low-oxygen environments.

Flatwoods soils are underlain by less permeable clay or organic hard pans, which severely restrict the downward movement of water and nitrate-N. The environment around an organic hardpan, characterized by the presence of C and absence of O, is favorable for denitrification, which appears to decrease the potential for nitrate to reach deep groundwater. If water contamination occurs in the Flatwoods, likely nitrates or phosphates in surface water or the surficial water table would cause it. Phosphate pollution of surface water is more likely where excess rainfall drains across the land surface to canals and lakes.

Best management practices (BMPs) for N fertilization (Figure 17.2) seek economically feasible methods to reduce N losses in drainage waters to an environmentally acceptable level. These practices include lower fertilizer rates, more precise application of fertilizer and split applications, application during the dry season, use of controlled-release or organic sources of N, alternative application techniques including foliar sprays, and improved irrigation management.

To decrease leaching potential, about two-thirds of the annual N fertilizer rate should be applied between January and May, the relatively dry Florida winter and spring, which coincides with the flowering and fruit-setting period. The rest of the N should be applied after mid-September to avoid the summer rainy period. Water-soluble N sources are the least expensive, but are also the most likely to leach if not managed properly. If only water-soluble, dry N fertilizer is used, the annual amount should be split into at least three applications for bearing trees and four to six applications for nonbearing trees. If nitrogen fertilizer solution is added to irrigation water, the application frequency should be increased. Any N-containing product applied should be considered in the total N budget.

ESTIMATED NITROGEN UTILIZATION BUDGET

A fertilizer management program should consider N removed by the crop, consumed in tree growth, and lost from volatilization, leaching, and denitrification (Plate 17.4). The total N removal, when subtracted from the total N applied, estimates the unaccounted-for N that is assumed available for leaching—an issue demanding increased attention. In addition to N applied as fertilizer, a relatively small amount of N becomes available through mineralization (breakdown of organic materials in the soil) and rainfall. The applied N rate should be adjusted for expected crop size to reduce the amount of N available for leaching during years of low production. A higher than normal leaf N concentration that often occurs during low crop years should be an added incentive to lower N input and reduces vegetative growth and leaching.

A nutrient BMP (Best Management Practice) is one that embraces source, rate, timing, frequency, and placement; it is designed to optimize fruit production and quality while producing acceptable economic returns and maintaining water quality standards (Figure 17.3). Thus, a nutrient BMP should be part of an integrated nutrient management scheme that considers all factors relating to fertilizer program development, nutrient application, monitoring, and evaluation.

Florida Citrus
Integrated Nutrient Management

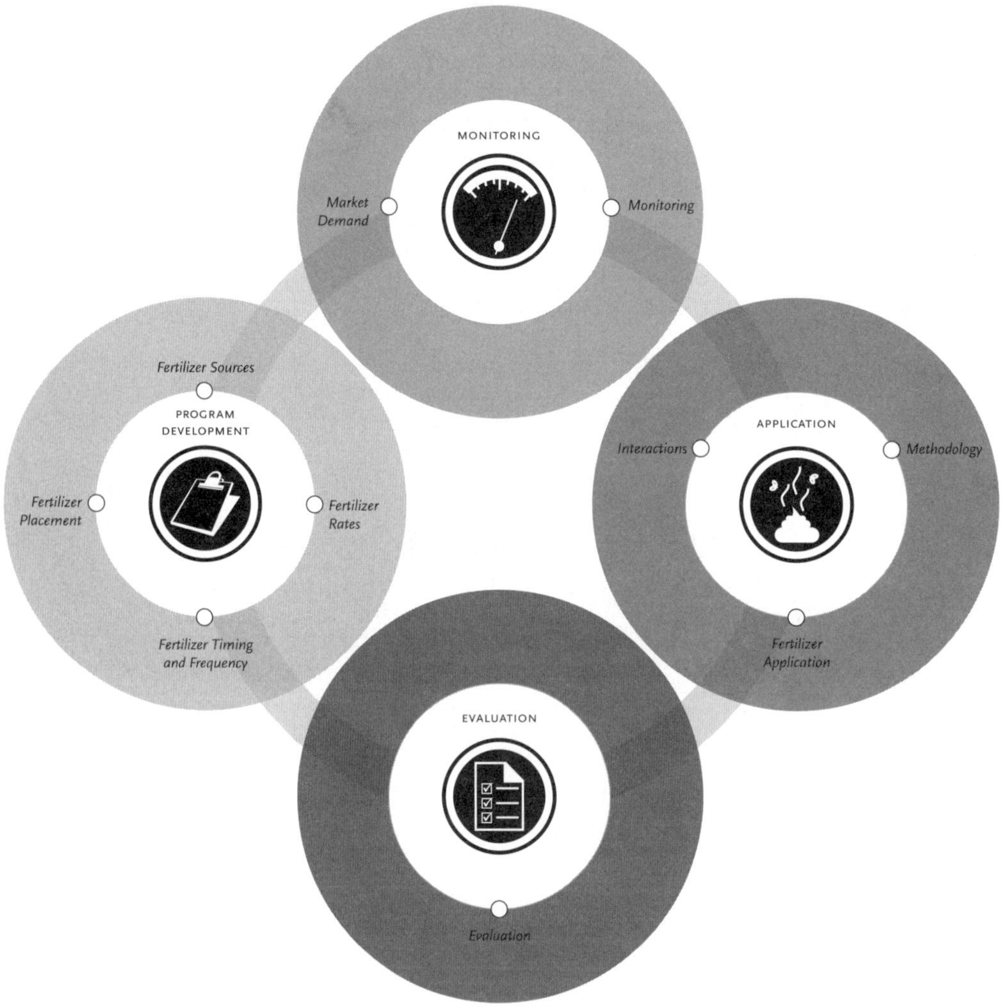

MONITORING

MARKET DEMAND

MONITORING
- Visual
- Leaf tissue analysis
- Soil analysis
- Irrigation and drainage water analysis

PROGRAM DEVELOPMENT

FERTILIZER SOURCES
- Economic situation
- Application method
- Tree age
- Soil pH
- Soil type (vulnerability to leaching)
- Salinity

FERTILIZER RATES
- Tree age
- Production history
- Variety/rootstock
- Soil type (organic matter)
- Organic soil amendments
- Destination of fruit
- Economic situation
- Diagnostic Information

FERTILIZATION TIMING AND FREQUENCY
- Climatic cycles
- Tree age
- Soil type (vulnerability to leaching)
- Fertilizer nutrients and source
- Time of year
- Irrigation method

FERTILIZER PLACEMENT
- Root zone location
- Application method

APPLICATION

INTERACTIONS
- Water management
- Weather variations
- Grove floor management
- Tree size control

METHODOLOGY
- Broadcast dry fertilizer
- Fluid suspension (under-tree boom)
- Fertigation
- Foliar

FERTILIZER APPLICATION
- Economics
- Fertilization timing and frequency
- Availability of equipment and labor
- Type of irrigation system
- Timing of other grove operations
- Environmental considerations

EVALUATION

EVALUATION
- Fruit yield and quality (fresh vs. processed)
- Tree size, foliage color, canopy density, growth habit
- Leaf tissue and soil analysis (visual and laboratory)
- Nutrient leaching and runoff
- Economics

Figure 17.3 Various factors influencing integrated nutrient management programs.

REFERENCES

Alva, A. K., A. Fares, and H. Dou. 2003. *Managing citrus trees to optimize dry mass and nutrient partitioning.* Journal of Plant Nutrition 26: 1541-1559.

Banuls, J., A. QuiZones, B. Martín, E. Primo-Millo, and F. Legaz. 2003. *Effects of the frequency of iron chelate supply by fertigation on iron chlorosis in citrus.* Journal of Plant Nutrition 26: 1985 -1996.

Syvertsen, J. P., and J. M. Dunlop. 2004. *Hydrophilic gel amendments to sand soil can increase growth and nitrogen uptake efficiency of citrus seedlings.* HortScience 39: 267-271.

Zekri, M. 2002. *Fertilizer management and nutrition of citrus trees.* Citrus Industry 83: 12, 14, 17, 29-30.

18

Grove Floor Vegetation Management

D.P.H. Tucker and C.G. Erickson

INTRODUCTION

Effective grove floor vegetation management reduces weed competition for resources and allows for efficient grove management and harvest operations with minimal environmental damage. It minimizes the impact of grove floor vegetation on tree health, growth, fruit production, and other grove practices, using a combination of control and suppression methods. To establish an integrated vegetation management program, the grower must first identify the principal plant species present in the rooting zone of the trees and in the row middles, understand their potential impact, and consider a variety of weed management options. The resulting integrated vegetation management program may include a variety of control methods including cultural, manual, mechanical, chemical, and biological techniques.

Effective grove floor vegetation management reduces weed competition for resources, allows for efficient grove management, and harvest operations with minimal environmental damage. In addition, weed control may contribute to control of pests and disease. A combination of control and suppression techniques is employed with limited chemical inputs.

A weed management BMP is one which minimizes the impact of grove floor vegetation on tree health, growth, fruit production, and other grove practices, using site selection and vegetation monitoring and a combination of control and suppression methods, while avoiding herbicide movement to ground and surface water.

VEGETATION MANAGEMENT

Modern citrus grove management typically involves maintaining tree rows that are largely free of competing vegetation (Plates 18.1a & b), and separated by mechanically and chemically controlled row middles. Ground cover is managed to assist equipment traffic, prevent erosion, and reduce invasion by problem weeds. To establish an integrated vegetation management program, the grower must identify the principal plant species present in the rooting zone of the trees and in the row middles, understand their potential impact, and consider a variety of ground management options. Weeds may compete with the trees for resources, thereby interfering with tree growth and productivity, or they may obstruct other cultural practices. Weeds may also have a significant impact on pest and disease control. The grower selects cost-effective, environmentally sound methods to control or suppress the growth of undesirable grove floor vegetation. The vegetation management program is integrated with other cultural practices to ultimately reduce reservoirs of weed propagules (seeds, tubers, etc.). In this way, the program is efficient, cost effective, and environmentally friendly.

IDENTIFICATION

Before embarking on a program of weed control, the weeds of concern should be identified. There are three categories of weeds: grasses, sedges, and broadleaf plants, including vines. Weeds may also be classified as annuals, biennials, or perennials. The spectrum of vegetation in a citrus grove varies with location, climate, soil type, and previous site history. Weed species are typically identified by a combination of characteristics, and several plants should be examined to confirm the identification. Management is easier if control begins while weeds are still seedlings. Species may also be controlled as though they were annuals or perennials, depending on the growing season and plant age; therefore, for accurate identification, aboveground rooting stems, stolons, and underground storage organs should be noted.

MONITORING

Regular monitoring of grove floor vegetation contributes to an effective weed control program. Weed species should be identified and their location (grove block and row, furrow, middle, etc.) and prevalence recorded regularly, so that population changes over time can be determined. Weed surveys should be conducted at least two to three times a year (e.g., in spring, summer, and fall) by scouts doing other monitoring work. As different locations may receive different treatment combinations and be subjected to various cultural practices, weed species vigor and abundance may vary (Plate 18.2).

Observations should be recorded on a weed survey form, with special attention paid to location of newly established species or difficult-to-control perennials. Vines, shrubs, and other species inhabiting ditch banks, fencerows, and other off-site areas should also be noted, as they may provide sources of future infestations. The objective is always to identify and correct small problems before they become too large and expensive to control.

Weed management should begin during site development, before planting. The site should be surveyed, particularly for hard-to-control perennials that can pose a problem for young trees, and a control program should be implemented.

SPECIES COMPETITION

Resources consumed by competitive species are unavailable to the citrus trees. Weeds compete for water, nutrients, and sunlight. Weed species are considered economic pests if they significantly reduce the survival rate of trees, negatively affect tree growth and health, delay bearing, or cause a significant decline in fruit quality or yield. Weed species vary in their level of competition and in the extent to which they may adversely affect citrus production. In addition, the risk presented by a particular species may depend on the location and conditions in which it is present. Grasses, especially sod-forming species, are more aggressive competitors than most broadleaf species, and mowed grass can be more competitive than unmowed grass because of the moisture demands of vigorous regrowth. Vines in tree canopies can become very competitive for sunlight, and may cover the tree. Relatively sparse weed growth on poor sandy soils may be more harmful than abundant growth on heavier soils with greater moisture and nutrient reserves.

Weeds may also interfere with grove management, reducing efficacy of low-volume irrigation systems, intercepting soil-applied pesticides, and lowering temperatures in tree canopies during freezing conditions by slowing heat transfer to and from the soil. In addition, weeds may increase insect and mite populations, disease incidence, and potentially damaging rodent activity.

Large numbers of seeds representing numerous weed species reside in the surface soil layers in which tree roots are established. Each species has its own period of emergence from these seeds, and weed competition is of concern for most of the citrus-growing season. Although total elimination of weed seed and reproductive plant material is neither economically practical nor environmentally desirable, appropriate control and suppression methods can serve to deplete these seed reservoirs over time. Nevertheless, they are quickly replenished in even one season of reduced weed management.

METHODS OF CONTROL

An integrated vegetation management program may utilize a variety of control methods including cultural, manual, mechanical, chemical, and biological techniques. The best method or combination of methods depends on: the grove location, age of the grove, planting system, weed species, location of the weeds (i.e., tree rows, row middles, fence rows, etc.), environmental considerations, equipment availability, and cost constraints.

Cultural Controls and Cover Crops

There are several cultural practices that may be implemented as part of a vegetation management program. Exclusion and sanitation practices can minimize the introduction, establishment, and spread of weed species. Carefully targeting irrigation and fertilizer application to an established vegetation-free zone (developed using herbicides or shading) limits the resources available for weed growth. Early shading of the tree row surface by the tree canopies reduces competing vegetation. It may also be beneficial to establish a leguminous cover crop to inhibit weed growth and supply some nitrogen.

There is an increasing awareness of the benefits of selected ground covers and their potential impact on management systems. Cover crops may consist of various plant species grown in monoculture or mixed stands and may provide an array of functions. Smother crops suppress weed growth; catch crops reduce nutrient leaching; green manure crops are selected for incorporation into soil, and insectary crops are used to attract beneficial arthropods and mites.

Cover crops offer a variety of benefits to the production of tree fruit crops such as citrus (Plate 18.3). They may supply or conserve nitrogen, reduce leaching and erosion, increase the organic content of the soil, suppress undesirable weed species, reduce maintenance and management inputs, provide an environment for desirable insects and mites, and suppress the establishment of insect pests, including certain nematode species. Cover crops, however, may present many of the same disadvantages as other grove floor vegetation. A cover crop may compete for limited resources, act as a reservoir for pathogens, provide a hospitable environment for rodents and insect pests, and serve to lower grove temperatures during freezing conditions. The grower should carefully evaluate the merits of a cover crop and consider the potential for the cover crop to become an economic pest if it cannot be controlled by cultural methods or herbicides currently available for use in citrus.

A relatively new cover crop candidate is perennial peanut (Arachis glabrata). This species has a number of desirable characteristics and should be of economic benefit in citrus production if it can be successfully integrated into grove management. It is a low-growing, deeply rooted ground cover with few maintenance requirements. It is quite drought tolerant once established and effectively displaces native weed species. The nitrogen-fixing capabilities are well documented and may contribute to nutrient management. It is readily controlled in the tree row by currently available herbicides. The variety that is currently available reproduces vegetatively and may prove difficult to establish economically. However, widespread commercial adoption should become economically feasible when seed for new selections becomes available. This cover crop merits careful grower evaluation as it may reduce vegetation management needs and increase available nitrogen.

Cultivation

Cultivation of row middles is largely restricted to citrus growing on deep sands but is seldom used on bedded groves where there is shallow rooting and where a good ground cover enhances equipment movement and decreases erosion. Cultivation in row middles kills annual weeds efficiently and economically by severing the stems from the roots while the plants are young. Each crop of weeds must be killed in order to prevent competition with the trees and production of seeds. Also, each cultivation brings seeds to the soil surface where they can germinate, so the number of cultivations depends on the number of flushes of weed emergence. The timing of the cultivation is critical to the success of this method. Cultivation at proper time intervals can theoretically be used to kill deep-rooted perennial weeds by exhaustion of their underground food reserves. Infrequent cultivation provides temporary control, but it spreads and invigorates perennials by increasing the number of buried seeds and by distributing rhizome and stolon cuttings, tubers, and bulbs. The result of infrequent tillage is the establishment of solid stands of resistant species that can greatly impact tree growth. Excessive cultivation results in the continual destruction of surface citrus fibrous roots that normally grow in the undisturbed portion of the soil.

Mowing

Mowing (between the tree rows) is routine in most bedded groves and presents a pleasing grove appearance, but requires high-energy input and high equipment maintenance costs. Weeds can also be spread by seed and vegetatively during mowing operations, reinfesting tree rows where herbicides have been applied (Plate 18.4).

It is important to mow before seedhead formation to reduce seed dissemination and reinfestation. While weed control by mowing is useful against tall-growing species, low-growing prostrate species are given an overwhelming competitive advantage. Mowed groves in many areas rapidly develop solid strands of perennial grasses that become sod cultures. Frequent mowing may actually increase water consumption by perennial grasses, although this is only a concern during dry, warm periods.

Chemical Control

In general, all susceptible weed species indicated on herbicide product labels will be controlled if the material is applied at the appropriate rate, season, and stage of weed development. Appropriate equipment must be used, and it must be calibrated and operated properly. Conditions during and following the application, including rainfall or irrigation, may also affect the herbicide action. Application of postemergent herbicides to weeds under stress can reduce control due to poor uptake and translocation of the herbicide. The timing and frequency of herbicide application are critical to effective vegetation management. More frequent application may be needed in young groves where the lack of shade may result in increased weed growth and where lower herbicide rates are required due to the risk of young tree herbicide damage.

There are various types of herbicides with different modes of action and types of application. In tree rows, pre-emergence herbicides are applied to the soil and absorbed through weed root systems and are most effective during germination and early seedling growth stages. Postemergence herbicides are applied to established weed foliage and may be characterized as having either systemic or contact activity. Systemic herbicides are those that are absorbed by either roots or aboveground plant parts and are translocated throughout the plant. Contact herbicides act as desiccants, damaging or killing all plant parts actually sprayed, with little, if any, translocation. For controlling established perennial weeds, a postemergence herbicide with systemic metabolic activity should be used often in conjunction with pre-emergence soil residual products.

The control of vegetation in row middles (middles management), is often achieved using a combination of mechanical mowing and chemical mowing, utilizing Low Rate Technology (LRT) postemergence herbicide spray applications, or wiping of herbicides so that only the top portions of groundcover plants are contacted (Plate 18.5). The cost of frequent mechanical mowing has led many growers to include chemical mowing or wiping with low rates of glyphosate. This method eliminates tall-growing species and aids in the establishment of more manageable sod type species, such as Bermudagrass and bahiagrass. Broadleaf weeds may sometimes proliferate under such programs, requiring modification of management practices. It is thought that chemical suppression of grass and broadleaf weed species in row middles reduces competition more than mechanical mowing, but there is no experimental evidence to support this hypothesis.

Rapid advances in herbicide application technology have resulted in the development of sophisticated equipment. Application equipment is now capable of selective delivery of multiple herbicide products; each directly injected into the booms and placed only where they are most economically efficacious. In a single application, tree rows and row middles may be treated with soil residual and postemergence products with selectivity for tree age, soil type, and vegetation type.

ENVIRONMENTAL CONSIDERATIONS

The objective of a sound vegetation management program is to reduce weed competition and interference while minimizing unintended environmental impact. Several options are available to control the weed species of concern. The grower must consider the soil type and rainfall distribution in the grove, the solubility of the product, and the possibility of leaching, runoff, or drift (Plate 18.6).

AQUATIC VEGETATION MANAGEMENT

An integrated vegetation management program must also address aquatic and ditch bank vegetation management (Plate 18.7). Aquatic and ditch bank vegetation influences water movement into or out of a grove, water quality, soil erosion, and other environmental or cultural elements. For efficient irrigation and drainage, canals and ditches should be maintained essentially free of floating, submerged, and emerged vegetation. Clogged systems will result in plugged pumping systems and culverts, resulting in drought stress or flooding. Low-growing vegetation and fibrous root systems can reduce erosion of ditch banks and filter soil particles that may carry fertilizer or pesticide residues. Proper aquatic vegetation management also minimizes the potential introduction of unwanted weed species. As with terrestrial vegetation management, an integrated approach to aquatic vegetation management consists of some combination of cultural, mechanical, chemical, and biological methods of suppression and control.

Since grove floor vegetation management interacts with many cultural practices, it is helpful to view an integrated vegetation management program through program development, application, evaluation, and monitoring.

REFERENCES

Futch, S. H. 2003. *Middles management in Florida citrus orchards.* International Society of Citriculture Congress. (9th 2000: Orlando, Florida), Orlando, Florida USA 1: 377-378.

Futch, S. H., and R. P. Muraro. 2002. *Weed control: An economical approach.* Citrus & Vegetable Magazine 66: 6-9.

Futch, S., and M. Singh. 2000. *Field evaluation of chemical weed control in Florida citrus.* Proceedings of the Florida State Horticultural Society 113: 68-74.

Noling, J. W., J. P. Gilreath, and P. Gilreath. 2003. *Weeds: The nemesis of Florida agriculture.* Citrus & Vegetable Magazine 67: 6-8.

Singh, M. 2003. *Weed management in citrus.* International Society of Citriculture Congress. (9th 2000: Orlando, Florida), Orlando, Florida USA 1: 370-376.

Tucker, D. P. H. 1999. *Grove floor vegetation management.* Citrus Industry 80: 28-29.

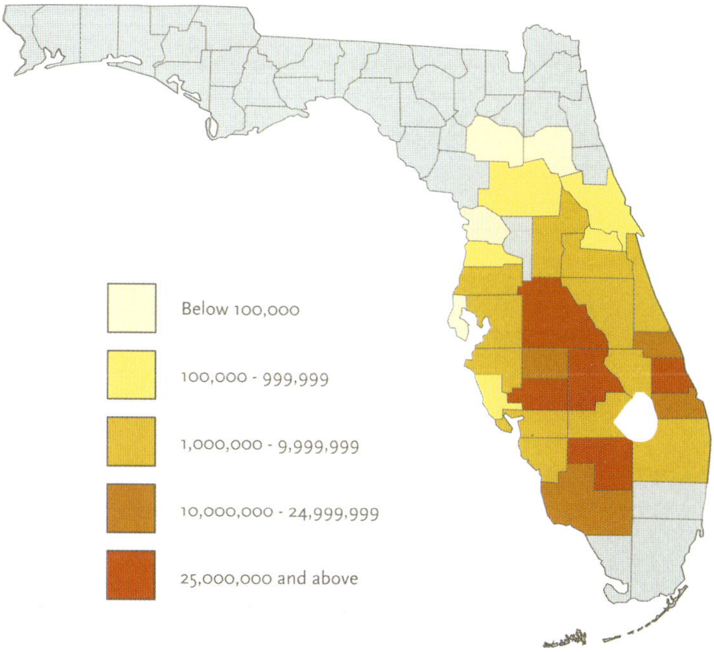

Plate 1.1a Citrus-producing counties in Florida showing yield production in boxes.

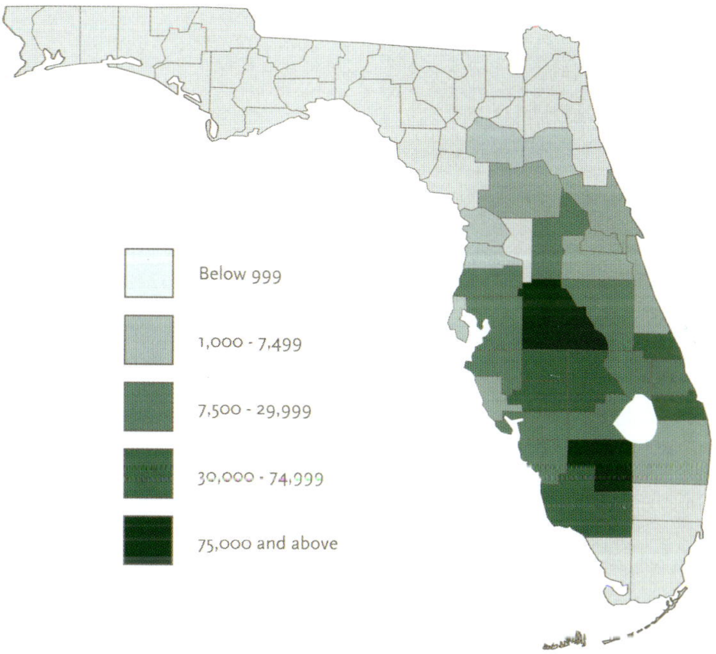

Plate 1.1b Citrus-producing counties in Florida showing total producing acreage per county.

Plate 2.1 A weather station and satellite dish link the CREC-Lake Alfred to the Florida Automated Weather Network (FAWN) and Web-based distance education programs throughout the UF system.

Plate 5.1a Secured greenhouse facility with scion trees for budwood production. Maintaining insect-free conditions and a suitable environment for good tree growth and health is management intensive. Such structures will become more important in the future as the industry strives to protect young nursery trees from organisms that cause problems such as canker and greening.

Plate 5.1b Potted trees inside greenhouse.

Plate 6.1 Close-up inspection of soil profiles is useful during the early site evaluation phase of grove development. A good understanding of the soils on which a grove is to be developed cannot be overemphasized.

Plate 7.1 Perforated tubing known as drainage tile is sometimes installed to enhance water removal.

Plate 7.2 Poorly drained coastal areas are most commonly developed by bedding soil to increase rooting depth. Pictured is the initial phase of land preparation and bed construction.

Plate 7.3 Young flatwoods citrus grove showing lateral and collector ditches.

Plate 7.4 A typical flatwoods grove development, with set-aside areas to protect wetlands and wildlife sanctuaries, with as much as 25 percent to 30 percent of acreage excluded from planting.

Plate 8.1 Typical field nursery with greenhouses in background. Here, seedling rootstocks are grown indoors and transplanted seedlings (liners) may be budded in the field. Such nursery systems will become less common in the future as the industry implements guidelines to reduce transmission of diseases such as canker, tristeza and greening.

Plate 8.2 Greenhouse facilities allow for total citrus nursery production indoors, offering better sanitation, pest and disease exclusion and freeze protection. insert: Enclosed tree production facilities like this one will become the industry standard in the future.

Plate 9.1 Citrus pollen is not easily transported by wind, so beehives are often brought into a grove to enhance pollination.

Plates 10.1a through 10.1f show a variety of equipment that may be used in citrus irrigation systems.

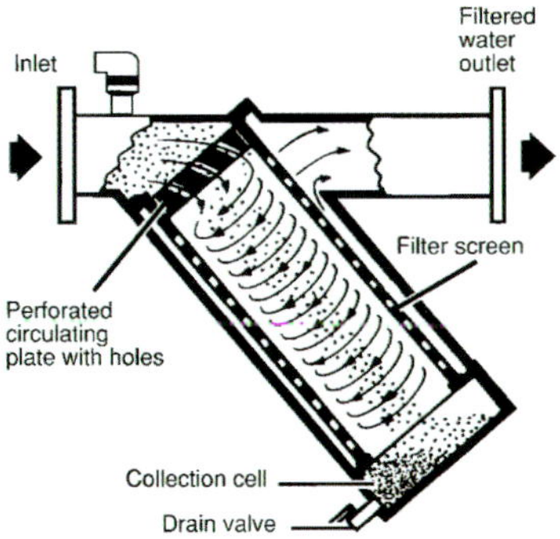

Plate 10.1a Screen filter.

Plate 10.1b Media filter system.

Plate 10.1c Turbine pump, which is standard for Ridge citrus.

Plate 10.1d Centrifugal pump, standard for flatwoods citrus.

Plate 10.1e Double backflow prevention for fertigation or chemigation.

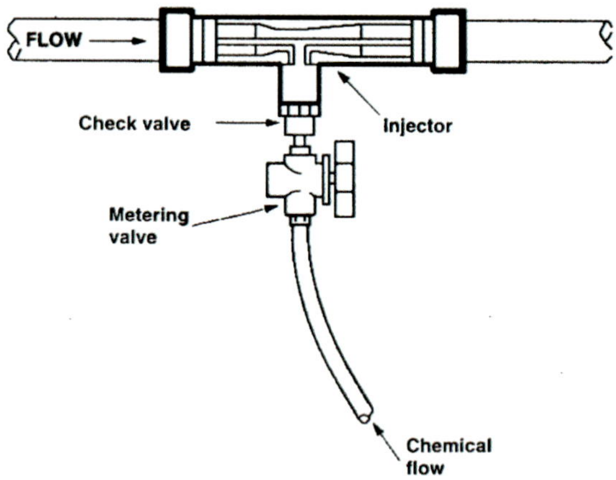

Plate 10.1f Venturi injector for fertigation or chemigation.

Plate 11.1 Well-conceived and maintained flatwoods citrus grove showing beds and furrows for drainage, microsprinkler irrigation system, weed control and ground cover management.

Plate 11.2 Improper planting practices can result in poorly developed and distributed root systems (shown here) that can lead to retarded tree growth, low production and lost income.

Plate 11.3 Young grove (3 to 4 years old) showing stunted trees that should be closely evaluated for citrus tristesa virus and if necessary removed.

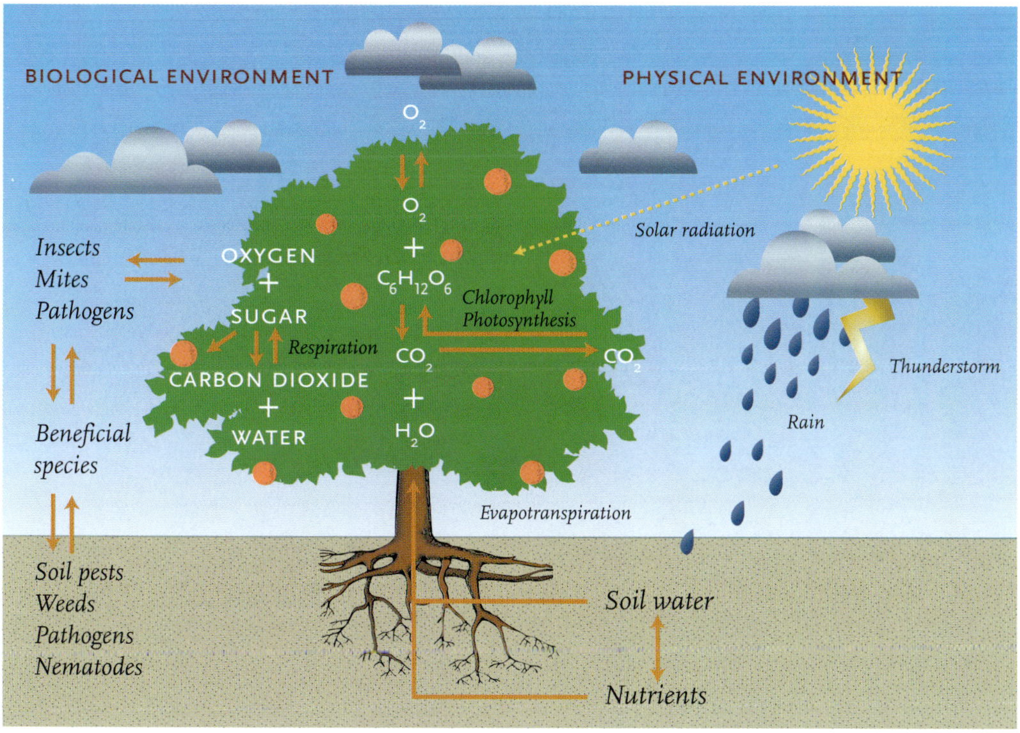

Plate 14.1 Biological and environmental factors influencing the growth and development of a citrus tree.

Plate 14.2 "Firing" of foliage in tree canopy due to combinations of stresses including moisture, wind, and spider mites. Such conditions can occur in the spring and fall.

Plate 14.3 Severely heat-stressed seedlings in greenhouse following failure of thermostat to activate cooling fans. Utility company power failures (brownouts) can also occur during winter allowing, plant damage to occur.

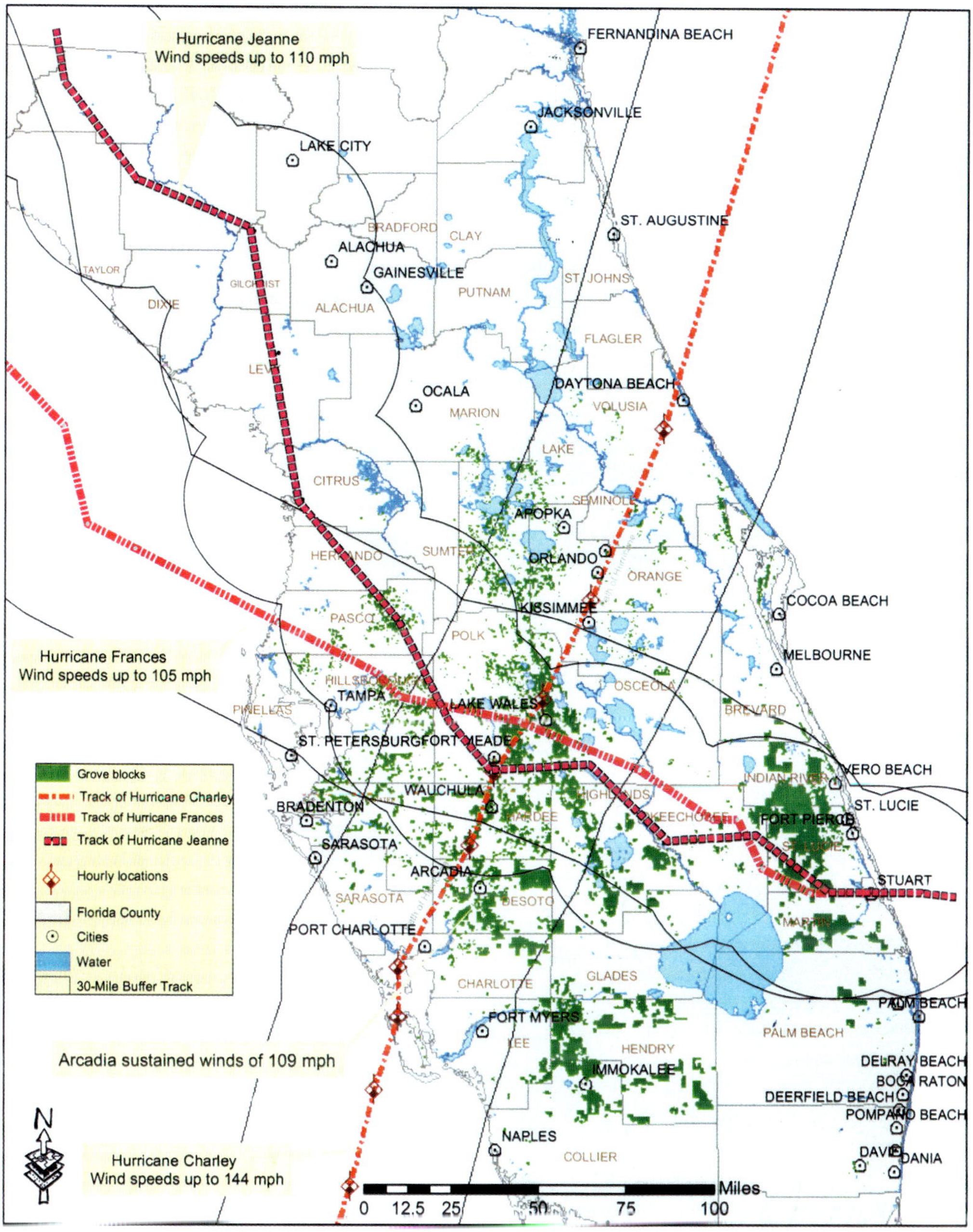

Plate 14.4 2004 Hurricane Event Map. The paths of Hurricanes Charlie, Frances, and Jeanne across the Florida citrus growing region.

(Created by CAPS—Andrea Chavez, Wayne Dixon and Xiaoan Sun 04 October 2004 Albers Coordinate System—NAD 1983 GRS80. Florida Cooperative Agriculture Pest Survey Program, Division of Plant Industry, Florida Department of Agriculture and Consumer Services, Gainesville, Florida.)

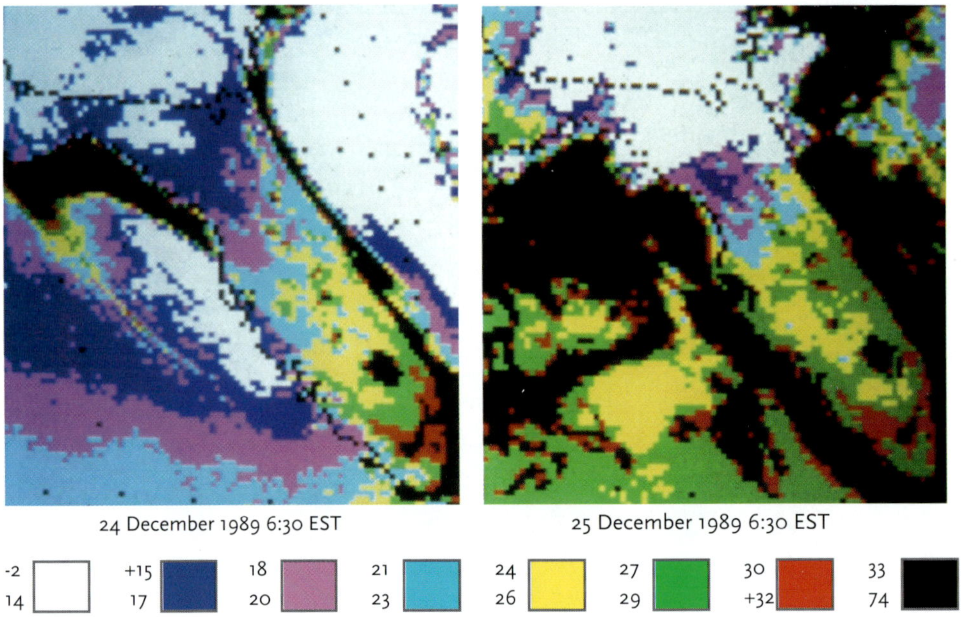

Plate 15.1 Satellite photograph showing temperature profile across Florida during the December 1989 freeze.

Plate 15.2 Young trees are more cold susceptible than older trees with larger canopies, thereby requiring greater management input during freeze events.

Plate 15.3 Microsprinklers are widely used for cold protection throughout the citrus industry.

Plate 15.4 Trees showing live lower canopy areas that were influenced by microsprinkler irrigation during a freeze event.

Plate 15.5 Difference in severity of freeze damage between Murcott trees with heavy vs. light crops. Trees under stress from heavy cropping have small fruit, are more susceptible to tree collapse and to freeze damage.

Plate 16. 1a Microsprinkler irrigation has become the standard of the industry because of low cost, high water-use efficiency, and cold protection.

Plate 16. 1b Although its use is declining, overhead irrigation is still used in some groves.

Plate 16. 1c Flood irrigation was standard in bedded groves and is still used in some situations.

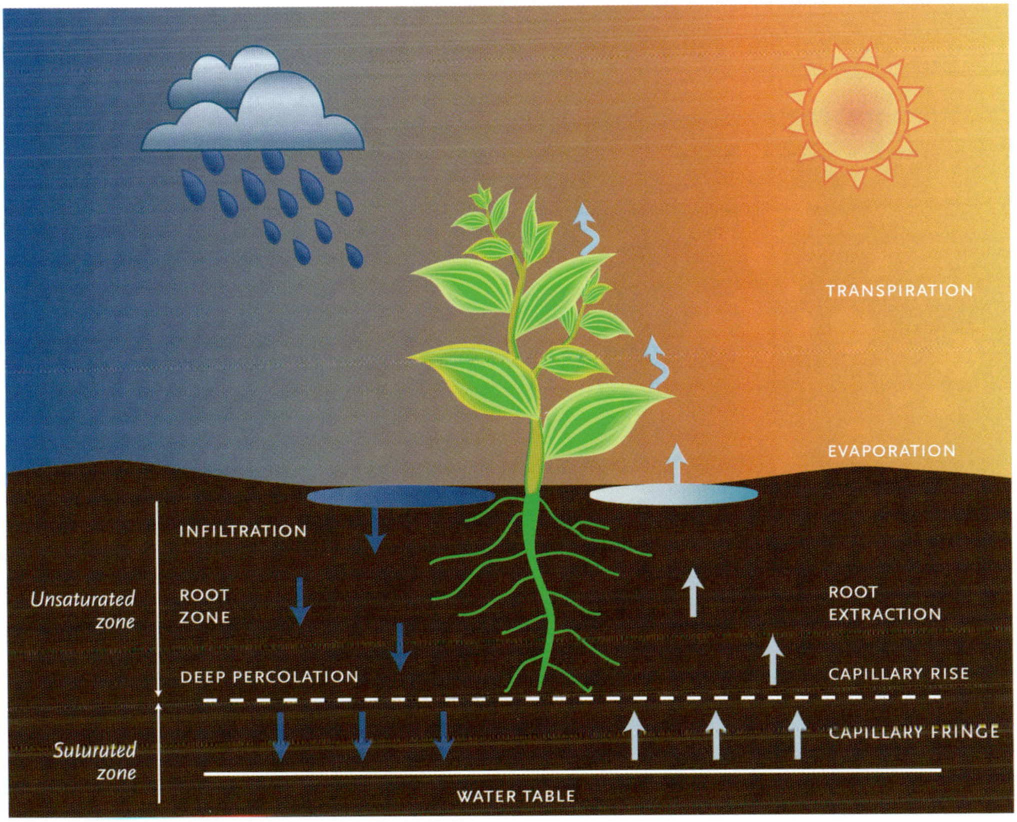

Plate 16.2 Understanding citrus tree moisture relationships is required to develop sound water management programs.

Plate 16.3 Proper placement of tensiometers under tree canopies and servicing of the instruments contributes to improved water management efficiency and tree productivity.

Plate 17.1 Typical soils on which Florida citrus groves are planted. A: Candler soil (Entisol). B: Riviera soil (Alfisol). C: Immokalee soil (Spodosol).

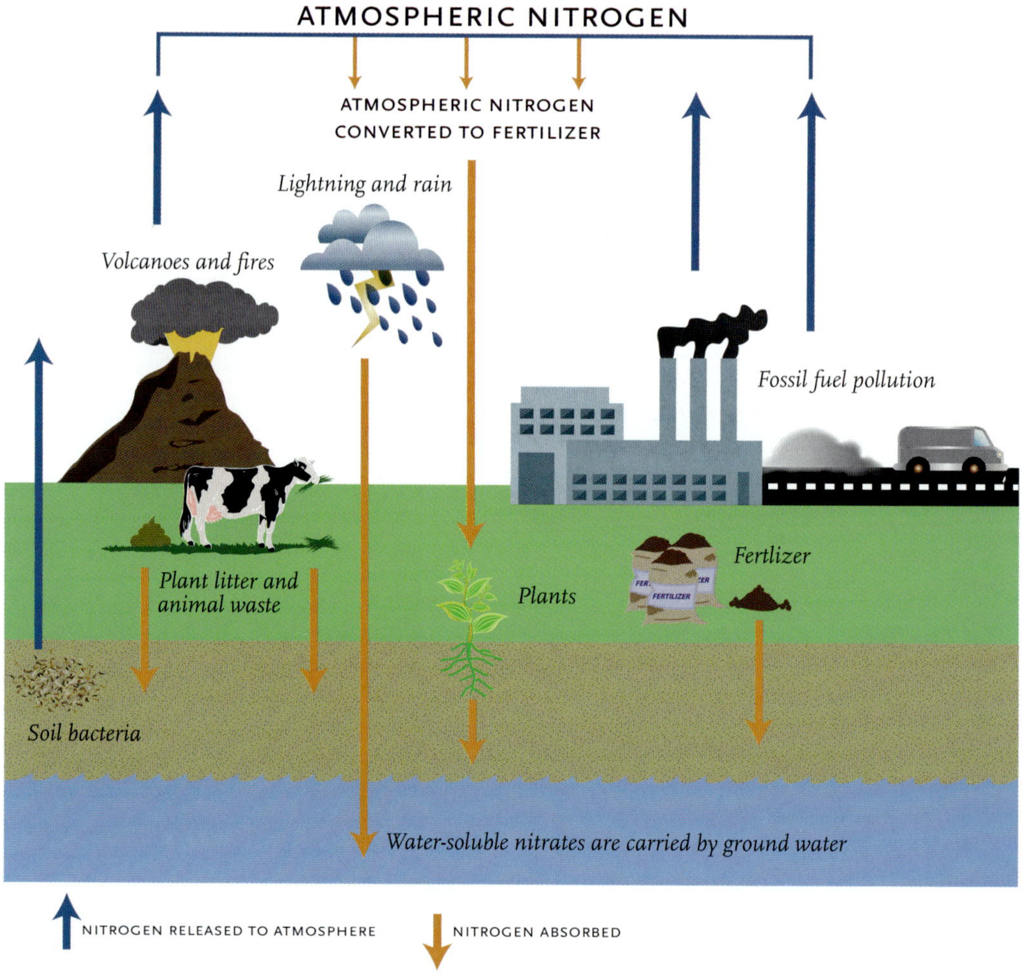

Plate 17.2 The nitrogen cycle. Nitrogen is vital to healthy plant growth. Abundant atmospheric nitrogen can only be utilized by plants if it is combined with other elements. Nitrogen may be combined with hydrogen to form ammonium ions, and with oxygen to form nitrate ions.

Plate 17.3 Injection equipment and liquid fertilizer storage tank installation allows for frequent and measured nutrient applications. Convenience, low application costs, and rapid young tree growth response are attractive aspects of fertigation.

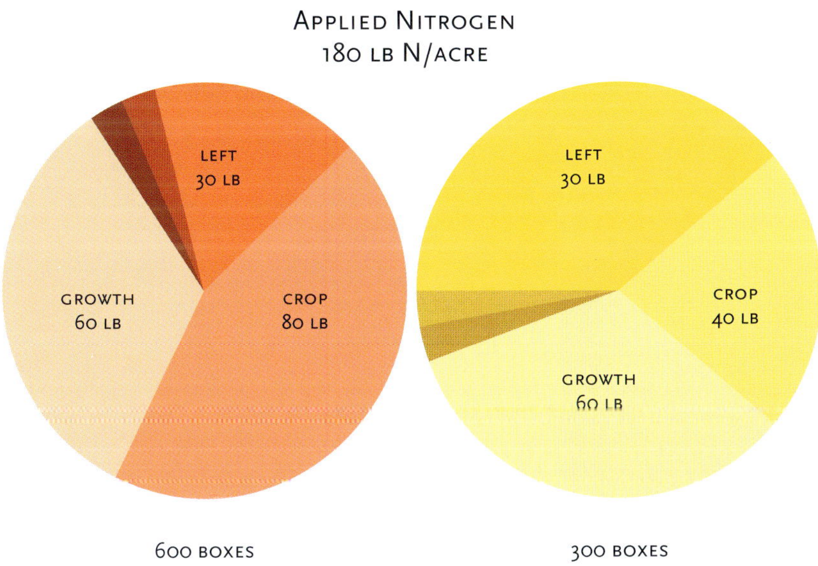

Plate 17.4 Nitrogen used for growth will be similar for mature citrus groves each year, but the amount of nitrogen removed by the crop will depend on the size of the crop. Fertilization should be adjusted to reflect crop size in order to reduce losses of nitrogen due to leaching.

Plate 18.1a Good vegetation management prior to planting provides an excellent start for this newly planted grove.

Plate 18.1b Inadequate weed control prior to planting may result in initial vegetation management problems and compromised tree development.

Plate 18.2 Shifting weed species are observed in tree rows treated repeatedly with one postemergence or residual herbicide. The original grass cover remains in the row middles, but has been essentially replaced by broadleaf weed species in tree rows. A suitable program of herbicide rotation will reduce this occurrence.

Plate 18.3 Experimental planting of the low-growing, rhizomatous perennial peanut cover crop. When established, it returns nitrogen to the soil and suppresses native weed growth, thereby reducing costly maintenance mowing operations. The benefits of suitable cover crop have not been exploited by Florida growers as has been the case in other citrus-growing areas.

Plate 18.4 Cultivation of row middles is used in deep sand to control weeds, but mowing is used in bedded groves.

Plate 18.5 Middles management utilizes low-rate technology (LRT) with postemergence herbicides and reduces the frequency of costly mechanical mowing operations. Fuel-efficient ATV-mounted equipment contributes to the energy efficiency of operations.

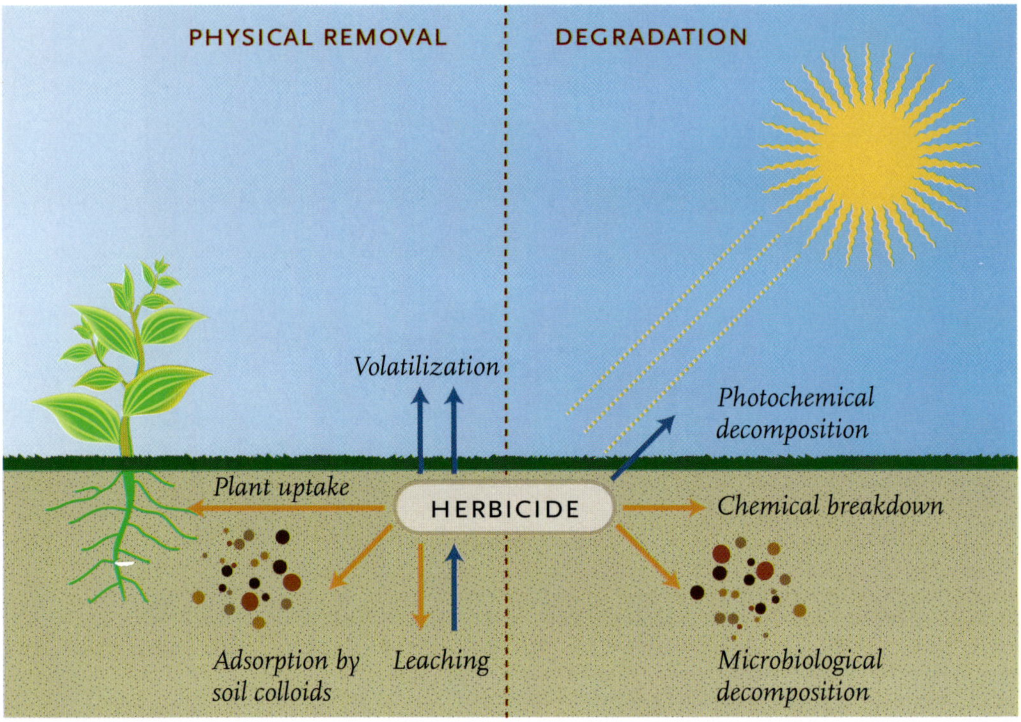

Plate 18.6 An understanding of degradation processes and other forms of dissipation following application of residual herbicides and soil pesticides to the grove floor is a part of informed agrichemical grove-floor management.

Plate 18.7 The controlled flow of water into and out of grove land is critical to good water management and is enhanced by integrated aquatic vegetation suppression/control programs.

Plate 19.1 Citrus rust mite damage on fruit. This insect can affect production and marketability, especially on fresh-fruit crops.

Plate 19.2 During scheduled tree removal operations, an opportunity exists to visually examine root systems for insect pests, diseases, water damage and other problems that can negatively impact root systems and tree growth and production.

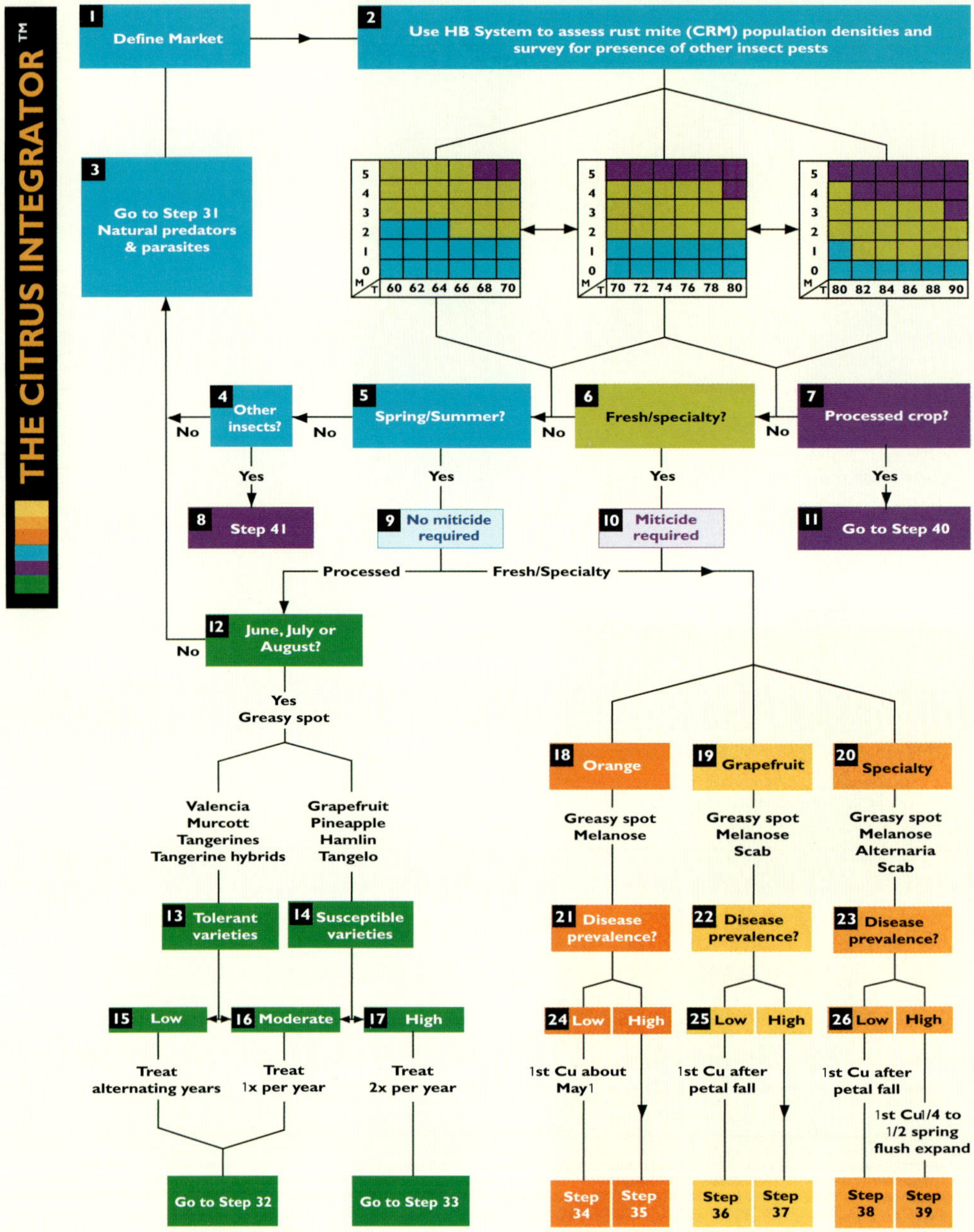
THE CITRUS INTEGRATOR™
1 Define Market
2 Use HB System to assess rust mite (CRM) population densities and survey for presence of other insect pests
3 Go to Step 31 Natural predators & parasites
4 Other insects?
5 Spring/Summer?
6 Fresh/specialty?
7 Processed crop?
No
Yes
8 Step 41
9 No miticide required
10 Miticide required
11 Go to Step 40
Processed
Fresh/Specialty
12 June, July or August?
Yes Greasy spot
Valencia Murcott Tangerines Tangerine hybrids
Grapefruit Pineapple Hamlin Tangelo
13 Tolerant varieties
14 Susceptible varieties
15 Low
16 Moderate
17 High
Treat alternating years
Treat 1x per year
Treat 2x per year
Go to Step 32
Go to Step 33
18 Orange
19 Grapefruit
20 Specialty
Greasy spot Melanose
Greasy spot Melanose Scab
Greasy spot Melanose Alternaria Scab
21 Disease prevalence?
22 Disease prevalence?
23 Disease prevalence?
24 Low High
25 Low High
26 Low High
1st Cu about May 1
1st Cu after petal fall
1st Cu after petal fall
1st Cu1/4 to 1/2 spring flush expand
Step 34
Step 35
Step 36
Step 37
Step 38
Step 39
M T 60 62 64 66 68 70
M T 70 72 74 76 78 80
M T 80 82 84 86 88 90
5 4 3 2 1 0

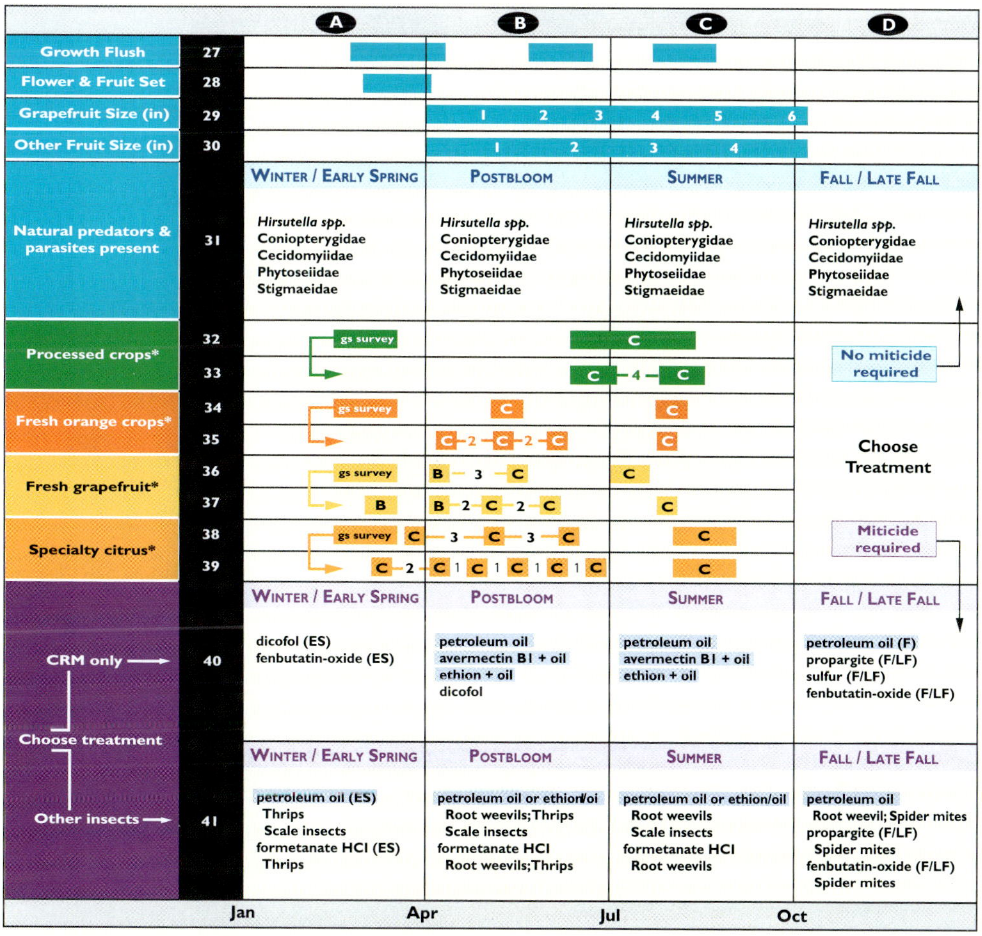

TREATMENT NOTES:

* Numbers between time bars represent the number of weeks between treatments.

1 - 30: See supplementary text for notes on steps 1 to 30.

31: Specific genera of predators and parasites vary from one geographical area to another. Others not listed here may also occur.

32: Processed Valencia, murcott, tangerine and tangerine hybrids will occur most often in this group.

33: Processed grapefruit, Hamlin, pineapple and tangelos will occur most often in this group.

34: Spray approximately May 1, then later in summer for greasy spot control.

35: Scab not usually a problem on oranges; melanose, copper at 2 to 3 wk. intervals.

36: Strobilurin or copper after petal fall with copper (4 to 6 lb. ai per acre) as the second application.

37: Strobilurin at 25% to 50% spring flush expansion. Alternate copper at early postbloom.

38: Spray at petal fall followed by 2 subsequent monthly applications. Specialty varieties not scab-susceptible may be treated using this program.

39: Spray at 25% to 50% spring flush expansion, after petal fall, then every 1.5 to 3 weeks depending on rain & disease severity.

40: Choose materials from this row if rust mites are the only problem insects. Blue highlighted materials may be mixed with copper

41: Choose materials from this row if there are insects other than rust mites present at damaging levels.

Plate 19.3 The Citrus Integrator™. For an example illustrating the use of the flowchart, see the description on page 202. (Reproduced by permission of Ecostat Inc.)

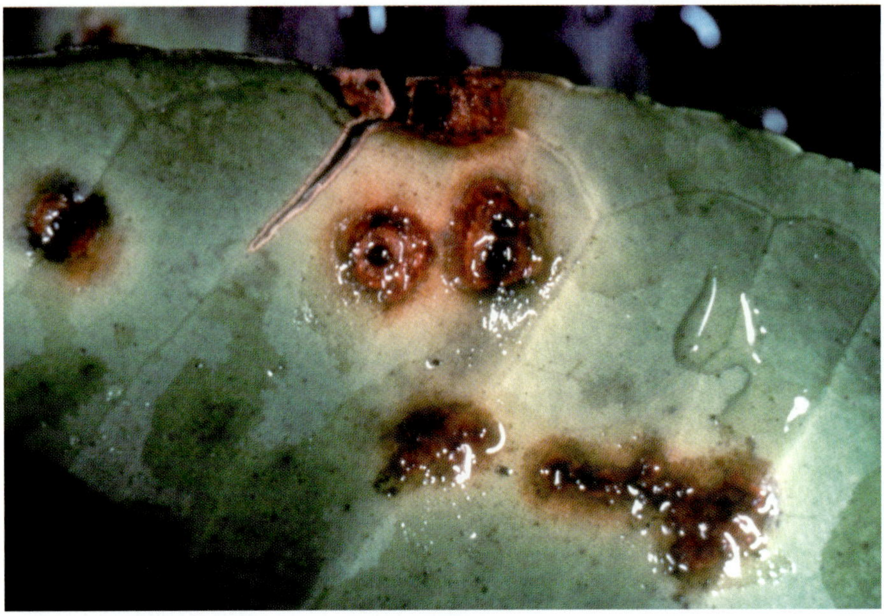

Plate 19.4a Citrus canker symptoms in citrus leaves.

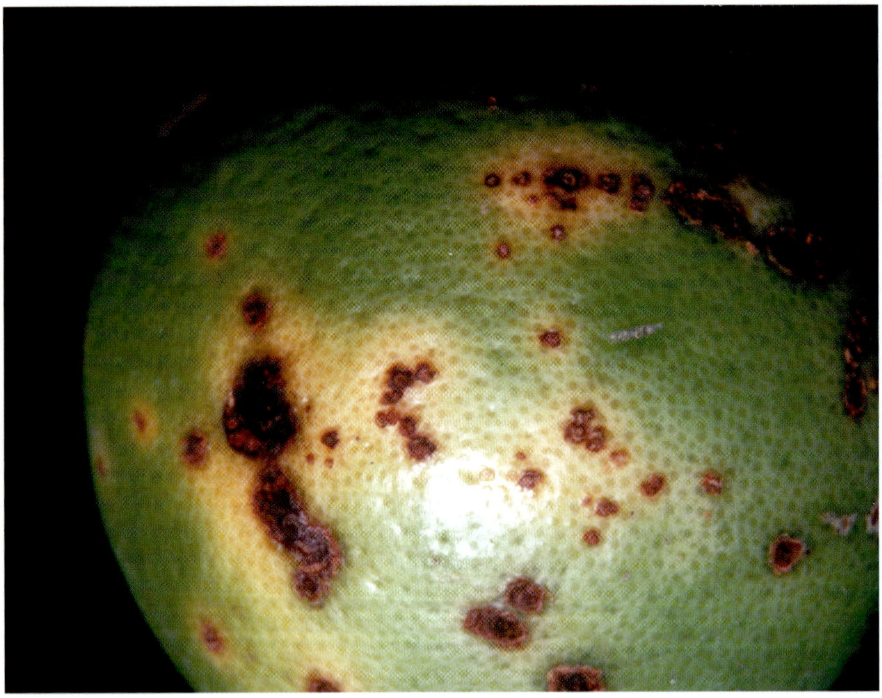

Plate 19.4b Citrus canker symptoms on citrus fruit. Citrus canker is a devastating disease of citrus that is currently under an eradication program in Florida until that program was terminated in January 2006.

Plate 19.5 Destruction of young trees in plantings found to be infected with the bacterial disease, citrus canker. Such actions were recently mandated by the State of Florida, but this practice was deregulated in January 2006

Plate 19.6 Inspection and sanitation stations are necessary precautions against the spread of diseases such as citrus canker.

Plate 19.7 Some unsightly fungi on citrus can actually be helpful. Here is an example of the "Friendly Fungus (*Aschersonia*)," which affects citrus whitefly populations.

Plate 19.8a Citrus scab symptoms on citrus leaves.

Plate 19.8b Citrus scab symptoms on citrus fruit. Citrus scab affects external quality, but is generally not a problem on oranges and grapefruit. It is found more commonly on specialty citrus varieties.

Plate 21.1 Ultrasonic sensor for detecting tree height in mixed plantings, to control sprayer nozzle output, conserve spray material, and reduce atmospheric pollution.

Plate 21.2 Equipment operating under high wind-speed conditions showing off-target spray drift.

Plate 21.3 Helicopter applications of agrichemicals can ensure timely management response to pest and disease outbreaks, particularly under difficult ground operating conditions.

Plate 21.4 Tower sprayer provides good spray coverage of pest populations in the tops of large trees.

Plate 22.1 Young, vigorously growing trees in a grove overdue for hedging with reduced sunlight penetration and impeded movement of equipment.

Plate 22.2 Hedging machine performing maintenance pruning in a young grove. Light cutting minimizes yield loss, vegetative regrowth and cost of debris disposal.

Plate 22.3 Removal of a tree within a hedgerow reveals loss of fruit bearing surface and structural framework of adjacent trees, which are devoid of internal fruit. Regrowth and fruiting of these surfaces will occur over the next several years until shaded by the replacement tree.

Plate 22.4 Trees recently maintenance-topped showing light-green colored regrowth. Light topping is preferable to severe cutting, which invariably results in excessive regrowth and reduced yield.

Plate 27.1 Conventional hand harvesting of grapefruit for fresh sale.

Plate 27.2 Fruit for the fresh market in pallet bins being loaded onto a flatbed trailer. Careful handling of fruit during all stages of harvesting and handling is essential in the fresh fruit business.

Plate 27.3 Fruit for processing being loaded from fruit handling equipment into a road trailer, part of the system for moving fruit to the processing facility.

Plate 27.4 Trunk shaker and catch frames for fruit removal and collection is a well-tested system that will play a larger role in harvesting future processed citrus crops.

Plate 27.5 Canopy shaker systems show promise for removal of fruit for processing as suitable harvesting labor becomes less available or too costly.

Plate 27.6 Trees with lower canopies skirted to allow unrestricted trunk shaker access.

Plate 28.1 Plastic pallet bins now in wide use throughout the industry are replacing standard wooden pallets because plastic bins are easier to keep sanitized.

Plate 28.2 Drenching fruits with fungicides on the truck prior to packing. Proper handling and treatment are important to maintain fruit quality after harvesting. Some buyers specifically request fruit that receive no postharvest fungicides, and this step is omitted.

Plate 28.3 Fruit pallet bins in a degreening room being subjected to ethylene gas under closely monitored environmental conditions.

Plate 28.4 Traditional grading line in a fresh fruit packinghouse. Adherence to USDA standards for grade size is essential in fresh citrus operations.

Plate 28.5 Electronic grading and sizing line in a packinghouse. Automation is a key element in identifying premium products and staying competitive.

Plate 28.6 Refrigerated vessel for the export shipment of fresh citrus. Refrigeration maintained immediately after packing has become increasingly important for long-term storage and international marketing.

Plate 29.1 View of an aseptic tank construction for the storage of not-from-concentrate (NFC) citrus juice.

Plate 29.2 Typical view of a fruit processing facility with waiting trailers.

Plate 29.3 Fruit trailers being unloaded at the processing plant. Trailers are hydraulically lifted so that fruit rolls onto a conveyer belt.

Plate 29.4 Fruit bins contain a load of fruit, where it is stored until the state quality tests are performed.

Plate 31.1 Regular training sessions for tractor drivers and other grove workers, including harvesting crews, along with wage and benefits incentives, are essential for attracting and retaining a competent work force with some pride in the profession.

Plate 31.2 Well-maintained housing is a positive factor in hiring and retaining workers.

Plate 32.1 This scenic view of central Florida citrus with rolling hills and lakes has been widely used in promotions of Florida citrus.

Plate 32.2 Large selections of citrus juice products are available to consumers in supermarkets and other retail outlets.

Plate 32.3 Appealing displays of fresh citrus fruits in both domestic and foreign retail outlets. Consumers need to be more aware that external appearance is appealing to the eye, but it has little direct relationship to internal quality.

Plate 33.1 A Florida Agricultural Statistics Service technician interprets aerial photography in the process of making tree counts and determining acreage changes.

Plate 33.2 Florida Agricultural Statistics Service technicians collect data in preparation for annual crop forecasting information which influences fruit pricing and marketing.

Plate 34.1 A change of scion, using a procedure known as topworking, is one option available to develop a more marketable variety. However, options must be carefully evaluated before decisions are made.

Plate 34.2 Severely pruned trees may be moved between sites using a tree spade. A number of factors should be considered when moving trees to ensure a satisfactory survival rate, including suitable irrigation, herbicide, nutrition and pest management programs, and sunburn and freeze protection.

Plate 34.3 Improved drainage can be an important aspect of grove rejuvenation. Excavated soil should be evenly distributed and not piled against tree trunks or thrown into tree canopies from furrows.

Plate 34.4 Aerial view of rejuvenation operation shows soil excavation for improved drainage. Such large projects should not be undertaken without a thorough understanding of the options available and costs involved.

19

Integrated Plant Protection

J.S. Rogers

INTRODUCTION

Citrus tree health may be compromised by many factors, including biological species such as insects, bacteria, viruses, fungi, nematodes, and weeds (collectively, "pests"). In fact, more than 100 species can attack or feed on citrus trees, resulting in typical symptoms of disease or decline. The physical environment, such as temperature, storms, drought, or flood, also influences citrus tree growth. Generally, five factors influence pest population development in citrus groves: (1) the host, (2) the pest, (3) time, (4) the environment, and (5) the activities of humans. These factors should be considered when developing an integrated pest management program for a citrus grove.

Professionals that observe citrus groves for pest populations are called "scouts." Scouts interpret their observations and produce a practical management program for pest problems they identify in the field. In smaller citrus operations, it is common for one individual to scout, design and even implement integrated plant protection programs. In larger operations, it is often common for the different tasks to be performed by different people. This chapter draws on the basic principles of ecology to discuss integrated strategic and tactical approaches to managing citrus pests.

RATIONALE

The plant protection business will be manifestly different in the future. The old images of farming technology such as tractors, spray machines, and mechanical harvesters are being supplemented by new symbols of progress: computer networks, satellite communications, and molecular genetics. The rate of technological change is accelerating and the time lag between the genesis of agricultural ideas and their application in the field is becoming shorter as production becomes more efficient. The underlying theme is greater emphasis on information technologies. To address these changes in corporate crop production, this chapter describes some basic principles of strategic plant protection for those with vested interests in agricultural crop production. This information should also be informative for those who do not necessarily work in the field on a daily basis.

Citrus trees can be attacked by a variety of biological species including insects, bacteria, viruses, fungi, nematodes, and weeds (collectively, "pests").

The practical application of generally agreed-upon optimum practices is referred to as a "best management practice" (BMP).

> An integrated plant protection BMP uses a combination of cultural, chemical, genetic and regulatory strategies and tactics to manage citrus insects and diseases. The BMP combines approaches from diverse scientific and practical disciplines to meet the economic, food, and environmental protection needs of farmers and society.

Table 19.1 introduces some common citrus pests and other problems: the common name of the problem and its general indications are presented. Other important pests not shown in the list may also be encountered in the field, but this list includes many that are encountered regularly. Common indicators of each problem are given, but field symptoms vary depending on location, environment and host. Many visible signs may occur that are not shown below. There are many excellent references describing the biology and ecology of the listed species. Readers are referred to Browning et al. (1995), which contains photographs and short descriptions of many important citrus pests and problems. Production managers responsible for on-going pest management duties should be familiar with the identification, symptoms and control for the pests listed in this table.

Table 19.1 Common citrus pest problems and their indicators.

Category and Pest Common Name	General Indications
Rust mites Citrus Pink Citrus	*Brown or rusty fruit*
Spider mites Citrus red Six spotted Texas citrus	*Scratched appearance on leaves; leaf drop if severe*
Scale Insects Florida red Purple Glove Citrus snow Fern Cottony cushion	*Raised, removable specks of various colors*
Mealybugs Citrus	*White, cottony masses on leaves, twigs and fruits*
Whiteflies Citrus Cloudy winged Woolly	*Small, white flying insects visible; honeydew facilitates fungal growth*
Aphids Melon Spire Black citrus Brown citrus	*Leaf distortion or curling; insects visible; may transmit disease*
Sharpshooters Glassy winged	*Multicolored insects visible; may transmit disease*
Plant bugs Leaf footed Southern green stinkbug	*Leaves cut, chewed or notched*

Leafminers Citrus	*Leaf distortion; insects visible*
Bacterial diseases	
Citrus canker	*Spots and lesions on leaves, fruits and stems*
Citrus greening	*Foliage yellowing; misshapen fruit; tree decline*
Fungal diseases	
Citrus scab	Scabby bumps on leaves and fruits
Greasy spot	Greasy spots on leaves; rind pitting on grapefruit
Melanose	Small, circular, black lesions on leaves and fruits
Phytophthora	Root rot; brown rot on fruits
Postbloom fruit drop	Brownish, water-soaked lesions on flowers; calyces remain attached to trees
Sooty mold	Black film covering leaves and fruits
Abiotic disorders	
Freezing injury	Injury to leaves and fruits; wood affected if severe
Chimeras	Genetic leaf and fruit growth abnormalities
Hail injury	Impact laceration of leaves, fruits and stems
Fruit splitting	Splitting of fruits
Wind scarring	Elongated lesions on leaves and fruits
Sunburn	Circular, brownish fruit scarring
Herbicide injury	Discoloration and disfiguration of leaves
Chemical injuries	Discoloration and disfiguration of leaves
Nutritional deficiencies	
Nitrogen	General yellowing of leaves
Phosphorus	Rare; dull bronzing in leaves
Potassium	Small, smooth fruit; thin peel, low acid content
Magnesium	Irregular, yellowish-green blotches on leaves
Zinc	Patterned leaf yellowing with green veins
Manganese	Patterned leaf yellowing
Iron	Patterned leaf yellowing with green veins
Orangedogs	*Leaves cut or notched; insect readily visible*
Root weevils	*Tree decline; leaves chewed or notched*
Grasshoppers Eastern lubber American	*Leaves cut, chewed or notched*
Katydids Broadwinged	*Leaves cut, chewed or notched*

Ants Red imported fire ant	*Girdling at base of young trees*
Decline problems Quick decline Slow decline Spreading decline Blight Citrus nematode Burrowing nematode	*Loss of tree vigor; sickly appearance; general decline*
Viruses and Viroids	
Tristeza	Loss of tree vigor; sickly appearance; decline
Psorosis complex	Symptoms complex, including bark scaling and gum exudation
Exocortis	Tree stunting; bark cracking
Cachexia; Xyloporosis	Symptoms complex, including bark scaling and stem pitting

FACTORS INFLUENCING PEST DEVELOPMENT

Generally, there are five factors influencing pest populations in citrus groves: (1) the host, (2) the pest, (3) time, (4) the environment, and (5) the activities of humans. The host is the species (i.e., the citrus tree) that is attacked or negatively influenced by the presence of the pest. The genetic and physiological composition of the host influences the spectrum of pests that attack and the rate at which infestations or diseases develop. In contrast, pests are the species that attack or negatively influence development of the host citrus tree. Pests may grow directly on the host (e.g., mites and scale insects), within the host (e.g., viruses and some nematodes) or may inhabit the same space and compete for essential light, water and nutrient resources (e.g., weeds). Problems develop over time after initial contact and recognition between the tree and pest. Sometimes this time period may be comparatively short as with leafminer development on young citrus tree foliage. Other times, it may take years for a pest to cause significant losses to a citrus tree, as in the case with root weevils.

Environmental conditions must be favorable to the pest for a disease or infestation to develop. A pest can come into contact with a susceptible host, but disease will not develop (or at least will develop much more slowly) if weather does not favor growth of the pest. For example, certain fungal diseases such as melanose are less of a problem during hot and dry periods with little or no rain because their aerial spread is limited during such times. However, during periods when wind and rain are prevalent, melanose fungi produce microscopic spores that spread to nearby branches and fruits resulting in infection and disease. Human activities influence pest populations and the extent to which they cause economic losses in groves. For example, even when weather and other factors favor development of insect or disease populations, a properly applied pesticide can often minimize economic damage. Conversely, lack of pest control may lead to greater-than-desired damage or disease.

COMMERCIAL CITRUS PEST MANAGEMENT PROGRAMS

For commercial citrus operations, pest management activities are usually a daily responsibility. Many pests occur at some level in all citrus groves, while other pest species may be found primarily in specific blocks or areas. Different pests also develop during different times of the year. For example, citrus rust mites (Plate 19.1) and the fungi that cause greasy spot and melanose are probably found at some level in every commercial grove in the state, but usually become bigger management problems during the spring and summer. Alternaria and scab are more restricted to susceptible cultivars and do not cause damage on as great a host range as do rust mites, melanose and greasy spot disease.

Corporate plant protection specialists should be able to foresee the development of potential future pest problems as part of their on-going field survey efforts. There are two primary duties associated with pest forecasts: scouting and production management. Scouting refers to the action of observing and collecting pest information from the field and the people who collect this information are often called "scouts." Scouts interpret their information and produce practical management programs for pest problems they identify in the field. Those responsible for carrying out these management tasks are usually the production managers. However, production managers also often scout and determine the actions needed to manage pest problems. In smaller citrus operations, one individual may scout, design, and implement pest management programs. In larger operations, it is more common for scouting and management tasks to be carried out by different people.

Once scouting information has been acquired (e.g., rust mite populations growing on fruits), experience often leads to a reasonable conclusion about the cause and treatment of a particular problem. Figure 19.1 shows a simplified integrated crop protection strategy. First, field scouting leads to the identification of potential crop problems. Diagnostic procedures may be used to confirm that the pest problem is real, after which possible control actions are proposed. The economic, environmental, and sociological impact of the proposed tactics is evaluated. If there is a strong possibility of a negative impact caused by the tactics, alternative management approaches are proposed, then reevaluated. Once negative impacts are minimized, growers carry out the actions. Finally, follow-up observations are carried out to help ensure the success of the management program. However, visual, microscopic or laboratory diagnostics, and printed or other professional sources of information may be consulted if there is uncertainty about the cause of a particular problem. Increasingly, computers are used to help interpret scouting information and produce forecasts regarding the development of insect and disease outbreaks. Confirmation of the causes of tree and crop damage is especially important for the often invisible and insidious problems caused by greening, nematodes, viruses, pest-related water stress, Phytophthora root rot fungi, or Diaprepes root weevils (Plate 19.2). Without verifying the cause, unnecessary chemical applications could be useless and extremely costly.

The intensity of a scouting effort is governed by a multitude of logistic and economic factors, but it is important to collect the correct type of information so that unnecessary management actions may be avoided. For example, citrus rust mites cause a brown to tan discoloration on the fruit surface at high populations making the fruit unpalatable for many consumers. However, the mere presence of rust mites at low levels is generally an insufficient reason for chemical treatment even on fresh and specialty citrus crops. Such low mite densities should not be used as the sole motivation to begin chemical treatment without first considering alternative cultural and ecological methods of control. Higher mite populations may be tolerated on processed crops, since external appearance is not as critical as with the fresh fruit crops.

If an agricultural chemical is applied to a confirmed pest problem, choices as to material, dosage, and timing of the application follow. Whether a chemical or natural management approach is used, the impact of the program on other aspects of the integrated crop management system must be considered. Finally, the economic, environmental, and sociological impact of a pest management action must be evaluated and weighed against potential crop or tree losses. For example, if a target planting is located near houses, schools, or ecologically sensitive areas, negative impacts resulting from chemical treatment must be considered and treatments modified accordingly with respect to people, materials, time, and space. Clearly, special professional qualifications are required to produce effective management decisions that take into consideration the complicated environmental and social practical factors that can influence the outcome of a pest control action.

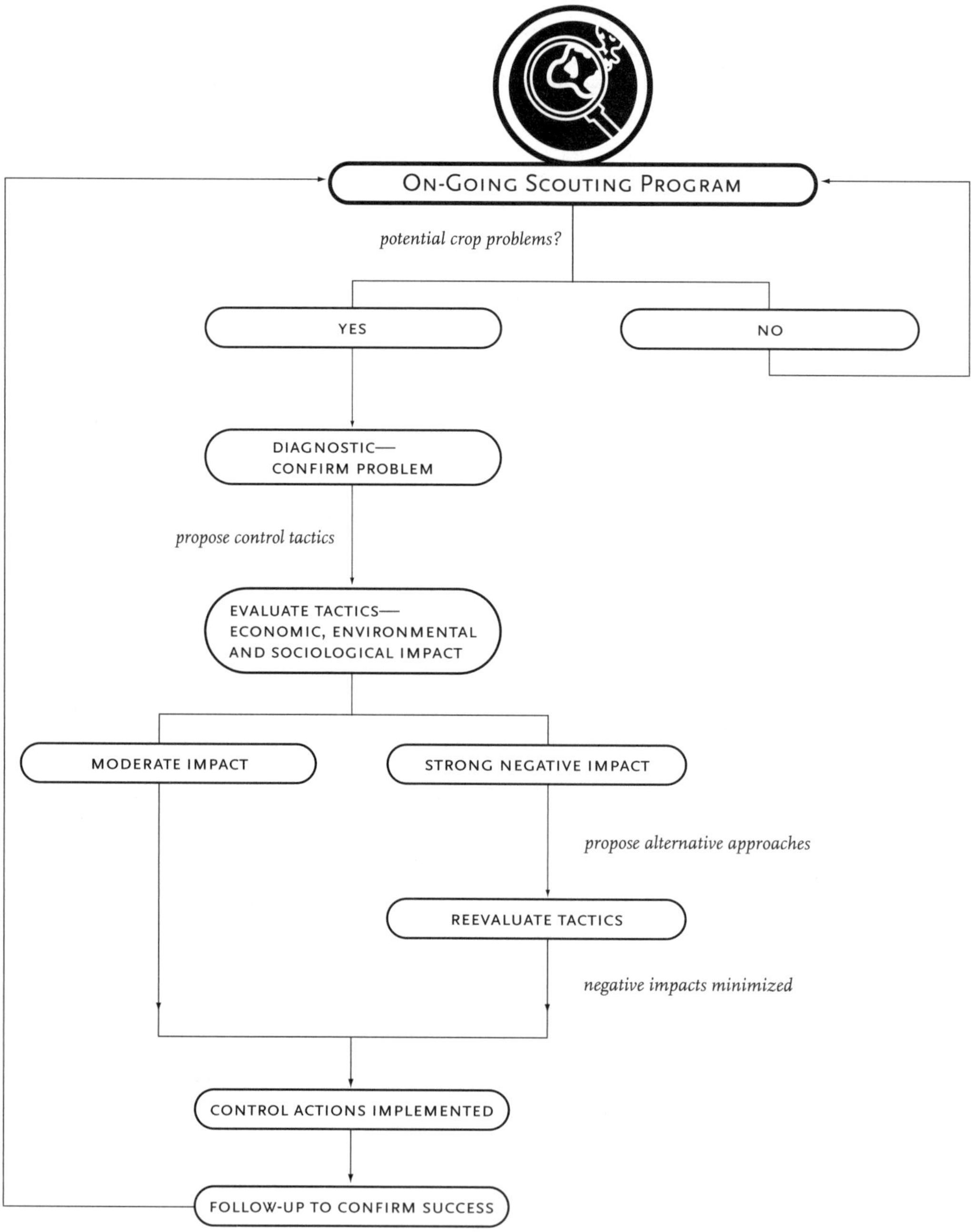

Figure 19.1 A simplified integrated crop protection strategy.

PLANT PROTECTION SPECIALISTS

Plant protection specialists should have expertise in three general technical areas: information technology (including the plant sciences), business management, and machinery operations. While different people naturally will have different levels of ability, those working in plant protection should have at least some working knowledge of each of these areas.

INFORMATION TECHNOLOGY

Information from different sources including regulations, biology, ecology, and handheld or desktop computing is increasingly required to render optimum crop management decisions. Written communication skills, public speaking, and mass media experience are becoming more important as the relatively poorly informed nonagrarian public becomes increasingly involved in agricultural policy and management, like when Mediterranean fruit fly or citrus canker issues arise. Private-sector consultants with expertise in the agricultural and environmental sciences can even be contracted as information specialists for general or specific problems that occur in the field.

Information skills required of plant protection specialists include: (1) An ability to identify insects, diseases, disorders, and agents of biological control with a reasonable degree of certainty. Ambiguity over problem causes and treatments is one reason unnecessary chemical applications in otherwise sound horticultural programs occur. (2) An understanding of pest development, disease progress, and death in quantitative specifics—not just intuitive generalities. Simply assuming by casual observation that pest populations are increasing at some general time may provide inadequate information for the most effective management decisions. Experience has shown that quantifying pest problems usually leads to better control decisions. (3) Knowledge of the strengths and weaknesses of various sampling procedures considered in relation to individual business goals. With rust mites, for example, percentage infestation measurements are of secondary value and are mostly useful after determinations of population density have been obtained. Although not always possible, rust mites should be measured according to some density rating scale, then decisions rendered accordingly. (4) An ability to integrate costs, returns and yield analyses into the pest management equation. It is an easy task simply to suppress a pest population. Verifying the cost benefits of such actions and optimizing the associated profits, however, requires that the biological, financial, and geographical nuances of production be scrutinized. Clearly, all of the above tasks require information, and computer-based technologies will by necessity play a more prominent role in future pest management decisions.

BUSINESS MANAGEMENT

Generally, there are five business resources that must be managed in commercial plant protection: (1) manpower, (2) logistics, (3) financial resources, (4) chemical inputs, and (5) environmental resources. Managing manpower is a key ability required by plant protection specialists that involves both business and personnel management. People have to be hired and fired, and their work must be efficiently scheduled. Time sheets must be maintained and payroll must be administrated. Costs and materials must be properly allocated to effectively control problem pests. Some of this knowledge can be gained through practical experience, but a basic education in the principles of business administration can help managers do a better job of managing these resources.

MACHINERY OPERATIONS

Intervention to interrupt the development of pest populations is required in many instances. For example, some form of control action (such as applying a miticide) is usually required to control rust mites on fresh fruit crops. Special equipment is needed for the application of such products. Disking, hedging, and topping are also accomplished by using specially designed equipment. A third requirement for people who work in plant protection is the ability to maintain and/or operate the special machinery used to control pest populations. Tractors, spray machines, and herbicide sprayers must be kept in good operating condition and calibrated regularly. Even if a scout or production manager does not personally operate this machinery, he or she often needs to manage or work with those who do.

DECISION LOGIC

In addition to some background in information science, business management, and machinery, plant protection managers should understand the logic underlying integrated plant protection decisions. This section introduces an example of the logic used by production managers to diagnose and treat five of the most common citrus problems. The Citrus Integrator (Plate 19.3) is a two-page decision flow chart that illustrates how citrus insect and disease strategies and tactics can be integrated. The first page is the Decision Guide and the second page is the Management Guide. The upper half of the Decision Guide holds the insect management model and the lower half contains disease management information. The top portion of the Management Guide outlines the general crop phenology and biological agents associated with changes in pest populations and the lower section details some integrated management information.

The Citrus Integrator presented in this section is an effective teaching guide, but certain field management details have been omitted due to space constraints. The five principal pest problems in the Citrus Integrator (citrus rust mites, greasy spot, melanose, Alternaria, and scab) are covered because the management actions and materials for these species overlap. Nevertheless, the general principles on which the Citrus Integrator is constructed apply to all pest management problems.

Use of the Citrus Integrator is best illustrated by example. Following the line arrows and responding to the queries at each junction box in the Decision Guide will lead to a management timeline on the opposite page. The numerical steps in the Citrus Integrator are not followed in sequence, but are given as positional references only.

All pest management decisions first involve market considerations, so it is appropriate to start with an economic evaluation. Beginning at Step 1, suppose that economic factors have been evaluated and it is desired to market a Valencia crop as a processed product. Pest populations usually begin at relatively low levels when their populations first become established. As time progresses and environmental conditions become more favorable, pest populations begin to increase slowly, followed by a rapid increase later in the season. For some pests, this cycle may occur within a single season (such as with rust mites), and for other pests, this cycle may occur over several seasons (such as with root weevils). With rust mites, for example, scouts usually begin scouting in Florida just before their numbers begin to increase rapidly (around May). Following the flow chart to Step 2, suppose mite densities are surveyed at about 3 mites per cm^2, while other economically important insects appear below damaging levels. Weather data suggest that average daily temperatures of about 76°F should be expected for the next several weeks. Referring to the middle of the three grids of Step 2, the production manager finds that the color of the squares at the intersection of 76°F on the "T" axis and 3 mites/cm^2 on the "M" axis is olive green. Moving directly below to the color that corresponds to the chart result, the production manager jumps to the olive-colored query box of Step 6 to check whether this crop is a fresh or processed crop. This is not a fresh market crop, so the "No" line is followed left to the Spring/Summer query box of Step 5. Assuming that it is summer, the "Yes" arrow is then tracked through Step 9 ("No miticide required"), then through the "Processed" option then down to the June/July/August query box of Step

12. Since it is June, the "Yes" arrow is followed down and to the left where Valencia is listed.

Next, an evaluation of greasy spot disease severity in the tolerant variety is made based on disease history, in Step 13. If disease severity is "Moderate," as in Step 16, the corresponding arrow is followed to the Go-To box at the bottom of the page, indicating that the crop is sprayed 1 time per year. The Go-To box refers the manager to the integrated protocol shown in Timeline 32 on the opposite page. Since a specific treatment for rust mites was not diagnosed, Timeline 32 shows that these processed Valencias need only the single copper plus oil spray for greasy spot, applied in June to early July. If a miticide were required, however, the production manager need only refer to Timeline 40 to find a list of materials that might be included in the June treatment.

The economic, environmental, and sociological impact of proposed treatment action must next be evaluated. If the treated grove is located near houses, schools, or ecologically sensitive areas, negative impacts must be carefully considered and treatments modified accordingly. Eventually, a treatment must be prepared for application to the crop. Pesticide applicators or other management specialists ensure that the appropriate machinery is correctly calibrated, that required materials are physically compatible, nonphytotoxic, in stock and that the water quality is suitable for mixture with the required ingredients. Weather conditions are usually checked prior to taking action to help ensure that wind or rain will not disrupt field operations. A follow-up survey of pest population densities should be conducted to verify that management operations are effective. If a heavy rain immediately followed a spray application, for example, an expensive material might be diluted or washed off the plant surface, resulting in reduced performance.

Other management scenarios are handled in similar fashion. For example, had the mite density or average daily temperature been much lower, the path could have tracked through the blue option in Step 2, also resulting in no recommended miticide. If it were early March instead of June, for example, the arrows would have led to the natural predators Go-To box of Step 3. (Over time recognizing natural predators by name will help citrus managers to better judge when pesticide applications can be delayed or avoided.)

Finally, the concept of the management cell is introduced. Associated with crop phenology, management cells are the grid boxes in the Management Guide located at the intersections of the rows and columns. For example, Management Cell 33-B and 33-C describe the development and protection of the spring growth flush from greasy spot on disease-sensitive processed crops.

INTERACTIONS AND CONSEQUENCES

It is important to remember that actions carried out in one area of integrated crop management result in responses in other areas of the program. Recall from previous chapters that weed management can improve crop quality, yields and water use efficiency. Such practices also alter the physical environment of the tree canopy by allowing more exposure to the sun. This exposure may influence the development of clustered rust mite populations, leading to a greater incidence of mite injury on some fruits. Interactions such as these are becoming more understood, but all production managers should be aware that such interactions are possible.

OVERVIEW OF PLANT PROTECTION

In addition to the straightforward -scout-and-treat approach described previously, a number of other strategies and tactics for controlling citrus pests are available. Figure 19.2 shows the four spheres of influential protection. Beginning from the outermost ring: 1) avoidance; 2) resistance; 3) scouting; and 4) direct intervention. The first line of defense for any nonintroduced pest is to prevent contact between the plant and the pest. This usually occurs at the regulatory level in which exotic diseases are checked at national borders. Second, if a pest has been introduced, the preferred line of defense is genetic resistance. Next, if a pest has been introduced, but the citrus trees do not possess resistance, then information must be obtained from the field regarding the incidence and severity of the attack. This is the "scouting" action. Finally, based on scouting information, direct intervention, such as initiating a biological control program or applying a chemical prophylactic are options. These other protection methods can be classified as regulatory, cultural, biological, physical, and chemical. The establishment and prevalence of a pathogen or pest usually dictate which methods are applied. Various methods are often combined to form an integrated strategy that is more effective than any single method alone. This section examines these plant protection methods in the context of how prevalent a pathogen or pest is in an area. For the following discussion, pest prevalence is broken into four levels: (1) not yet introduced, (2) newly introduced and/or in low incidence, (3) established but pest abundance is low, and (4) pest is widespread and variable.

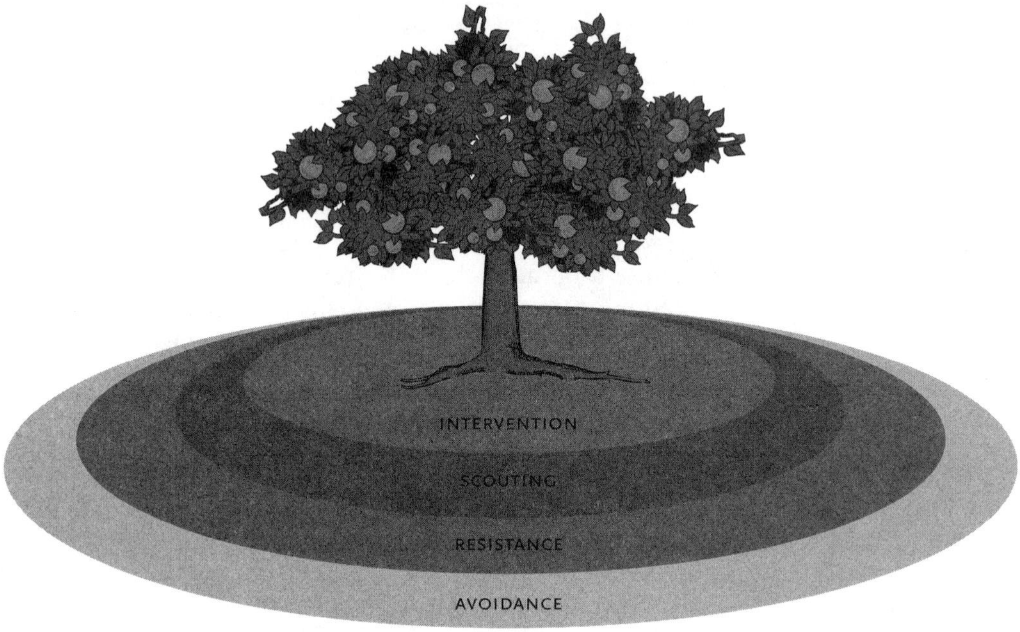

Figure 19.2 Spheres of plant protection for Florida citrus.

Pest not yet introduced

The first line of defense is to avoid or exclude the pest. No outbreak can occur if a pest can be kept out of a citrus area. In many areas of the world, citrus crops are grown free of certain devastating insects and diseases. Even if the pest does exist in a certain area, it's extent may be limited, meaning that exclusion is still a viable option for new strains of the species. For example, depending on the severity of the viral strain or isolate, citrus tristeza virus (CTV) causes a decline of citrus trees such that affected trees appear sickly and unproductive. If CTV is restricted to a certain area, often a limited number of strains of the virus may be present and exclusion still works for more severe strains of the virus that have not yet appeared in the area. In Central California, for example, CTV is present at a low incidence and only a limited number of isolates exist. In Florida and Spain, however, more isolates have been introduced over time and so CTV is more prevalent. Some of the worst CTV isolates (e.g., those causing stem-pitting) are not common in Florida, but this situation is likely to change over time. The greatest diversity of CTV strains is found in Southeast Asia—the geographical origin of citrus. So, with the exception of some areas of Southeast Asia, exclusion of CTV is a viable and necessary crop protection method to prevent establishment of the virus in the area. It should be noted that CTV exclusion will probably become less of an option in Florida over time as CTV losses continue to increase and other strains develop or are introduced.

Citrus cultivars that have developed in the absence of a pest have no opportunity to develop resistance to it. Catastrophic results can thus occur when a pest is introduced to citrus cultivars that have no general resistance or tolerance to the introduced pest. In Florida citrus, the primary avoidance or exclusion methods are quarantines and budwood registration programs. Quarantines are effective at restricting entry or passage of infected plant materials, pests and diseases across political borders. Unfortunately, modern transportation and increasing human commerce and tourism have resulted in quarantine inspectors at the borders becoming overwhelmed. The USDA inspection of produce, including citrus arriving at international airports and other ports of entry, is critical to reducing the introduction of foreign pests and diseases to Florida agriculture. The importance of this first line of defense, along with public education cannot be overemphasized considering the prohibitive costs associated with control and containment. In 1996, for example, the USDA, APHIS, and PPQ reported about 1,500,000 interceptions of plant materials at major U.S. ports of entry such as the Miami and Los Angeles International airports. They also reported about 50,000 interceptions of diseases or insects occurring on those same plant materials. Therefore, around 3 percent of all interceptions of plant material had detectable levels of a potentially harmful disease or insect. Unfortunately, these detections probably represented a fraction of the total pest introductions into the country. Few of these introductions are likely to cause problems by coming into direct contact with commercial citrus, but even a single introduction of a bacterial disease such as greening or citrus canker (Plates 19.4a & b) can be devastating.

Budwood Registration. Budwood registration programs effectively exclude pathogen introductions into geographical areas. Further, approved budwood brought into Florida under quarantine can be propagated in quarantine greenhouses and tested for viruses and viruslike diseases to expose known susceptible cultivars in special nurseries. Plants propagated at budwood nurseries are periodically inspected for the presence of disease. Such procedures help ensure that pathogen introduction is avoided. Unfortunately, it is sometimes the citrus growers themselves who find cultivars or propagation material in other countries and import them illegally that compromise budwood registration efforts. The recent discovery of citrus greening may result in citrus tree production in Florida moving to screenhouses some distance away from commercial groves.

Newly introduced or in low incidence

If a pest is introduced into an area, the second line of defense is to reduce the pest population. This method is a form of sanitation. For pest reduction to be effective, the disease or pest must be in low incidence and readily detectable soon after it invades a certain area. Eradication can be achieved if the rate of infected or infested host plant removal is greater than the rate of increase of the pathogen or pest. As disease or pest progression continues, however, this becomes more difficult since greater numbers of plants must be removed to offset new infections or infestations. If an infestation continues until it grows exponentially, sufficient tree removal to offset new infections becomes economically unfeasible and logistically impossible.

Eradication. The most extreme form of pest reduction is eradication (Plate 19.5). For eradication to succeed, there must be sufficient manpower and funds to maintain the program at highly active levels. Surveys must continually be directed to surveying new areas at the fringes of infested areas while also conducting repeat surveys in known infested areas to find the pest and aggressively eliminate it. Successful eradication programs for plant diseases are rare, however. In Florida from the 1900s to 1930s, and again from the 1980s to 1990s, and then recently from 1995 to 2006, eradication programs were conducted to suppress or eradicate citrus canker. Extensive funds and manpower were used to locate and remove all known infected trees. Over several years, apparent eradication was achieved in the cases of the first two outbreaks. However, some people question whether future outbreaks can be contained. In Argentina, citrus canker eradication by infected tree removal has not been successful. Eventually, Argentine citrus growers felt that tree removal was costing too much in crop losses, damaging the integrity of their groves, and not supplying sufficient disease suppression. Thus, their eradication program was discontinued and the country was divided into canker-free and canker-infected zones, with quarantine regulations imposed to avoid transporting the disease between the different zones. Florida terminated their canker eradication program in January 2006 due to the spread of the disease over large areas by the hurricanes of 2004 and 2005. Canker management recommendations in Florida will undergo changes as the ecology and epidemiology of the disease in the state become better understood.

Since 1929, more than twenty outbreaks of Mediterranean fruit fly (Medfly) have occurred in Florida. In contrast to plant disease eradication programs, the Medfly eradication program (which relies on continual monitoring of pheromone attractant traps) appears to be highly effective. The insect has been eliminated on these occasions by combinations of bait sprays and release of sterile male flies each time it has been reintroduced.

Sanitation. Eradication programs are often combined with sanitation (Plate 19.6). Sanitation is the disinfection of people and equipment to reduce spreading a pest or pathogen from grove to grove. For example, citrus canker bacteria can be found in or on plant parts, clothes, shoes, tools, equipment, and elsewhere. Therefore, containment of the spread of citrus canker is dependent on an effective sanitation program. Several chemicals are effective for use in sanitation programs. Among the most effective are quarternary ammonium sprays and bleach (for equipment), and various formulations of surgical soaps (for people). These products kill bacteria by disintegrating bacterial cell walls. Although the bacteria are sensitive to sunlight and desiccation, vehicles and equipment should be reasonably cleaned of leaves, twigs, and fruit before leaving a grove.

Chemical Control. Chemical control is one of the most frequently used methods to reduce or eradicate pests. Soil fumigation has traditionally been used to control nematodes and certain soil fungi. In many places however, including the U.S., soil fumigation has become limited since methyl bromide and other soil fumigation compounds are being phased out in response to environmental concerns. Effort is now being placed on finding less environmentally damaging alternatives. Insecticides, fungicides, and bactericides still form a main defense for other citrus pests, however. These pesticides reduce infection by foliar and fruit pests and controlling insect vectors of citrus diseases. Nevertheless, the array of compounds registered for use is dwindling in many countries because of environmental concerns and lack of profit-base in many countries.

Biological Control. Biological control is yet another method of pest suppression, such as the "friendly fungus" shown in Plate 19.7, which attacks and helps control citrus whitefly. Another successful example of biological control is suppression of insect vectors of citrus greening disease by a parasitic wasp. (Citrus greening is a disease that causes trees to become yellow and unproductive. An insect vector can transmit the disease by feeding on diseased plants and transferring the pathogen to healthy plants in later feedings.) On the island of Reunion, heavy damage to citrus plantations occurred due to citrus greening. Two insect vector species were also present. Researchers there imported an Asian wasp species that parasitized both vector species. As a result, the insect vector population was decimated and continues to be kept at low levels by the wasps. Today, citrus greening is just a minor propagation problem in that area because of this beneficial wasp (and the use of budwood registration programs). Greening was identified in Florida in 2005, and researchers are currently determining the best approaches to controlling this disease.

Low pathogen or pest viability

Crop losses can increase significantly when a pathogen or pest becomes established. Many of the crop protection strategies discussed above (especially the use of quarantines, registration programs, and pest population reduction techniques) should be continued or expanded. However, eradication programs are often abandoned after some period of time, as occurred with canker in Florida in January 2006. When this occurs, additional long-term crop protection strategies must be considered. If insects transmit the disease, control of the insects by chemical and biological means may be intensified to help reduce new infections. Cultural considerations can also help reduce the impact of a disease or pest that must now be lived with and managed. Some examples of cultural management are (1) modified irrigation schedules and reduction of overhead watering to reduce bacterial and fungal infection by citrus canker, citrus bacterial spot, scab (Plates 19.8a & b), and greasy spot; (2) care in pruning to reduce disease transmission either by avoiding wet conditions or by periodically sterilizing pruning equipment; and (3) the effective use of windbreaks to limit wind transmission of citrus canker cells.

Resistant Cultivars. Disease or pest resistance programs increase in importance and are often expanded as higher investments in crop protection are required to offset crop losses. For insect-vectored diseases, tolerance to disease is often sufficient when no vectors are present whereas resistance is preferred after vectors become established. For example, when CTV is firmly established but the variability of virus strains is limited, tree losses become significant usually due to the presence of virulent strains that cause decline of trees on sour orange rootstock. Over time, growers may be forced to convert their groves to rootstocks tolerant to the virus (such as rough lemon, trifoliate orange, Rangpur lime, and the citrange rootstocks). When stem-pitting CTV strains are present, emphasis can shift to a search for resistance within cultivars. Unfortunately, this can be dangerous to surrounding citrus plantations. CTV strains that cause no symptoms in mandarins can be devastating to adjacent sweet orange and grapefruit cultivars. The mandarins act as a symptomless reservoir for severe types of CTV that may then spread to surrounding susceptible cultivars. In South America, the importation of mandarins carrying severe CTV has led to destructive epidemics in several countries.

Widespread and variable

The ultimate prognosis for most citrus-producing areas is that most exotic pests will eventually become established. Over time, exotic diseases and pests will be imported and battles will be fought and lost to keep their incidence low. Integrated plant protection techniques such as those discussed in this chapter must be aggressively implemented to reduce the impact of virulent pathogens. Crop and tree losses are expected and must be factored into the cost of doing business.

The diversity of established pest problems requires a careful analysis so that insect and disease populations can be suppressed to economically feasible levels. For example, greasy spot is a disease that causes leaf spots and rind blemishes. Greasy spot was poorly understood in Florida for many years and was controlled with only one or two midsummer copper sprays. The citrus industry in Florida has moved south because of a series of freezes in recent years, and greasy spot has become a much more prevalent and damaging disease in South Florida. Epidemiological studies of greasy spot spores released from decaying citrus leaves have revealed that the disease now occurs earlier in the year and for longer durations in South Florida than in more northern areas. Hence, control measures must be applied earlier and spaced out over a longer time frame to achieve adequate control of greasy spot in the southern part of the state.

Sometimes even more exotic plant protection solutions must be considered as with the CTV example introduced previously. Tree immunization is one such example. This is called "cross protection" because less virulent strains of CTV are inoculated into trees to help prevent more severe strains from developing and causing disease. Although not an exact analogy, this process is similar to the vaccination (immunization) process in humans. Such cross protection is complex because one mild strain may only inhibit infection of one or a few severe strains. Careful site-specific analyses and trials must be conducted to assess the diversity of the severe isolates present and to determine which mild strains would be most effective for use in cross protection. Cross protection can slow down a CTV epidemic, improve crop quality and quantity, and extend productive tree life. Unfortunately, cross protection is not a permanent solution and usually breaks down over time.

Resistance Management. Citrus pests can respond to stresses in ways that enable them to survive adverse environments. These stresses include physical factors (e.g., temperature or humidity), biological factors (e.g., predators, parasites, or pathogens), and environmental contaminants. In commercial agriculture, chemical sprays to control pests create severe environmental stresses for the pests. After treatment, a small percentage of pest individuals survive because of genetic characteristics they possess that allow them to resist the chemical's effects. Consequently, those individuals survive and reproduce, while weaker individuals of the same species die. Resistant pests develop the ability to tolerate amounts of the chemical that would kill the majority of other individuals in the population. Many pests have even developed resistance to more than one chemical.

Over time, repeated applications of the same chemical can result in the number of chemical-resistant pests increasing. Production managers recognize this as a decrease in the effectiveness of an applied chemical. Practically speaking, the effective way to delay the onset of resistance is to limit the number of applications of the same chemical to one per season. Rotating the use of specific chemicals within or between years may also reduce the likelihood of developing resistance. If resistance to a particular chemical develops in a pest population, use of that material and materials in the same class should be discontinued.

FUTURE OF INTEGRATED PLANT PROTECTION

In the future, integrated plant protection will be increasingly technology driven. National and international companies often have more abundant capital than smaller firms; therefore, they will be among the first to incorporate new technologies into their production systems. However, small- to medium-sized farms still stand to benefit by taking advantage of new technologies to increase efficiency. In the case of technologies such as classical biological control or computer-based systems, larger corporations may provide these as services to smaller operations for a fee or as value-added services as a way to establish influence over a larger geographical area.

Increasingly, citrus pest management will involve the use of computers in routine commercial management. One commercial system has been developed and consists of a field-data collection network that permits georeferencing insect and disease incidence and severity data into handheld computers.

Data collected by scouts into the handheld computers are transmitted to a central location through the Internet that makes the data available for real-time analysis or mapping. Efficient collection of such data is critical if certain exotic pest species are introduced. Such technology eliminates administrative tasks of keying in and analyzing or interpreting extensive field data. Data export functions enable researchers to update their datasets at any time to facilitate epidemiological analysis and modeling. Worker administration and scout accountability functions enable time tracking of personnel. Growers may similarly access disease data for mapping or to relate back to their company information on yields, profits, and other grove conditions. In other computer-based systems, disease diagnosticians have access to a network to track the development of invasive pests and diseases to help improve early detection and control of disease outbreaks.

It is important to understand that technology does not necessarily imply gadgetry. New machines, integrated circuitry, and information that improve business operations will continue to appear. But there must be no confusion between the tools of technology (devices and databases) and the intellectual techniques (problem-solving abilities) that will empower plant protection professionals in the future. Methodological strategies for problem solving and decision-making can increase profit margins as significantly as the most ingenious new devices. The greatest predictor of future success in plant protection will be the ability of production teams to exploit field techniques and information technology tools that increase profits and efficiency, reduce costs, and protect the health of people and the environment.

REFERENCES

Browning, H. W. 1994. Biological control of the citrus snow scale, *Unaspis citri*, in Florida: evaluation of aphytis and other natural enemy species, p. 119-142, ed. D. Rosen. *Advances in the Study of Aphytis*. Andover: Intercept.

Browning, H. W. 1996. *Early classical biological control on citrus*, p. 27-46, eds. D. Rosen, F. D. Bennett, and J. L. Capinera. *Pest Management in the Subtropics: Integrated Pest Management, A Florida Perspective*. Andover: Intercept.

Browning, H. W., and R. P. Muraro. 1993. *Approaches to integrated pest management on citrus fruit destined for processing.* Citrus Industry 74: 49-56.

Browning, Harold W., Robert J. McGovern, Larry K. Jackson, David V. Calvert and Wilfred F. Wardowski. 1995. *Florida Citrus Diagnostic Guide.* Florida Science Source, Lake Alfred, Florida. 244 pp.

Childers, C. C. 2003. *Integrated mite control on Florida citrus.* International Society of Citriculture Congress. (9th 2000: Orlando, Florida), Orlando, Florida USA 2: 740-750.

McCoy, C. W. 1999. Arthropod pests of citrus roots, 149-156, eds. L. W. Timmer and L. W. Duncan. *Citrus Health Management.* APS Press, St. Paul, Minn.

McCoy, C. W., R. J. Stuart, L. W. Duncan, J. H. Graham, and H. N. Nigg. 2003. *Soil arthropod pest management for citrus: Future solutions to a growing problem.* International Society of Citriculture Congress. (9th 2000: Orlando, Florida), Orlando, Florida USA 2: 733-735.

Ploetz, R. C., L. W. Timmer, and S. M. Garnsey. 2003. Management of tropical fruit diseases: Current overview and future outlook, 465-481, ed. R. C. Ploetz. *Diseases of Tropical and Subtropical Crops.* CABI Publishing, Wallingford, U.K.

Rogers, J. S., C. W. McCoy and M. M. Manners. 1994. Standardized visual comparison keys for rapid estimations of citrus rust mite (Acari: Eriophyidae) populations. *Journal of Economic Entomology* 87:1507-1512.

20

Integrated Crop Scouting

J. S. ROGERS

INTRODUCTION

All farm decisions are ultimately based on some type of scouting or survey information. Scouting information is derived from the field aspect of agricultural production. Even financial data from the caretaking and sale of the crop also represent important types of information strongly tied to field scouting data. In an agricultural context, Phase I and Phase II environmental audits represent a special category of scouting information. Sometimes scouting information may be quantitative, as in the maturity tests collected prior to, during, and immediately after harvest. Sometimes the information may be more qualitative, as in the general assessments of soil moisture for irrigation purposes used by many growers. Spring begins an active period of scouting devoted to assessing grove condition, meeting irrigation needs, and starting up survey programs for insects and diseases. Summer includes the period when other pests and diseases are aggressively scouted. With the onset of canker and greening in a citrus area, systematic scouting and surveying become required practices. This chapter explains why scouting is performed and how to apply that information to improve the citrus ICM program.

All farm decisions are ultimately based on some type of scouting information. Scouting information is derived from the field aspect of agricultural production. Examples of scouting information include pest population surveys, crop quality tests, soil and tissue test results, weed population surveys, and crop yield estimates. Even financial data from the caretaking and sale of the crop also represent important types of information strongly tied to field scouting data. In an agricultural context, even Phase I and Phase II environmental audits represent a special category of scouting information. Sometimes scouting information may be quantitative, as in the maturity tests collected prior to, during, and immediately after harvest. Sometimes the information may be more qualitative, as in the general assessments of soil moisture for irrigation purposes used by many growers. Nevertheless, scouting information is the ultimate foundation of all farm management decisions.

Historically, scouting protocols are not sufficiently broad in scope, and the data are not properly used for optimizing crop yields and profits. In the citrus industry, this is due primarily to the lack of a clear definition of scouting, as well as a lack of technology that supports scouting efforts. Scouting information should not be limited to pest and disease information, but there clearly has been a tendency to think of scouting information primarily in those terms. Such a constrained approach limits the use of scouting information to a "scout and treat" model for insect and disease suppression. Unfortunately, practices supporting just this limited insect and disease application of scouting information result in the acquired information being useless for almost any other purpose, such as for yield and profit optimization.

A clear understanding of the information needs that can be met through scouting is required to properly use the information collected and extract its extended value. Specifically, the use of scouting information must go beyond merely making determinations for pest suppression practices. Scouting information must also advance to yield and profit optimization. Therefore, scouting is more correctly thought of as a year-round process involving the acquisition of a broad base of diversified information that can be quantitatively related back to yields and profits.

Optimization is the process of fine-tuning what we are able to accomplish through approximation. However, optimization of a technical program is impossible without record keeping. Grove-scouting information should be collected in such a way that data can be accumulated over time and analytical results can be presented in real time. Subsequently, retrospective analyses can be performed and optimal production practices more easily evaluated.

Remarkably, there are more than 900 billion measurably distinguishable grove conditions. The question becomes which of these billions of possibilities are the most productive or profitable? On a practical level, the human mind can probably differentiate between no more than about 50 or so of those billions of possibilities. The task becomes insurmountably more difficult as productivity differences become subtler. However, if optimal approaches to productivity are to be clearly identified, there must be a systematic approach to managing scouting information that can help differentiate and prioritize the most profitable production scenarios. Further, data so acquired should be broadly inclusive of a diversity of parameters, and those data should be centralized so that the appropriate real-time analyses can be applied.

Centralization of scouting data within a single company may provide a temporary competitive advantage for that company, but centralization across the industry will ultimately be required to identify the most profitable production practices, especially with the increasing distribution of problems such as citrus canker, post bloom fruitdrop, and citrus tristeza virus complex. Data centralization is the only way that subtle influences of genetic background, soil type, etc., can best be understood. Segregating scouting efforts into separate public and private efforts is a most inefficient manner in which to carry out a scouting effort. Integrated pest and crop management should be a region-wide effort, with financial and logistic support for the scouting effort justifiably coming from the public, private, and nonprofit sectors. This approach spreads the cost of the development and maintenance of centralization across the industry, so the incremental cost per acre over time becomes miniscule—much less than one dollar per acre per year. Region-wide integrated plant protection will become increasingly important in the future as pest and disease profiles become more exotic and environmental and human health issues become more entwined with product marketing.

Unfortunately, in most commercial operations, the collection of scouting information is often haphazard and nonsystematic. Until centralization of yield, maturity, cost, return, profit, and pest population data is achieved, we will be approximate, but we will be unable to verify what are the most profitable production practices in a given region. Nevertheless, the first step to true optimization is to acquire scouting information in a way that all information collected from one location can be simultaneously related to all other information from the same and other locations. This is the conceptual basis for economic ecology. For confidentiality reasons, georeferenced data domains can be encrypted. However, the information needs and the associated acquisition and reporting methods across all locations should become thoroughly integrated. The remainder of this chapter addresses the management of scouting information to meet those needs.

SCOUTING THROUGH THE YEAR

The first step to developing an integrated scouting approach is to map the various information needs that can be addressed by scouting and order them chronologically throughout the year. Table 20.1 shows the type of information collected by scouting and surveillance throughout the year in commercial citrus grove operations. This list does not differentiate between bearing and nonbearing trees, although scouting needs for the two differ somewhat. Variable information includes those information needs that vary from season to season, while nonvariable information includes those information needs that do not change from season to season. Inclusion of a category does not preclude it from being important in other seasons, only that it is generally a priority in the listed season.

Winter Scouting

January through March. There has been a general tendency for systematic scouting and data documentation to be overlooked during the cooler winter months, but valuable information can be collected from the field during this time. Information needs during the winter include greasy spot disease levels (examining fallen leaves), crop maturity, weed populations, certain mite and other insect problems, and possibly postbloom fruit drop. Preparations for hedging and topping can also be accomplished noting tree development and productivity and scheduling the work accordingly. Scouting for citrus canker disease can also be accomplished during this period, although the lesions will likely be rather aged.

Spring Scouting

April through June. Spring begins an active period of scouting devoted to assessing grove condition, meeting irrigation needs, and starting up scouting programs for insects, diseases and weeds. It is during spring and continuing into summer that on-going surveys for insects and diseases become most critical. In seasons following fairly warm winters, early scouting for rust mites is important, because early population increases on leaves can serve as predictors of early population increases on fruit. If conditions become too dry, scouting for spider mites could become an issue, especially on younger trees. Such problems may be exacerbated by the windy conditions that sometimes occur during the spring. Tree responses to irrigation practices, leafminers, and crop set also are important information records to acquire during this time, along with postbloom fruit drop. The earlier information from this scouting period is documented, the earlier it can be used to forecast probable conditions likely to develop in groves later in the season. Follow-up scouting for postbloom spray applications is important, including measuring the effectiveness of postbloom fruit drop control applications and tree responses to micronutrient applications applied to the foliage.

Table 20.1 Type of information collected by scouting and surveillance throughout the year in commercial citrus grove operations. PFD = postbloom fruit drop; PDC = Phytophthora-Diaprepes complex.

Winter	Spring	Summer	Fall
Weather	Weather	Weather	Weather
Weeds	Weeds	Weeds	Weeds
Greasy Spot	PFD	Rust Mites	Soil Nutrition
New Flushes	Crop Growth	Greasy Spot	Tissue Nutrition
Frost Damage	Melanose	Plant Bugs	Crop Maturity
	Diaprepes	Diaprepes	Diaprepes
	Canker	Canker	Canker
	Aphids	Aphids	Spider Mites
	Leafminer	Leafminer	Brown Rot
	Irrigation	Fruit Drop	Irrigation
	Tree Stress	Crop	Fruit Drop
	Alternaria	Destination	Crop
	Scab		Destination
	Scales		

Summer scouting

July through September. Summer includes the period when pests and diseases are heavily scouted, primarily citrus rust mite populations. Monitoring for effectiveness of spring weed control operations is important. Fruit drop on some cultivars, e.g., Navels, may manifest during this time, and subsequent applications of growth regulators may also be planned and prioritized. Follow-up surveys for mite control and assessments of crop quality for marketing purposes are important to document. If certain mite control programs were more or less effective than expected, changes in the market destination of some crops may be considered.

Fall scouting

October through December. Slightly cooler temperatures often prevail in fall, so rust mite scoutings tend to slow to moderate diligence, but they cannot be ignored. Most scoutings during this time relate to obtaining information needed for harvest and sale of the crop. Also, leaf flushes from previous seasons become fully expanded, so soil and fall flush tissue sampling are important. Greasy spot disease assessments may be made during this period as well. As early fresh grapefruit crops are harvested, the number of groves that require frequent scouting for rust mites and other crop-damaging pests declines, but information reconnaissance regarding fruit quality and yield should be maintained.

ACQUISITION, CENTRALIZATION, AND ANALYSIS OF SCOUTING INFORMATION

Specially trained personnel who visually survey each grove several times per month may observationally acquire scouting data. Data may also be acquired using electronic or other automated technologies, such as weather stations or computerized soil moisture detection devices. Traditionally, citrus scouts record their observations in more or less haphazard form on paper or similar media. While systematic, the paper approach does little to improve the overall problem of integrated scouting logistics and information management. The law of diminishing returns applies to the acquisition and interpretation of paper-based scouting data for optimization purposes. Labor costs are simply too high to administrate and analyze all data that are necessary for quality yield and profit optimization. However, automated technologies permit the acquisition of data through computerized or similar methods. Due to the inherent differences in ease of data management between the paper and computerized approaches, automated methods are currently receiving the greatest attention.

Recent innovations in digital scouting technologies are permitting scouts to record their observations into handheld computers, which can be directly linked to computers and integrated decision support networks for improved data management. One important advance in the utilization of this information is the improved accuracy in handheld GPS (Global Positioning Satellite) units, which permit georeferencing of scouting data to latitude and longitude. Accuracies to within about 10 to 12 feet are now obtainable in many instances.

Once all of this integrated crop information is collected, the next task deals with utilizing the information for optimizing yield and profits in citrus groves. As mentioned above, it is first important to centralize the information so that trends and relationships between the different observational parameters may be identified. Centralization does not necessary imply that all information be physically located in the same place; rather, centralization is a virtual concept. As long as critical data are accessible for analysis through a centralized diagnostic server, the centralization caveat can be met. From these centrally accessible data, real-time reports may readily be made available. For example, it would be a simple matter to map the extent of tree damage after a freeze if scouts are georeferencing measurements of leaf drop, crop damage, or tree damage and posting it to an accessible database. This would assist greatly in the management of markets after the freeze. If a more robust use of the data were desired, it is a simple extension, for example, to subsequently determine what profile of damage on freeze-damaged trees is most susceptible to, say, citrus canker infections.

Such management of commercially acquired information would also serve to validate much research. Even the best-designed research can only provide general advice on how to improve production. It requires site-specific accuracy in data collection to validate research findings across increasingly large geographical regions. For example, with insect/plant pathogen complexes, the commercial database itself could serve as a valuable research tool. Population distributions could easily be analyzed with respect to weed profiles, natural predators, diseases, scion, soil type, etc. The relationships within and between species and their environments could be much better characterized. The future will eventually bring the industry a centralized, integrated, scouting information database, and it is important to note that the technology to create such a system already exists. The design and implementation of a industry-wide, centralized data system should be a concerted and cooperative effort, and it would be for the benefit of all if it were sooner, rather than later, that it emerged.

REFERENCE

Clemenzie, J. R., G. F. Fairchild, and S. Thornsbury. 2001. *Internship in Florida grapefruit utilizing ENTONET™ through Citrus Solutions™ marketing initiative*. Teaching and Learning Paper Series TLP 01-5: 12 pp.

Rogers, J. S., H. N. Nigg, I. Clark, D. S. Hart, R. Ehsani, D. L. Brock and E. W. Schuster. 2006. Internet-based survey and modeling systems for agricultural insects and diseases: Applications in Florida citrus. Annual Meeting of the Florida State Horticultural Society, June 5, 2006, Tampa, Florida.

21

Materials Application Technology

J.S. Rogers and P.D. Spyke

INTRODUCTION

Application of chemical and other inputs is a major component in citrus management. Various materials are applied to citrus in an effort to enhance growth, production, fruit quality, and to achieve other desired outcomes. The basic consideration when choosing a method for applying these materials is to consider the effectiveness with which the inputs will be delivered to the target. However, this is not the only consideration. Inputs must be applied in a manner that is economically viable, within acceptable risk limits, and safe for operators, grove workers, the general public, and the environment.

Agrichemical product recommendations for crop production, nutrition, and other practices may be found in annually updated UF/IFAS publications. These include the Florida Citrus Pest Management Guide and the Florida Citrus Nutrition Management Guide.

The application of materials affects the profitability of the operation as much as any other single factor. For example, the proper material improperly applied can be ineffective at the least or can even lead to damage of some kind. Economic losses from improper applications can range from the value of the material and application itself up to the entire value of the treated crops or even the trees themselves. A great deal of effort and capital investment is required to achieve optimal applications. For example, growers may choose to make applications themselves or contract the work to companies that specialize in such operations. Regardless of this choice, the method and timing of an application can be just as important as the actual material and dose. Therefore, the grower and production manager should regard application methods as significantly as any other management function. Records should be kept regarding dates, times, and composition of specific applications (Table 21.1).

Any method that applies the material to the target at the proper concentration is suitable for the task. This requires that the machinery is properly calibrated and operated. There is usually a wide range of choices for any particular application, but issues such as initial cost of the machinery, operation costs, support needs, and other logistical considerations affect individual choices of machinery.

In general, however, there are four broad application target categories: (1) the aboveground portion of the trees, that is, the fruit, leaves, limbs, and branches; (2) the surface area beneath the trees; (3) the row middles between the spaced trees; and (4) the soil within the root zone. Within these four broad categories are subcategories, based on specific target area, the delivery method, or the purpose of the application. For those unfamiliar with the insects and diseases discussion that follows, Table 19.1 provides a general background on the symptoms caused by the various organisms.

ABOVEGROUND PORTION OF THE TREE

Generally, airblast sprayers are used to apply materials to the tree itself (Plates 21.1 & 21.2). Airblast machines operate on the principle that a combination of water and moving air is used to deliver the pesticide to the target. Airblast machines are manufactured with a wide range of delivery capabilities, from machines with high air and water volume to small machines with relatively low air and water capacity (Plates 21.3 & 21.4). Prior to the development of airblast machinery, water alone was the standard carrier for canopy applications, and applications were performed with high-pressure water sprayers of various kinds. This method is still quite effective and is actually a low-cost alternative. The drawback is that the machinery involved is normally less reliable than modern airblast machines, and the volumes of water involved require frequent sprayer refilling. Consequently, modern airblast sprayers that use moving air as a component of the delivery system are generally a better choice for commercial operations.

Foliar sprays are applied to the aboveground leaves and fruit on the tree for controlling diseases, insects, and for nutritional purposes. The specific target is a major factor in the choice of the machinery and calibration.

Table 21.1 Suggested record keeping for spray operations.

Item	Description
Date and Time:	State the date and time of the application.
Block:	State the farm, field and block name of the treated area.
Variety:	State the varieties covered.
Product Applied:	State the full trade name of the product and list the EPA registration number, if available.
Rate per Acre:	Indicate the rate or dose per acre of he applied product.
Application Volume:	Enter the volume of spray per unit area applied.
Operator Name:	State the name of the operator of the application machinery.
Target Pest:	State the target pest(s) of the application, e.g., rust mites, greasy spot, etc.
Justification:	Enter the treatment threshold or justification for the pesticide application.
Temperature:	Enter the temperature maxima and minima for the day of the treatment.
Wind:	Record wind speed.
Rainfall:	Indicate whether or not there has been rain near the time of the application.
Re-entry Times:	Specify time of posting and removal based on product labels.
Days to Harvest:	Indicate preharvest intervals as specified on label and relationship to expected date of harvest.

FOLIAR FUNGICIDES

Greasy spot

Greasy spot is generally considered to be the primary cause of premature leaf drop in Florida. Therefore, greasy spot control is a critical component in commercial citrus management programs. To control greasy spot on foliage, the material must be applied to the underside of the leaves. No other target site has any effect on reducing disease. Since citrus leaves are usually oriented with the underside facing downward and inward, applications machinery must turn the leaves upwards or sideways and apply the material to the leaves on the side closest to the sprayer (or penetrating the canopy to reach the undersides of leaves on the other side of the tree). The best assurance for the control of greasy spot on foliage in large, mature trees requires the use of high air volume conventional airblast machinery. Ground speeds must be matched to air volume to assure proper penetration of the air stream into the tree canopy; sprayers with lower air volumes must be operated at lower ground speeds to achieve the same amount of leaf disturbance necessary for a suitable application. Tree size and canopy characteristics affect the specific calibration combination.

Lower-volume applications using aircraft or low-volume ground machinery have been used successfully to control greasy spot on foliage under certain conditions. Where disease pressures are low, and with cultivars that are less sensitive to the disease, low-volume applications can achieve commercial efficiency. Another consideration is that the greasy spot fungus usually grows for a period of time on the underside of the leaf before penetrating stomata, so imperfect coverage is less likely to fail than is the case with other kinds of disease control. While it is still an absolute requirement that the material be applied to the underside of the leaf, the density of the droplets as well as some other factors can be compromised somewhat and still achieve commercial control.

Citrus Scab

Scab sprays are applied to foliage and fruit in an effort to reduce infections on fresh fruit. With most commercial cultivars, scab infections on the leaves themselves rarely cause enough damage to warrant special applications. The major consideration is to reduce spores that can be produced by the disease on leaves that can later spread to the fruit. Therefore, the application target for scab control must be young leaves that are located on the outside of the tree canopy. Generally, foliar applications for scab are applied only to the spring flush.

Alternaria

Alternaria is very difficult to control on foliage with current fungicide choices. The situation could change if new fungicides are registered for Alternaria control, but copper is an effective material currently available. Copper is a protectant, and germinating fungal spores must come into actual contact with the copper chemical for control to be achieved. In contrast to greasy spot control, distribution of copper on the leaves and fruit must be maintained evenly on both sides of young leaves throughout the period of susceptibility. However, the percentage of the leaf surface covered by copper is reduced through weathering, growth, and enlargement of the leaf, so applications must be frequent. New decision aids are under development to assist in determining the proper timing of applications. However, any spray method that delivers an even coating of copper to both sides of young leaves should be sufficient for achieving optimum levels of control.

FOLIAR INSECTICIDES AND MITICIDES

In general, the insects and mites that affect the leaves and fruit of the tree are mobile. Any application method that applies the material evenly to the tree should be sufficient for control of many insect and mite pests. Timing is generally the major consideration for these kinds of applications, so machinery that can cover a grove quickly allows the manager or grower more flexibility in control programs. An exception to this general statement would be spray intended to control Diaprepes root weevil egg masses. Diaprepes lays its eggs between two leaves and attaches the leaves together with a sticky substance to form a protective sheath around the egg mass. When citrus spray oils are applied at the correct time, the leaves will slip apart, exposing the egg mass to solar desiccation. Therefore, sufficient water and air volume must be used to allow the oil to penetrate the leaf sheath.

FOLIAR NUTRITIONAL MATERIALS

Most nutritional materials are translocated within the tree, so thorough coverage is not usually necessary. The effectiveness of most applications of nutritional materials can be improved if they coincide with the presence of young leaves, which have lower wax deposits and absorb materials better than older leaves. Nutritional applications should be designed to apply the materials evenly throughout the tree, especially with materials that are not readily absorbed or translocated. Caution should be used when concentrating applications of materials with high salt content. Foliar and fruit damage can occur if concentrated solutions of salts are applied in a nonuniform manner, such as with low-volume sprayers or aircraft.

FOLIAR GROWTH REGULATORS

Growth regulator sprays (PGRs) can be translocated within the tree. Therefore, while uniform coverage to the tree is preferable, the main consideration is the concentration of the PGR material. Unfortunately, most research has not included the effect of growth regulator concentration, so PGRs have often been applied using thorough coverage with a specific material-to-water ratio or concentration. If applicators are not careful, PGRs can harm trees if applied improperly. Consequently, every effort should be made to rigorously follow PGR label rates and usage recommendations. Low-volume concentrate sprays should not be used unless the recommendations specifically address this option. Calibration of machinery should be precise, and machine operators should be briefed on the need for accuracy.

APPLICATIONS TO YOUNG FLUSH AND FLOWERS

Flowers on older trees often appear primarily on the outside of the tree. With young, rapidly growing trees, flowers are usually produced deeper within the canopy, and it may be more difficult to achieve coverage on a four-year-old tree than on an older tree that is less vegetative. Sprayer configurations should reflect the position of the flowers within the canopy of the trees. The major product applications to flowers in Florida citrus are for the problems discussed below.

Citrus Scab

Scab can infect the young fruit from the time they are formed within the flower, so where infections are severe and/or cultivars are susceptible a spray for scab control is frequently applied at 2/3-petal fall. The spray must be applied so that the fungicide is deposited on the young fruit, which are often 1/8 to ¼ inches in diameter at the time.

Postbloom Fruit Drop (PFD)

PFD first infects the petals of the flowers themselves, so materials must be applied to the flower petals. The flower petals are somewhat waxy, so thorough coverage is required. Low-volume applications can be successful, provided the droplet size is small enough to result in a uniform coating of material on the flowers, particularly when contact materials are being used. Some fungicides used for PFD control are inefficiently systemic, so thorough coverage is quite important.

APPLICATIONS DIRECTED AT THE FRUIT

Many materials are applied to the fruit throughout the year. The choice of application method is largely related to the maturity of the fruit on the tree. The fruit does not have uniform target characteristics throughout the season. From flowering until the diameter reaches about 0.75 inches, the fruit is oriented upwards in the same manner as the flower. As the fruit increases in size, the weight bends fruiting twigs downward. By summer, the fruit is hanging downward. Coverage of fruit with sprays is a difficult challenge, especially as the size of the fruit increases later in the season. Linear airflows from low-volume sprayers generally do not achieve thorough coverage on the inside of the fruit. Similarly, concentrate sprays from conventional air-blast machines can result in uneven coverage. This uneven coverage, particularly when highly concentrated spray mixtures are being applied, can result in fruit burn on the outer portion of the fruit and concurrent lack of control on the surface of the fruit facing the tree trunk.

The best assurance of complete coverage of the fruit surface of larger fruit is achieved by using high water volumes applied with air-blast machines with sufficient turbulent airflow to cause the liquid and air mass to swirl within the canopy of the tree. Where complete coverage of fruit in the later stages of size growth and development is important, the machinery choice and calibration should be adjusted accordingly. Discussion of some specific applications to fruit follow.

Melanose and citrus scab

The target for a melanose spray is generally the small, developing fruit in the spring of the year. Coverage on the leaves or other parts of the tree is less important, unless other materials such as miticides or nutritional materials are added to the spray mixture. The materials used to control melanose on fruit generally are contact protectants (such as copper), so a uniform, thorough coverage of the fruit surface is important. The fungicides are not systemic and generally do not redistribute themselves through rainfall; therefore, the material must be placed in the proper target area at the time of application.

The size of the fruit during the melanose control period is generally from ¼ to 1 ½ inches in diameter. It is usually oriented upward in the early stages of the control period, but the fruiting twigs start to bend downward toward the end of the control period, which could dictate a change in the calibration of the machinery. Lower water volumes and faster ground speeds are possible when applying fungicides for melanose control. Low-volume machinery also can perform satisfactorily since: (1) the small fruit are generally in exposed locations or behind leaves that can be easily moved by the air-blast of the machine; and (2) miticides, insecticides, and nutrients do not usually require thorough, even coverage to perform effectively. In the postbloom period, however, sprays intended for melanose will also control citrus scab. The requirements for both are essentially the same.

Greasy spot rind blotch

Rind blotch caused by the greasy spot fungus is occasionally a problem on grapefruit and certain other cultivars, particularly when disease pressure on leaves is high. The control period is in the summer, after the fruit has tilted to its downward orientation. The fruit is usually about 75 percent of its final size, which presents a challenge when trying to achieve even coverage on the entire surface. Attempts to concentrate spray materials for control of greasy spot rind blotch applications (which are distinctly different from applications designed to control greasy spot on leaves) have resulted in burn and reduced growth on the outer surface of the fruit and a failure to control rind blotch on the surface of the fruit facing the tree trunk. The problem has occurred particularly when copper fungicides and citrus spray oil have been mixed and applied in very hot conditions, but the problem has also been noted when oil alone was applied. In general, better results can be expected when spray water volumes are higher and material concentration in the mixture is lower. This factor has to be weighed against the higher costs and greater time required for a high-volume application, and if disease pressure is low, the choice to sacrifice coverage could be acceptable.

Mites

Mites are mobile, so coverage of the fruit is less important than it is with fungicides. The exception to this rule is materials that are incorporated into the cuticle of the peel by combining them with citrus spray oil. In this case, it has been noted that a loss of control can occur when the material is not applied uniformly to the fruit surface.

Plant Growth Regulators

Gibberellic acid (GA) sprays. Gibberellic acid applications made to delay the aging of the rind on approved cultivars is an exception to the discussion on growth regulator sprays previously mentioned. GA must be applied uniformly over the surface of the fruit since it does not translocate within the peel. Also, since GA can be toxic to the peel, uneven coverage greatly increases the risk of spray burn.

TRUNK AND BRANCHES

Occasionally, it is necessary to apply materials to the trunk and/or the branches of the tree. The normal circumstances for these kinds of applications are for problems such as scale control. Natural enemies normally control scale insects, but unbalanced predator and pest relationships can result in the need for supplemental spray applications. Snow scale can build up to damaging levels on older trees, and black or brown soft scales are common problems on younger, rapidly growing trees in newly established groves. Scale control applications are normally performed with airblast machines on older trees, while younger trees are given spot treatment with hand sprayers. Sufficient water must be used to penetrate the canopy and thoroughly coat the infested areas. Timing is also critical since the effectiveness of some sprays is reduced if not applied during the correct life stage of the developing scale insects. Normally, these sprays are applied to a point where some runoff onto the ground occurs.

Trunk Paints

Some materials are applied by painting them on the trunk or main branch of the tree. These materials may be systemic materials that can be absorbed through the bark of the tree, such as treatments for *Phytophthora*. When trunk painting with highly concentrated hazardous materials, precautions must be taken to prevent worker exposure. Workers must wear the correct protective clothing and be trained to apply the proper amounts of material to the trees.

UNDER TREE AREA

Many different kinds of materials are applied beneath the trees. The technology used must deliver them in the proper amount to the intended target, while protecting the trees, nontarget areas, and machinery operators from unintentional exposure. As with tree canopy applications, the various kinds of applications can be segregated by method, purpose, or material applied.

BANDED HERBICIDE APPLICATIONS

With the exception of organic citrus production, almost all commercial citrus grove programs include under-tree weed control using herbicides. The machinery used to apply materials under the trees must be capable of making uniform applications while controlling nontarget exposure. In general, spray booms that extend under the tree are used for these types of applications. The booms must be designed to reduce interference from low-hanging branches and reduce nontarget exposure to applied materials. Considerations when choosing or designing machinery for under-tree herbicide applications include nozzles, shields, hydraulics, and filtration. Wider-angle nozzles, spaced to achieve full overlap at a low height, allow booms to be operated closer to the target area. Further, herbicide spray booms must be shielded to prevent exposure of tree branches to herbicides and to reduce damage to fruit by drift of materials. Generally, rounded boom designs are superior to boom shields with sharp points or angles. Hydraulic devices operated by the machinery operator control the height of the spray booms. Hydraulic designs that depend on wheels to maintain the proper boom height have limited usefulness in commercial situations. Finally, filter choices must be based on nozzle size and material used. When using wettable powders as the suggested material, a series of filters is often more effective than a single filter unit.

HERBICIDE TREATMENTS

Herbicide materials applied under the trees can be divided into two broad categories: (1) contact or systemic materials (foliar absorbed), and (2) soil residual herbicides (root absorbed). Contact or systemic materials must be applied directly to the foliar surface of the weeds with sufficient coverage for the active ingredient to initiate control. It is usually not important in this situation that the material reaches the soil. Water volume is also less important than the spacing of nozzles, boom design, and operation of the machinery. However, booms must be capable of applying material uniformly to the target weeds under the trees if weeds are too tall. Control will be reduced if booms are incapable of achieving adequate coverage. Also, timing of applications is often as important as specific boom design to avoid the necessity of treating weeds that have reached into the tree canopy. In contrast to materials applied directly to weeds, soil residual materials must be applied to achieve a uniform layer of material atop the soil surface. Booms that apply materials unevenly or the presence of emerged weeds that interfere with the application pattern can reduce the effectiveness of soil residual applications. The nozzle spacing must be matched to the height of the boom above the soil surface to assure a uniform application. Non-uniform applications can result in an inadequate or premature breakdown of weed control in areas that receive less than the desired amount.

Some herbicide materials have both weed surface contact and soil residual activity. In this case, the intention should be to maintain a uniform application rate across the boom swath, even though emerged weeds might intercept the pattern to some extent. The same is true when contact materials are tank-mixed with soil residuals.

HERBICIDE SPOT TREATMENTS

Herbicide materials are applied to the tree row as spot treatments for various reasons. Spot treatments are applications that target only specific areas of the groves. For example, tall weeds that cannot be treated with a boom, occasional weeds that have escaped soil residual control, or weeds within reset sites all might be treated on a case-by-case basis using an herbicide spot treatment. Generally, high-volume handguns used to apply foliar materials are not well suited to this type of application. Lower-volume handguns can apply sufficient material to achieve control, although handguns with extremely small nozzles can result in excessive drifts and time-consuming applications. Soil residual materials should never be applied with handguns since calibration is impossible. Under-tree booms can be used for spot treatments where soil residual materials are needed (such as around resets). With this type of application, the normal calibration methods are used, but the liquid flow is turned on and off as needed while the herbicide machine moves down the row at a uniform speed to maintain an even application rate. Various methods of mechanizing this operation have been tried with some success.

MULTIPURPOSE HERBICIDE APPLICATIONS

Recently, herbicide machines have been built with multiple tanks and either separate nozzle manifolds on the boom or direct-injection capability. These configurations allow the option of applying different mixtures of materials as the herbicide machine moves down tree rows. Normally, the machine carries two mixtures, one for mature trees and the other for resets, and has the ability to apply chemical mowing materials using the nozzles as well. With such machinery, various other options are possible, such as applying contact materials on a spot-treat basis while applying soil residual materials in uniform bands. Another long-term possibility is the variable-rate application of soil residual materials, based on differences in soil types or other factors.

FERTILIZER MATERIALS

In Florida, most macronutrients (N, P, K, Ca, Mg, and S) are applied to the soil under the trees. The general definitions for the various types of fertilizers are defined in Chapter 17. The type of fertilizer material applied determines the application method for the various fertilizer types. Generally, there are three types of fertilizer materials: (1) dry bulk, (2) liquid solutions, and (3) liquid suspensions.

Dry bulk

Dry bulk fertilizers normally are delivered in large trailers that contain up to 22 tons of materials in bins. Broadcast spreaders are used to apply the material primarily and under the trees. The spreaders are either tractor-drawn or truck-mounted. Calibration is achieved by weighing the material caught in a bin as it is discharged by the mechanism while the machine moves down a measured distance. Dry bulk materials are usually premixed combinations of nutrient materials, and this will likely remain the standard for the foreseeable future. New innovations in application machinery have been attempted in row-crop farming where the various materials are kept segregated and mixed “on-the-go.” If research supports such a practice, it may be used in Florida citrus in the future as well.

Liquid solutions

Liquid solutions of fertilizer materials can be applied with herbicide machinery, but the practice has not gained great favor in Florida for a number of reasons. Most importantly, cost-competitive solutions generally must be formulated using materials with high volatility, so gaseous losses after application become unacceptable. Also, the percentage of materials in the solution must be fairly low to avoid precipitation or other unwanted chemical reactions. Limited attempts have been made to combine stabilized nitrogen solutions with herbicides in an attempt to achieve contact activity without using a systemic or contact material. While high concentrations of the material may cause salt burn on the weeds, vigorous regrowth from surviving weeds stimulated by the fertilizer material can offset any gain from such applications. Liquid solutions may also be applied to the foliage as a means of supplying both macro- and micronutrients that can be rapidly utilized by the plant.

Liquid suspensions

A liquid suspension is a highly concentrated mixture of fertilizer materials and water, where water is used as the carrier for the material. Such suspensions must be constantly agitated from the time they are mixed at the fertilizer plant until they are applied. The application machinery resembles a normal herbicide machine but has the capability of applying higher volumes of abrasive, more corrosive, materials. Some soil residual herbicides can be mixed with the suspension to achieve weed control at the same time the fertilizer is applied. Use of suspensions has advantages and disadvantages over conventional dry bulk applications. The advantages are that some weed control can be achieved at the same time, and material placement can be precise and accurate. The primary disadvantage of liquid suspensions is that the machinery requirement is more complex than a dry bulk spreader.

REFERENCES

Miller, W. M., and J. D. Whitney. 2003. *A test program to assess VRT granular fertilizer applications for citrus.* Presentation at the 2003 ASAE Annual International Meeting Paper Number 031126: 11 pp.

Schueller, J. K. 1992. *A review and integrating analysis of spatially-variable control of crop production.* Fertilizer Research 33: 1-34.

22

Management of Tree Size, Shape and Crop Load

D.P.H. Tucker, T.A. Wheaton, and E.W. Stover

INTRODUCTION

A tree size control program integrates site characteristics, variety, rootstock vigor, tree spacing, nutritional and water inputs, and mechanical hedging and topping to maintain optimal production with balanced vegetative regrowth. An ideal balance between fruiting and vegetative growth provides sufficient shoot growth for excellent cropping without aggressive pruning. Selecting a tree spacing and scion/rootstock combination appropriate for the soil type is critical in establishing a balance between vegetative growth and fruiting. In the field, mechanical hedging and topping are the most common forms of tree size control, but there are several other effective techniques that should be considered. The need for pruning is in large part determined by decisions made in site selection, grove design, and planting, which should integrate the vigor of the scion/rootstock combination, tree spacing, and soil type. At first, these decisions might seem simple, but once economics are factored into the equation, the process can become quite involved. This chapter presents basic information related to the factors that go into making hedging, topping, and pruning decisions.

There are a number of factors that can affect tree size and crop load management, such as whether tree allocation to vegetative growth or yield depend on if inputs are moderate or excessive (Figure 22.1). Best management practices for tree size control and crop load management require an integration of grove design and production inputs with application of a pruning program that is continually evaluated and monitored. Tree size control strategies should integrate all factors to provide a balance between fruiting and vegetative growth to provide sufficient shoot growth for continued cropping without the need for aggressive pruning. The benefits of maintaining a proper balance between vegetative growth and reproductive crop development include good tree health, consistent high production, better fruit quality, and more efficient harvest. The selection of a tree spacing and scion/rootstock combination appropriate for the soil type is critical to establishing a balance between vegetative growth, and fruiting. A tree size control BMP integrates site characteristics, variety, and rootstock vigor, tree spacing, nutritional and water inputs, and mechanical hedging and topping to maintain optimal production with minimal vegetative regrowth. Maintaining this balance requires knowledgeable grove management, even when these components have been carefully chosen (Figure 22.2).

TREE SIZE CONTROL

Tree size control must be considered in the layout and design of a new grove. The ideal citrus tree is one that grows rapidly to fill its allocated fruiting surfaces and then stops, or greatly reduces its rate of growth. Thus, tree size control begins with selecting a tree of appropriate vigor for its allocated space. At present, the selection of dwarfing rootstocks for citrus is limited. However, there are significant differences in vigor among existing rootstocks, and new rootstocks are being developed. Swingle rootstock has been used extensively for the past decade and results in a smaller tree than other rootstocks widely used in Florida, but has performed poorly on some flatwoods soils.

Mechanical hedging and topping are the most common forms of tree size control, but there are several other effective techniques that should be considered. Heavy croploads reduce the amount of vegetative growth and, thus, effectively reduce tree size. Limiting fertilization and irrigation will slow tree growth, so management strategies that avoid excesses in nutrition and water assist in tree size management. In addition, there are selected strains of viroids that can be introduced into citrus that reduce tree size in a predictable manner, but these have not been thoroughly tested in Florida.

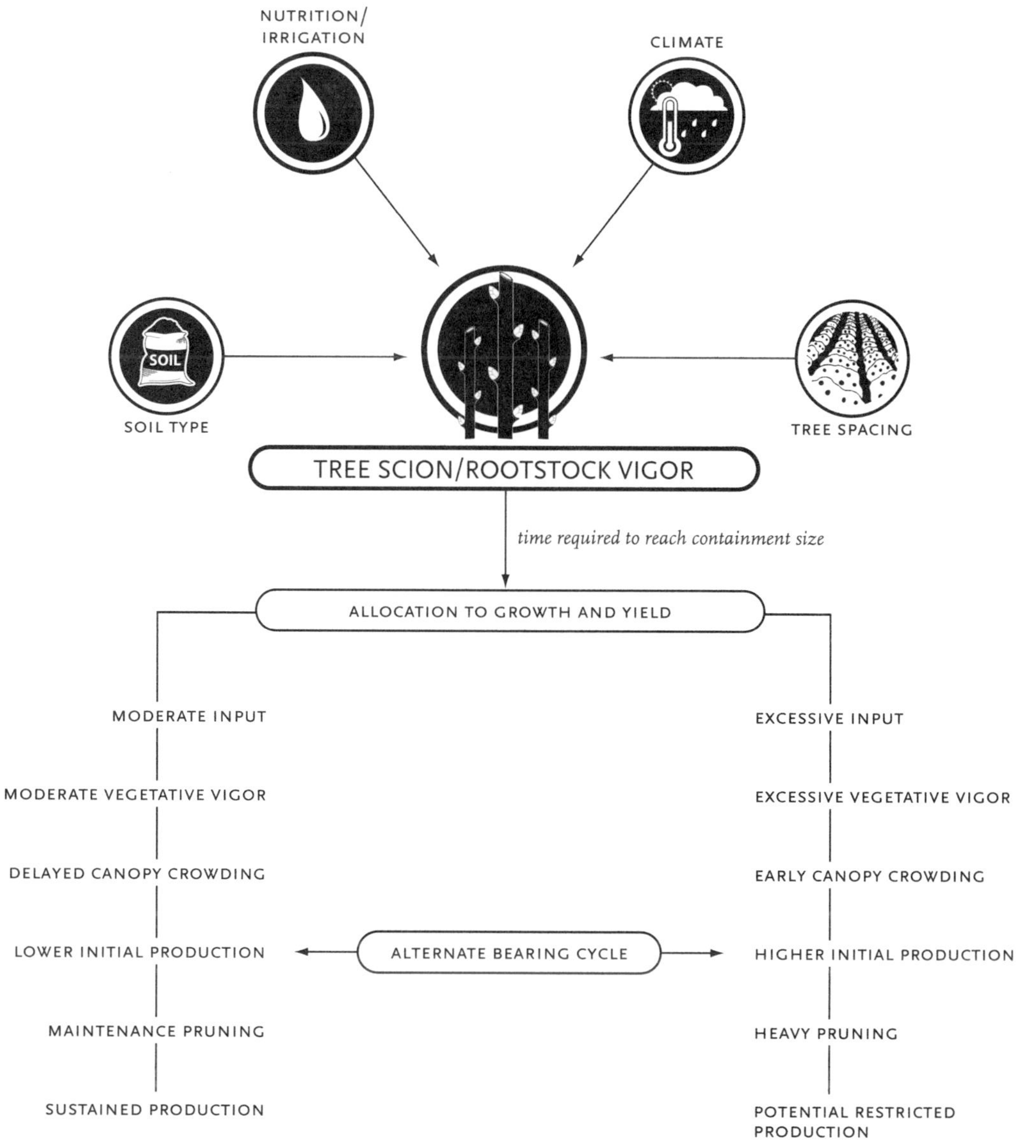

Figure 22.1 Early, high production from groves with closely spaced trees of high vigor, followed by stabilized or declining production after 10 to 12 years. Best management practices include careful evaluation of scion/rootstock compatibility, tree spacing, soil type, and nutrient/water inputs to ensure more sustained production.

Tree Size and Crop Load Management

GROVE DESIGN AND PROGRAM DEVELOPMENT

GROVE DESIGN

Row spacing
In-row spacing
Row orientation
Tree age
Equipment access
Harvesting

SCION/ROOTSTOCK

Vigor
Final tree size
Adaptability

SITE CHARACTERISTICS

Soil pH, CEC, OM
Soil water holding capacity
Climate limitations (freeze)

MONITORING

YOUNG TREES

Growth rate
Vigor
Canopy closure
Cropping
Irrigation and nutrition
Inputs
Leaf analysis

MATURE TREES

Pruning regrowth
Yield trends
Alternate bearing
Fruit distribution and quality
Lower canopy shading
Irrigation and nutrition inputs

APPLICATION

HEDGING

Containment size and light interception
Frequency
Timing
Width
Angle
Crop load adjustment

TOPPING

Containment size and light penetration
Frequency
Timing
Height
Angle
Crop load adjustment

SKIRTING

Timing
Height
Frequency
Fruit quality
Mechanical harvesting
Irrigation maintenance
Fertilizer and pesticide application

RECONSTRUCTIVE AND REHABILITATIVE PRUNING

Freeze damage
Recovery
Rejuvenation
Timing
Exposure and sunscald
Pest and freeze susceptibility and protection

SELECTIVE TREE REMOVAL

Canopy crowding
Tree age
Yield trends
Timing

EVALUATION

YIELD TRENDS AND TREE VIGOR ASSESSMENT

Crop load and fruit size
Regrowth
Topping height
Tree removal
Alternate bearing
Leaf nutritional status
Irrigation and nutrition
Management

Figure 22.2 Various factors that can affect tree size and crop load management.

Tree Size Control

Tree size control should begin with selection of spacing appropriate for site, scion, and rootstock.

Mechanical hedging and topping are implemented as soon as trees exceed desired width and height.

When trees reach containment size, fertilizer, and irrigation should be reduced to limit tree growth.

Selected strains of viroids are used to dwarf citrus in some areas, but they are not currently used in Florida.

PRUNING

Trees require little pruning until they exceed containment size, but minor selective pruning may be necessary during the first several years in the grove. Removal of sprouts from the trunk is important, and sprouting may be controlled using trunk wraps or sprout inhibitors on young trees. Shading by foliage makes trunk sprouting uncommon on trees older than three to five years. Vigorous branches extending beyond the rest of the canopy are frequently cut off, but there is no evidence that this cutting is necessary or beneficial. Young trees may produce a few fruit during the first two years in the grove. In some cases, this fruit is removed so it will not compete with vegetative growth.

Pruning needs are in large part determined by decisions made about site selection, grove design, and planting, including the vigor of the scion/rootstock combination, tree spacing, and soil type. These factors along with certain cultural inputs, especially water and fertilizer, determine how rapidly trees attain containment size and how quickly the ratio of tree height to middle width hinders adequate sunlight interception by lower canopy areas.

Once trees have grown to containment size, regular pruning becomes necessary. It is important to begin hedging and topping as soon as trees achieve their allocated space and before severe cuts are necessary. Pruning is most effective when the cuts are made in wood that is still small in diameter. When large wood is cut, the regrowth is very vigorous and rapidly extends to or beyond the original canopy. Good production of quality fruit may not resume for two or three years after severe pruning.

In the past, it was not uncommon for growers to also remove dead wood from mature trees every few years in order to reduce melanose damage. While stringent quality demands may make dead wood removal desirable, it is rarely done anymore. Experienced labor is not available and/or the investment is perceived as excessive for expected returns.

HEDGING

Hedging consists of cutting back the sides of trees to maintain a clear alley for equipment access and to provide good light distribution to the lower parts of the tree. Hedging is usually done using a large machine with several saws on a fixed or rotating boom designed especially for this purpose. Hedging operations cut off all the twigs or branches extending beyond the allocated space. New sprouts form at the cut surface and form a new layer of canopy that becomes a significant fruiting zone of the tree the year following the new growth.

Hedging should be started early so that only small branches are cut and crop reduction is avoided or minimized (Plates 22.1 & 22.2). In some groves, hedging may be required as soon as five years after planting if trees are on a vigorous rootstock and tree growth is rapid. Hedging is also less expensive when started early because brush removal is not required. Small branches and twigs may be left on the ground, but if large wood is cut, removal is necessary.

The hedging cut is usually made at an angle so that the tree is narrower at the top than at the bottom, providing better light distribution to the lower parts of the tree and helping avoid the snagging of debris in trees. An angle of approximately ten degrees is recommended, with angles between five and fifteen degrees commonly used. Tree height and row spacing may influence the hedging angle to some extent. With closer row spacing and taller trees, hedging at greater angles cuts more severely into the top of the tree, which may have a negative impact on yield.

Hedging operations may be performed annually for vigorous trees or less frequently for trees of reduced vigor. Annual hedging of every row-middle is popular because it avoids large pruning cuts. There is no brush disposal problem, it does not stimulate excessive vegetative growth, and good fruit quality is maintained.

Groves should be hedged at about the same time each year, and hedging shortly after the harvest of early and midseason fruit is common. Choosing a time to hedge Valencia (a citrus variety bearing both a new and maturing crop simultaneously) is more difficult, since there is an overlap between the mature crop and the crop developing for the next year. Thus, fruit at some stage of development may be lost no matter when hedging is conducted. In practice, there is little yield loss associated with hedging if it is practiced annually at approximately the same time of year. In colder locations, however, trees may be more susceptible to cold damage if hedging or topping is carried out during the fall or winter.

Cross hedging between trees in the row is usually not warranted, as continuous hedgerows produce large crops. However, a tree removal option may be considered where crowding reduces yields (Plate 22.3).

TOPPING

Topping consists of trimming the tops of trees to control height and should be started as soon as trees grow to a desired height. Topping maintains tree height for economic harvest and effective pest control. It also allows adequate light to reach the lower parts of the tree. Topping, however, usually causes reduced yield. With vigorous trees, annual topping may be required to avoid excessive tree height. On the other hand, if trees of less vigor are planted using a grove design that provides adequate space and height, topping may be required less frequently, or not at all. Trees that grow rapidly to the desired height and then stop growth are ideal because canopy development is rapid, and tree size is easily managed at maturity (Plate 22.4).

Trees may either be flat-topped or topped at an angle of fifteen to thirty degrees from horizontal, resulting in a peak about two feet or more above the shoulders. Rows that are closely spaced and hedged with a sufficient angle are relatively narrow at the top, and can be flat-topped with a single pass of the topping boom. The cost of topping may be reduced if a machine with booms on each side can top two rows in one pass. As with hedging, disposal of small brush during maintenance topping does not require much removal. Severe topping results in excessive vegetative growth and substantial yield reduction. Many growers with bedded groves feel that severe topping may sustain production of trees where rootstocks are poorly or marginally adapted to the site (for example, Swingle on calcareous solids), but this belief has not been verified through research.

The conventional wisdom regarding topping suggests an ideal ratio of tree height to middle width of 2:1. This ratio, in combination with a ten-degree hedging angle, provides good light distribution throughout the tree and maintains good fruit production in the lower canopy. Topping heights of fourteen to eighteen feet are common. Trees at these heights are easily managed and can be harvested with standard ladders. The practice of topping trees at twelve feet or less is of questionable value. Topping heights in this range are sometimes observed in young plantings.

Fresh fruit growers may also top trees below a height needed for other purposes, since reducing cropping in this way will sometimes enhance fruit size. Topping height affects harvesting efficiency. Most of the fruit from short trees can be harvested from the ground. Because harvesting from the ground is more rapid, there may be a small reduction in picking costs associated with shorter trees. However, topping to maintain trees of standard vigor at heights of twelve feet or less causes a significant reduction in yield.

Because of the yield reduction associated with topping, it is difficult to make a general recommendation about timing and frequency of topping. Timing of topping, like timing of hedging, is not critical, as long as it is done at the same time every year. Here again, it is common to top shortly after harvest of early and midseason cultivars, but before major bloom development. Valencia may be topped either before or after harvest. Vigorous trees benefit from annual topping, or possibly from topping more than once per year. With less vigorous trees, it may be possible to top only every two, three, or four years and avoid some of the yield losses caused by topping every year. In these cases, the year to top may be chosen based on expecting heavier than normal flowering.

Hedging and Topping

Effects are least deleterious when cuts are made into small-diameter shoots.

Trees should be topped at a height that is about twice the width of the clear alley between rows to provide good light to tree bottoms.

Hedging and topping are least disruptive to cropping when they are done at the same time each year.

Early and midseason cultivars are best hedged after harvest.

In colder areas, pruning is usually delayed until after freeze danger, since pruning can increase cold tenderness.

Valencia and late-harvested grapefruit, which have fruit present year round, should be pruned at around the same time each year: when pruned annually at the same time, research suggests that pruning time doesn't affect cumulative yield.

SKIRTING

Skirting is the practice of pruning lower branches to remove low-hanging limbs and foliage and fruit in direct contact with the ground. A skirting machine is used to remove the lower eighteen to thirty inches of the canopy. Because it improves fruit quality, skirting is becoming an accepted practice with citrus grown for the fresh fruit market. Lifting the skirts means less mechanical and chemical damage to fruit from fertilizer and herbicide application, and less disease from direct contact of fruit with the ground. Skirting is also required for mechanical harvesting of fruit, which requires visibility of the trunk for shaking and room for the catch frame. Additional benefits include better distribution of fertilizer under the tree and easier maintenance of undertree irrigation systems.

CROP LOAD, FRUIT SIZE, AND QUALITY

Hedging and topping may also be used as a means of reducing crop load in some cultivars. Murcotts and some other mandarins tend to set excessive crops some years, which results in poor fruit size and quality, and may even cause death of the tree. In addition, a very heavy crop may encourage alternate bearing; that is, the tree will alternate between very heavy and very light crops in successive years. When performed early in a heavy crop year, both hedging and topping are effective in reducing the crop load and consequently discouraging alternate bearing by cutting off a significant portion of the newly formed fruit. In contrast, pruning during a light crop year may increase the problem of alternate bearing by further reducing the crop load. If an excessively heavy crop is expected, hedging and topping to reduce crop load should be done before fruit set. Delaying pruning until the amount of fruit set can be more accurately determined is still likely to reduce alternate bearing, but may be less effective in increasing fruit size. Hedging or topping later in the season is less desirable, but may help avoid tree collapse of Murcotts or other cultivars with excessive crop loads.

Any pruning that reduces crop load generally increases fruit size and lowers soluble solids content. The increase in fruit size is usually desirable for fresh fruit, but the loss of solids reduces the value of the fruit for processing. Severe pruning can create a juvenile growth habit and result in the poor-quality fruit typically associated with young trees. Internal quality, external color, and peel characteristics may all be diminished, and juice content may be reduced.

FREEZE DAMAGE AND PRUNING

Decisions about pruning groves with severe freeze damage are best delayed until the full extent of the freeze damage can be assessed. New growth will appear, but dieback may continue for some time. Several months may be required to determine the full extent of freeze damage. Once the extent of damage is known, pruning to remove damaged wood should begin. Trees should be cut back into live wood either by hand or through mechanical pruning. Selective pruning by hand to remove only dead wood is ideal, but costly. Mechanical hedging and topping are frequently used as alternatives to hand pruning. If freeze damage extends into the primary branches down to the trunk, the trees should be removed and the grove replanted.

REFERENCES

Barry, G. H. 1993. *Rootstocks for tree size control.* Proceedings of the World Congress of the International Society of Citrus Nurserymen 4: 367-369.

Eissenstat, D. M., and L. W. Duncan. 1992. *Root growth and carbohydrate responses in bearing citrus trees following partial canopy removal.* Tree Physiology 10: 245-257.

Wheaton, T. A., J. D. Whitney, W. S. Castle, R. P. Muraro, H. W. Browning, and D. P. H. Tucker. 1995. *Citrus scion and rootstock, topping height, and tree spacing affect tree size, yield, fruit quality, and economic return.* Journal of the American Society for Horticultural Science. 120: 861-870.

23

Plant Growth Regulators

E.W. Stover, L.G. Albrigo, and F.S. Davies

INTRODUCTION

Plant growth regulators (PGRs) may be used to improve fruit set of bouquet bloomers, increase fruit set after bloom, increase fruit size by reducing crop load, extend the harvest season by delaying rind aging, reduce preharvest fruit drop, increase juice extraction, increase fruit Brix, and control trunk sprout growth in young trees. Yet, it is easy to cause damage to trees and fruit if PGRs are not applied correctly. There are many factors that must be considered when making the decision to apply PGRs to a crop, including weather, dosage, variety, economics, and more. This chapter presents information related to how and when PGRs should be applied to citrus crops. However, the information here should not be used to render a final decision to apply a PGR. Such a decision should be made on a case-by-case basis.

Plant growth regulators (PGRs) are widely used in other tree fruits and in citrus production in California, Spain, South Africa, Australia, and Israel. In these citrus-producing regions they provide significant economic advantages. Similar benefits are likely to be realized in Florida as Florida citrus producers incorporate the appropriate use of PGRs into crop management. Citrus producers should be aware, that unlike most pest control agents, response to PGRs might not always be obvious unless nontreated plots are used for comparison. Results after PGR use are usually within the range of variability observed between different groves or years. PGR response also tends to be more variable than with other spray materials, with little effect in a significant proportion of cases. Extension recommendations are made when research indicates that there is a substantial likelihood that PGR use will provide a benefit to the grower.

The three PGRs currently available for Florida citrus are NAA, GA_3 (gibberellins), and 2,4-D. The judicious use of these PGRs may significantly increase the profitability of citrus production. The response of citrus to PGRs depends on the citrus variety, physiological state of trees, environmental conditions, and the timing of the application. The potential benefits and hazards of plant growth regulators are discussed in terms of each PGR.

PGRs may be used to reduce excessive flowering (GA_3), improve fruit set (GA_3), improve rind quality (GA_3), increase juice extraction (GA_3), reduce preharvest fruit drop (2,4-D), increase fruit size by reducing crop load (NAA), and control trunk sprout growth in young trees (NAA).

EFFECTS OF GA_3

GA_3 is used to enhance juice extraction, improve fruit set, improve fruit quality, extend the harvest, and reduce excessive flowering to enhance cropping in bouquet blooming cultivars.

Reducing Floral Initiation with GA_3

Citrus that blooms excessively, particularly with a lot of bouquet bloom, may actually set more fruit if the number of flowers is decreased through use of winter GA_3. Thus far, research has only shown this benefit with 'Navel' and 'Ambersweet.' In other cultivars GA_3 can also sometimes be used to increase fruit size by spraying during the winter to reduce flowering, although this practice remains experimental. The relationship between numbers of flowers and resulting fruit is important in determining whether these treatments should be used. In general, any reduction in floral initiation is more likely to reduce fruit production in blocks with average flowering and may reduce crop value. However, in blocks that bloom excessively, reducing resources wasted on excess flowering may result in increased fruit set. In such cases, reducing floral initiation may substantially increase yield (typically 15 percent to 25 percent).

Increasing Fruit Set

GA_3 application during bloom can significantly increase fruit set in some light cropping cultivars, especially Minneola and Orlando tangelos. It can also be useful for blocks of tangerines with inadequate pollenizer placement. Similar benefits for 'Star Ruby' grapefruit are reported from Texas, but have not been tested in Florida. Essentially, the applied GA_3 substitutes for gibberellins produced by developing seeds and enhance the retention of seedless fruit. Resultant fruit size may be somewhat reduced. While much of this reduction is the direct effect of increased cropping, in some cases fruit size is smaller than would be expected without GA_3 use. Citrus that blooms excessively, particularly with a lot of bouquet bloom, may actually set more fruit if the number of flowers is decreased through use of winter GA_3.

Fruit Quality and Market Season

Rind quality factors are important in determining fresh fruit marketability. Many citrus cultivars can be harvested with reasonable quality over a fairly broad period of time. Generally, premium prices are achieved for very early- and late-harvested fruit of good quality when there is less competition from other producers. However, it should be noted that substantial delay of harvest in many cultivars might reduce cropping in the subsequent season. The internal quality of grapefruit, in particular, remains high throughout much of the season. The quality of the peel, however, slowly declines, and this is the primary limitation on late sales, as fruit becomes more susceptible to decay and peel disorders and shelf life is lost. In some studies, including trials on Florida grapefruit, applications of GA_3 have helped maintain peel quality of late season citrus. GA_3 is routinely used on California Navel for maintaining peel quality and in many grapefruit-producing areas as well, but is rarely used in Florida, except on Minneola. Most Florida Navel growers prefer to market their crop prior to Christmas and before the California crop, minimizing the value of GA_3 treatments. Use of GA_3 on Florida grapefruit has been limited in part by frequent reports of significant spray burn, which can cause damage that exceeds the potential benefits of GA_3. The degree of burn is increased with higher concentrations of GA_3, application in water volumes less than

250 gallons per acre, and the use of potent surfactants. Applications of GA_3 prior to color break in Minneola tangelo can significantly delay stem rind deterioration and enhance fruit quality received at the market. Studies also indicate that GA_3 greatly reduces susceptibility of white grapefruit and Fallglo tangerine to postharvest pitting, a peel disorder that has caused extensive losses of exported Florida fruit. Using GA_3 to reduce postharvest pitting could offer considerable economic advantage, but again deleterious effects such as poor degreening of Fallglo and spray burn in grapefruit preclude recommendations for this purpose. Overall, it is recommended that use of GA_3 to delay rind aging should be largely restricted to Minneola tangelo in Florida.

Enhancing Juice Extraction

Studies indicate that GA_3 enhances juice extraction in the processing of oranges. It appears that GA_3 either increases fruit rind firmness so that standard extraction yields more juice, or decreases peel growth. Juice production has been increased by as much as 9 percent following GA_3 treatment, with an average increase of 4 percent. In addition, processors report much lower losses of late-season fruit from crushing in trailer bottoms, and less time is required to clear filters of fruit particles. Juice yield, however is not increased from late-harvested fruit. If such improvements are routinely realized, juice production and profits will significantly increase, and profitability should be enhanced. It is particularly attractive that this increase in juice occurs with no additional cost beyond GA_3 application, and environmental impact is negligible. Some growers report, however, that application of GA_3 to Valencia for increased juice yields sometimes reduces flowering and may affect yields.

GA3 APPLICATION

Timing and Rates

The timing and rate of GA_3 application are very important in achieving the desired effect and minimizing unwanted effects. Some studies have shown a substantial benefit to including surfactants when using GA_3 for some purposes.

To reduce flowering and increase cropping if excessive blooming is expected, GA_3 is applied during the winter at 30 ppm (see Table 23.1) using a surfactant at 0.05 percent. Navel has shown the best response when treated in mid to late December, while Ambersweet has responded best from mid to late January. Ideal timing is likely to vary with variety and with environmental conditions that influence bloom initiation.

Updates are available during the crucial winter period through the University of Florida at http://www.lal.ufl.edu/crechome/groweraids.htm.

Table 23.1 Calculating GA3 needed for desired ppm.

Desired ppm GA3	Ounces of 4% GA3 per 100 gallons	Ounces of 4% GA3 per 500 gallons
10	3.8	18.9
15	5.7	28.4
20	7.6	37.8
25	9.5	47.3
30	11.4	56.8

Application of GA_3 to enhance fruit set is recommended at full bloom, but no careful study has been made to determine optimal timing. A good response is unlikely when applied postbloom. Studies suggest that rates of 15 to 25 ppm (Table 23.1) are optimal, and surfactants are not recommended for this use. Some growers report excellent response when two 10-ppm applications are made at early and late bloom.

Rind quality enhancement is typically greatest when the treatment is applied just before color break. Yet, application at this time can significantly delay loss of chlorophyll from the fruit, both in the field and following degreening with ethylene. Application of GA_3 during color break can result in irregular green spotting. Many growers choose to apply GA_3 after color break (December), accepting a reduced efficiency in order to protect color development. Late winter application of GA_3 can also hasten degreening, which may outweigh the benefits of greater rind firmness for fresh fruit. GA_3 applications to enhance rind firmness are made at 20 ounces per acre. Use of a surfactant enhances GA_3 effectiveness, but also significantly increases the risk of peel blemish in grapefruit. Use of an organosilicone surfactant at 0.05 percent in spray formulations is common for all cultivars except grapefruit.

Applications to increase juice extraction are most effective near color break for the variety of interest, using 20 ounces of 4 percent GA_3 per acre. The importance of using surfactants is still under debate, but most growers use an organosilicone surfactant at 0.05 percent.

The timing and rate of GA_3 applications are very important so that the desired effect is achieved and unwanted effects minimized.

EFFECTS OF NAA

NAA (naphthalene acetic acid) is used primarily to control crop load by increasing drop of fruitlets shortly after set (fruit thinning). It may also be used to control trunk sprouting.

CHEMICAL THINNING WITH NAA

There is considerable economic incentive to produce larger fruit in fresh market citrus, especially in specialty fruits. Fruit thinning increases the average fruit size by reducing the number of fruits competing for available photosynthates. Thinning to increase fruit size can substantially increase profitability in blocks that otherwise bear an excessively heavy crop load. Controlling cropping may also result in more regular production by discouraging alternate bearing. Thinning may also reduce limb breakage. In some situations, controlled cropping may also enhance the concentration of soluble solids. Only groves that are expected to have excessively heavy crop loads should be thinned with NAA. Because results are unpredictable, NAA is best used where the danger from overcropping is greater than the potential harm from overthinning.

Hedging and topping are the chief means of controlled cropping in Florida citrus. Removal of the outer canopy and the associated fruit does reduce the number of fruit per tree and may increase fruit size, but the reduction in crop load may not be adequate and often produces little increase in fruit size. In addition, hedging to control excessive cropping is inappropriate in a young planting since it delays canopy development. Since hand removal of excess fruit is too expensive to consider, PGRs provide an attractive alternative for controlled cropping.

The plant growth regulator NAA is widely used in apples and pears to reduce the number of fruit on trees with excessive fruit set. It is also registered for use in tangerines (Mandarins), oranges, tangors (Temple, Ortanique, King, etc.), and tangelos, but has received little use in Florida citrus. NAA is thought to enhance early fruit drop by increasing competition between fruitlets and increasing production of abscission-inducing ethylene, both of which serve to encourage "June drop," the natural drop of citrus that usually occurs in May in Florida. In many citrus cultivars, use of NAA or other PGRs for thinning appears to enhance fruit size beyond the effect of reducing competition among fruits, and in some cases actually increases overall yield.

Only groves that are expected to have excessively heavy crop loads should be thinned with NAA. Alternate bearing cultivars such as Sunburst, Murcott, and Dancy are obvious candidates for chemical thinning in a year when exceptionally high yields are expected (on-year), but it may also be profitable to thin other cultivars.

Young plantings that have already developed an alternate bearing habit are also well suited to chemical thinning with NAA, since hedging small trees is generally not advisable. Thinning these young blocks may be especially beneficial because excessive crop loads not only result in many small, unmarketable fruit, but may also delay development of the canopy volume needed to maximize production. In addition, increased tree growth and earlier development of mature canopy volume may offset losses possible from overthinning (excessive reduction in crop load) in these blocks. Lower spray volumes are required because of the small size of young trees, so the cost of treatment per acre is generally reduced.

CONTROL OF TRUNK SUCKERS

NAA can be applied to the trunks of nonbearing citrus to reduce sucker formation. This treatment can significantly reduce the hand labor required to remove unwanted shoots. The undiluted formulated product is applied as a thorough spray or wipe to tree trunks only. Application from September through February is not recommended since this treatment may slow recovery following a freeze. The grower should be aware that excessive application might cause tree damage, especially where material puddles around trunk bases. In the one study on use of NAA to reduce trunk sprouts in citrus, tree wraps (even aluminum foil) were more effective than NAA. The best use of this product appears to be on problem spots, which repeatedly sprout even when trunk wraps are used.

APPLICATION OF NAA FOR THINNING

Rate of Application

The amount of active material absorbed and moved to the target tissue is of great importance when NAA, or any PGR, is applied. The absorption depends on a number of factors, including the concentration of material, the volume applied, and the uniformity of application. Since the concentration of active material is so important, PGR treatments are usually described on a concentration basis (parts per million or ppm), rather than an application rate (grams per acre). Concentrations in the range of 250 to 500 ppm NAA have been most effective in thinning citrus cultivars, not only in Florida, but also around the world.

Most citrus-thinning experiments have applied NAA in high volumes of water (400 to 600 gallons per acre) for uniform absorption; however, NAA only needs to be adequately absorbed by leaves in the fruiting zone to thin effectively. Recent experiments in Florida indicate that application in lower volumes of water is still sufficient to thoroughly wet the canopy. For mature groves of large trees, 125 gallons per acre is probably adequate. Even lower volumes should be used for smaller trees, and this can be accomplished by simply closing some of the sprayer nozzles. Some growers may not be comfortable calculating their applications on a ppm basis. They can instead use the ounces of NAA per 125 gallons, at appropriate ppm, as a rate per acre when applying at 125 gallons per acre. Note that the results are likely to be poor if the application is in 250 gallons per acre instead of 125 gallons per acre, unless the amount of NAA used is also doubled to maintain the recommended concentration.

NAA application to Florida citrus has been studied too little to recommend broad use, and at present it is recommended only on blocks where it would be better to overthin than permit excessive cropping. With this strategy in mind, it is advisable to use 500 ppm in thinning vigorous, well-watered blocks of all cultivars except Murcott. Studies show that Murcott usually thins effectively at lower rates of NAA, and 250 ppm is suitable for application to healthy, unstressed trees. Trees that are weak, freeze damaged, or under nutrient or drought stress may be overthinned by NAA and should be treated at a lower rate or left untreated. All NAA applications should include a nonionic surfactant at 0.05 percent and should not be tank mixed with other materials. The selection of surfactant should be discussed with an agrichemical dealer.

Timing of NAA Application for Thinning

Worldwide, thinning studies have shown that on a wide range of cultivars, response to NAA is greatest when the average fruitlet diameter is 8 to 15 mm, typically at six to eight weeks postbloom. Work in Florida suggests that best results are achieved when NAA is applied at the beginning of or just prior to the period of natural fruit drop that normally occurs in May.

REDUCING FRUIT DROP WITH 2,4-D

The plant growth regulator 2,4-D (2,4-Dichlorophenoxyacetic acid) can be used to reduce preharvest fruit drop in citrus. In Navel orange, application six to eight weeks after bloom can reduce early drop, while application in August or September can substantially reduce late season drop. Early drop in Navel orange frequently begins with abscission of the navel, which is anatomically a small fruit enclosed within the primary fruit. The navel then turns yellow and often begins to decay, making it appear that disease is causing fruit drop. Grapefruit or other citrus (for example, Pineapple orange) can be treated with 2,4-D during November or December to reduce losses from preharvest drop. When used on grapefruit, 2,4-D is often tank mixed with GA_3 and applied to blocks scheduled for late harvest. The GA_3 maintains rind quality while the 2,4-D reduces loss of fruit through abscission. To avoid drift onto susceptible crops in surrounding areas, products containing 2,4-D have stringent requirements for application conditions (see label for requirements). Since some phytotoxicity may occur, treatment of trees less than 6 years old or with substantial young flush should be avoided.

CAUTIONS IN PGR USE

It is vitally important that PGRs be applied at the proper concentration and with sufficient water volume, at the proper time of year or stage of development, under appropriate environmental conditions, and only in tested surfactant combinations and tank mixes. Even when properly applied, some PGRs may cause leaf curling, especially when sprayed on young leaves. Improper application of plant growth regulators can produce erratic, perhaps highly undesirable results, including significant phytotoxicity and greatly reduced cropping. Growers are urged to become familiar with PGR use and evaluate applications on small plots before treating significant acreage.

CONCENTRATION AND SPRAY VOLUME

Most registered pesticides are effective over a fairly broad range of concentrations, with little likelihood of phytotoxicity. In contrast, since PGRs function by directly influencing plant metabolism, plant response can vary considerably with concentration, making sprayer calibration and accurate material measurement especially important. Studies show that spray deposit becomes more variable when spray volume is reduced below 250 gallons per acre in mature citrus groves. At lower water rates, canopy areas closest to the sprayer manifold tend to retain much more material than other plant surfaces. Since material concentration is especially important in PGR use, water volumes below 125 gallons per acre (to a mature planting) are not recommended for most applications.

> Growers are urged to become familiar with PGR use and evaluate applications on small plots before treating significant acreage.

PGR UPTAKE

Unlike most agrichemicals applied to crop plants, the efficacy of PGRs depends on entry of the material into plant tissues. Uptake of a PGR is influenced by a number of factors, including the amount of the PGR applied, the concentration of the PGR, the presence of surfactants in the application, the pH of the PGR solution, environmental conditions during and after the application, and the plant stress level.

SURFACTANTS AND TANK MIXES

Surfactants and other spray adjuvants can affect uptake in several ways. Surfactants and oils spread spray materials over leaf surfaces and increase uptake by enhancing the total area contacted by the spray solution. Many surfactants, urea, ammonium salts, and oils can also directly enhance uptake by helping materials penetrate the plant cuticle. Organosilicone surfactants and some oils can result in very rapid uptake by carrying material through plant pores know as stomates. Clearly, surfactants can significantly enhance entry of PGRs into plant tissues; nevertheless, most PGR studies in citrus were conducted without surfactants, or with milder surfactants than many of those currently available. Use of untested surfactants may significantly enhance uptake, resulting in excessive plant response and possible increased phytotoxicity. Tank mixing with other spray materials may enhance PGR uptake due to surfactants or oils present in the material formulation; or other materials may deactivate PGR molecules, thereby reducing plant response.

ENVIRONMENTAL CONDITIONS

Growers should consider the likely influence of environmental factors in timing their PGR sprays. Studies with crops other than citrus have shown that weather conditions greatly influence PGR uptake. Uptake generally increases as temperature and drying time increase. Application at night or in the early morning often enhances uptake because the greater drying time more than compensates for the somewhat lower temperature. Dew following an application is likely to enhance PGR uptake by prolonging drying. Considerable uptake often occurs after spray has dried; therefore, rain within a few hours of an application may significantly reduce PGR effectiveness. In addition, many PGRs degrade rapidly in sunlight. Growers should also be aware of regulations regarding PGR application. In particular, it is illegal to apply 2,4-D when the wind speed is above 10 mph, and the required distance to susceptible crops downwind is specified for lower wind speeds.

PLANT STRESS

Trees under significant drought, cold, or pest stress may respond excessively to PGR treatments. Application of PGRs is recommended only on healthy citrus blocks.

It is clear that PGRs offer both tremendous potential benefits and considerable risk. Of course, the use of PGRs is only one of the many crop management resources available to the citrus grower. Careful site selection and selection of appropriate cultivars, followed by sound nutrition, weed, and pest management will continue to be the grower's principal strategy to achieve efficient, profitable production of high quality citrus.

REFERENCES

Agusti, M. 2003. *Regulation of citrus cropping and improvement of fruit quality using exogenous plant growth regulators.* International Society of Citriculture Congress (9th 2000: Orlando, Florida), Orlando, Florida USA. 1: 351-356.

Fidelibus, M., F. S. Davies, and C. A. Campbell. 2002. *Gibberellic acid application timing affects fruit quality of processing oranges.* HortScience 37: 353-357.

Grierson, W. 2000. Fruit development, maturation and ripening, ed. M. Pessarankli in *Handbook of Plant and Crop Physiology*, Marcel Dekker, New York.

Guardiola, J. L., and A. Garcia-Luis. 2000. *Increasing fruit size in Citrus. Thinning and stimulation of fruit growth.* Plant Growth Regulation 31: 121-132.

Kender, W. J., U. Hartmond, J. K. Burns, R. Yuan, and L. Pozo. 2001. *Methyl jasmonate and CMN-pyrazole applied alone and in combination can cause mature orange abscission.* Scientia Horticulturae 88: 107-120.

Whitney, J. D., U. Hartmond, W. Kender, M. Salyani, and J. Burns. 2000. *Abscission chemicals affect trunk shaker orange removal.* Proceedings of the Florida State Horticultural Society 113: 93-96.

24

Precision Agriculture and Crop Management

J.S. Rogers and J.D. Whitney

INTRODUCTION

Precision agriculture refers to an emerging trend in crop management in which the grower endeavors to identify and respond to the specific needs and requirements of small areas in a planting. Precision agriculture reduces unnecessary applications of pesticides, herbicides, and fertilizers in areas with lesser needs and consequently reduces the economic and environmental costs of citrus production. It is a highly computerized approach that usually involves the use of satellite-based spatial information. Geostatistical methods may be applied that enable citrus managers to make more precise determinations of the relationships between the various factors that are being measured in a field, such as nutrition, yield, insects, and diseases. This chapter introduces the basic concepts of precision agriculture and provides some examples of how the information is used to render site-specific assessments for improving citrus production.

Precision agriculture refers to an emerging trend in crop management in which the grower endeavors to identify and respond to the specific needs and requirements of small areas in a planting. Precision agriculture reduces unnecessary applications of pesticides, herbicides, and fertilizers in areas with lesser needs, and consequently reduces the economic and environmental costs of citrus production. It can provide the grower with greater resource control, which is especially valuable in blocks that are not yielding their full potential. It also aids in identification and correction of limiting or complicating factors in underperforming blocks. The benefits of precision agriculture have long been apparent; however, specific economic assessments of these benefits in citrus are still somewhat elusive. Early efforts to manually identify and resolve highly localized problems were labor intensive and seldom cost effective.

The implementation of precision agriculture has accelerated dramatically in recent years as sophisticated new computer technologies have become available and affordable. Most significantly, the release of satellite-based geographical referencing systems (Global Positioning Systems, or GPS) to the public has tremendously increased the practical application of precision agriculture. The general public now has access to handheld GPS equipment that provides latitude and longitude coordinates that are accurate to within a few meters. More sophisticated (and expensive) devices can identify locations within several inches. These systems allow growers to precisely and automatically record variations in yield and grove conditions by location. This capability has encouraged the development of many variable-rate application devices that allow the grower to tailor grove maintenance activities to the specific needs identified in each location. Although these technologies are newly emerging, they promise to revolutionize agricultural production. Many of the proposed applications are, as yet, untested, but their development and implementation seem virtually inevitable.

Precision methods can increase agricultural productivity and profits by helping the grower better understand where financial, nutritional, and other resources are best allocated on a farm. The challenge is to identify the specific field locations where nutritional deficiencies or disease and insect pressure indicate a need for active intervention and to target the response to precisely those locations. Florida citrus groves are generally managed according to units referred to as "blocks," the size of which may vary from less than one acre to more than 100 acres. Grove conditions may vary tremendously within a single block. A technique called geo-referenced sampling is gaining momentum in a variety of agricultural crops and holds promise for Florida citrus production. This technique consists of defining a spatial grid on a farm or planting and collecting representative soil or leaf samples from each of the individual grid blocks. The grid blocks are delineated and identified by GPS coordinates. Field samples collected are analyzed for various soil characteristics, concentrations of nutrients, or populations of pathogens using standard laboratory procedures. Readily available computer software then translates the geo-referenced data into color 2-D or 3-D maps that display variations and trends in the conditions or characteristics that have been measured and recorded.

This same technology may be used to map yield or fruit quality information throughout the grove. To map yield, GPS coordinates are used to locate and identify individual field containers during harvest. These containers are tracked through the processing plant or packinghouse, where measurements of fruit quality and maturity are recorded. The information obtained by geo-referenced sampling is overlaid by a GPS map of yield and fruit quality. This overlay allows the grower to correlate unsatisfactory or suboptimal production with specific grove conditions. Growers use these overlay maps to determine appropriate crop production practices, observe crop response to those practices, judge the suitability of scion/rootstock combinations to particular grove sites, measure the degree of alternate bearing, or identify trees in decline that should be reset.

There are many factors that account for variations in yield and fruit quality, including scion/rootstock interactions, tree age and health, tree size and spacing, canopy density, soil type and pH, soil fertility and water-holding capacity, water quality, and pest pressure. Any of these factors can be indicated on a grid map once the technology to measure and record that parameter becomes affordable. Aerial photographs already provide some site-specific information, and various remote-sensing devices are in development that will communicate with GPS equipment to automatically measure and record many of these factors.

Effective precision agriculture further involves the development of affordable variable-rate application equipment to allow the grower to customize the cultural practices in each identified area. Many such devices are already available, especially for applications of materials to the soil. With this equipment, soil pH and nutritional deficiencies that have been mapped on a grid can be corrected as needed at precise locations by tailoring application rates to each GPS coordinate. Equipment is also currently available that uses ultrasonic or laser sensors to determine tree height or size. These sensors are used primarily to control discharge patterns of foliar pesticides or to vary the application rate of dry fertilizer. Again, it is reasonable to assume that a GPS map may soon control these devices so that trees benefiting from higher fertilization will receive such treatment. The market for precision agriculture equipment is small, but likely to expand rapidly.

Precision agriculture may also improve the efficiency of general grove management. Cultural practices and grove maintenance activities may be automatically recorded, reducing the need for supervisors to physically visit and observe widely separated grove locations. Irregularities will also be recorded, and any necessary mitigation may be promptly initiated. This may be especially important if environmental damage or tree injury is potentially associated with the error. Yield and fruit quality maps can also be used to record and rate the performance of individual harvesters or teams of harvesters. This information may be used to identify both exemplary harvesters and those in need of further training or greater supervision. Ultimately, this information may be used to automate payroll transactions.

It should be noted that although the value of precision agriculture has been demonstrated in many cropping systems, it remains relatively unproven in Florida citrus. Such systems are likely to be cost prohibitive to all but the largest citrus growers in the early stages of development. However, the potential benefit resulting from widespread implementation of precision agriculture is so extraordinary that it seems virtually certain that these techniques will soon become standard Aerial, topographical, and interpolated maps are commonly used in precision agriculture, and technologies to produce these maps are rapidly advancing.

REFERENCES

Balsari, P., P. Marucco, M. Tamagnone, C. Bellissima, E. Cerruto, and S. Failla. 2003. *Evaluation of a tree sensing device to reduce the use of pesticides in citrus.* International Society of Citriculture Congress. (9th 2000: Orlando, Florida), Orlando, Florida USA 1: 252-256.

Clark, I. and W. Harper. 2000. *Practical Geostatistics 2000.* Greyden Press, Columbus, Ohio. 415 pp.

25

Meteorological Data Collection and Application

K.T. MORGAN AND J.L. JACKSON

INTRODUCTION

For more than sixty years the National Weather Service provided weather forecasts and meteorological information for Florida citrus and other agricultural producers. When that service was discontinued in 1996, citrus growers began to seek and develop new sources of weather data. Several public and private organizations collect meteorological data using sophisticated, automated equipment. Information from these organizations is available to growers on the Internet, or as a call-up service. Site-specific weather information and forecasting is extremely valuable to the citrus grower. Improved weather information will eventually lead to more effective control of fungal diseases and other pest problems. The importance of site-specific weather data cannot be underestimated. This is true especially during freeze nights. Good weather records are also useful if a grower must recover crop losses due to unusual weather events such as wind or hail. It is to every grower's advantage to collect site-specific weather information to guide individual grove management and to support broad research into the effect of weather on citrus production.

For more than sixty years the National Weather Service provided weather forecasts and meteorological information for Florida citrus and other agricultural producers. When that service was discontinued in 1996, citrus growers began to seek and develop new sources of weather data. Several public and private organizations collect meteorological data using sophisticated, automated equipment. Information from these organizations is available to growers on the Internet, or as a call-up service. The Florida Automated Weather Network (FAWN), which is operated by the University of Florida's Institute of Food and Agricultural Sciences, provides information at its Internet site: http://fawn.ifas.ufl.edu.

Sophisticated weather stations historically have been operated primarily for research purposes. However, falling equipment costs and technological improvements have made site-specific weather data collection practical for most citrus growers. Several manufacturers offer weather stations with a wide range of sensors and features; the price depends on the model and the manufacturer. Relatively inexpensive devices measure air temperature, relative humidity, wind speed and direction, water evaporative demand, and rainfall. More elaborate models also measure soil temperature, solar radiation, and barometric pressure. Weather stations can be customized to contain only the sensors a particular grower needs or wants.

Site-specific weather information and forecasting is extremely valuable to the citrus grower. The grower can use this weather data for various aspects of crop management. Growers may be able to confidently anticipate and prepare for severe weather, evaluate if freeze protection is needed, determine when to turn off irrigation systems used for cold protection, or estimate potential damage after severe weather. Growers also use this information to manage insects and diseases, schedule irrigation, and anticipate floral bud initiation. Computer systems are under development that will greatly improve the ability of citrus growers to manage all aspects of their citrus operations.

DECISION INFORMATION SYSTEM FOR CITRUS

There are several computer-based expert systems currently developed or deployed in the citrus industry that provide a valuable service by reducing the management burden on citrus managers. Some of these are from the private sector, and there are many being developed by the University of Florida under a project called DISC. The Decision Information System for Citrus is intended to provide specific management recommendations to individual growers in a variety of situations. The system reflects the ICM philosophy and attempts to integrate the many interrelationships in the citrus grove environment, of which weather is a crucial component. The success of this project depends in large part on the ready availability of site-specific weather information. Current elements of this project focus on the following important aspects of citrus management.

FLOWERING

Intensity and date of flowering are predicted from winter and spring temperatures. Bloom dates are being predicted within one week, sixty days in advance.

FUNGAL DISEASE CONTROL WITH COPPER

Control of fungal diseases in citrus depends on the growth and development of the damaging fungi, the placement and persistence of fungicides on the tree, the stage of fruit development, and perhaps other factors not yet identified. Copper is the primary material used to control these fungal diseases, but copper coverage is compromised through weathering and growth of fruit and/or foliage. Fungal growth and the development of fruit and foliage are also weather dependent. With the input of accurate weather information, especially rainfall, humidity, and temperature data, the DISC program can help the grower develop an effective and efficient schedule of copper applications to control these fungal diseases. DISC provides guidance to growers concerning the need for reapplication if plant material is susceptible and fungi are present. The copper weathering DISC module for control of melanose on fresh grapefruit was grower tested for several years and is now available for general use.

INSECT AND DISEASE MANAGEMENT

Predictive, weather-based models for insect and disease development are currently under development and will be extremely useful to citrus growers in the timing of pesticide applications. One such model has been developed for postbloom fruit drop in Florida. It is helpful in predicting disease outbreaks and in determining the most effective timing of fungicide applications. Similar models could be developed for most of the foliar and fruit fungal diseases, including melanose, scab, greasy spot, and Phytophthora brown rot. Control of these diseases can be achieved with minimal fungicide applications if the applications can be timed precisely. To use these programs, the citrus manager will need to input current, accurate weather data. Further, scientists rely on accurate weather information in the development of these programs as they study and quantify the relationships between weather conditions and the severity of insect and disease pressure.

One model has also been developed to evaluate the effect of temperature, wind speed, and humidity on the deposition of spray materials on the tree. This program will help growers to apply pesticides safely and effectively and to determine the need for applications of material.

IRRIGATION AND COLD PROTECTION

Programs have been developed to calculate evapotranspiration (ET) rates with the input of weather data. Sophisticated follow-on programs then use this information to calculate when and for how long a grower should irrigate. These programs can be used to check for rainfall, turn on irrigation pumps as needed, and turn off the irrigation system when the appropriate amount of water has been applied.

A simple program has also been developed to monitor wet bulb temperature and to predict when wet bulb temperature will reach 32°F. This is the temperature at which it is safe to turn off irrigation for freeze protection. This program can be used to automatically turn off these freeze protection systems.

Yet another straightforward program uses air temperature and dew point temperature at sunset to predict minimum temperature the following morning to advise growers when freeze protection is necessary. This program works well during a radiation freeze when the movement of cold air into the area is not a factor.

Another program monitors temperature and records the number of hours below specific critical temperatures to predict flowering. The program provides recommendations to the grower concerning the need for measures to increase flowering and/or attempts to achieve a single period of flowering rather than a series of overlapping blooms resulting from several induction events.

DAMAGE ASSESSMENT

This program can be used by the grower to evaluate the impact of severe weather, such as freeze, drought, or flood on the crop and the trees. After damaging weather, the grower is faced with many important decisions concerning salvage and sale of the damaged fruit, management or removal of the injured trees, and possible recovery of losses through tree crop insurance. The information provided by this program can help the grower quickly assess the situation and respond to the damage.

An understanding of the direct and indirect relationship between weather conditions and the countless factors that affect the productivity and profitability of citrus production is vital to ICM. This research will require routine weather data collection at a large number of grove sites and more sophisticated data collection at a few appropriately located field sites. The more elaborate weather stations will probably be maintained publicly, and individual growers will also be asked to share information collected at their own smaller weather stations.

REFERENCES

Gerber, J. F. 2003. *Florida freeze forecasts.* International Society of Citriculture Congress. (9th 2000: Orlando, Florida), Orlando, Florida USA. 1: 387-389.

Jackson, J. L., and T. W. Oswalt. 1977. *Dissemination of weather information to citrus growers at the local level.* International Citrus Congress (2nd 1977, Orlando, Florida), Orlando, Florida USA, 1: 190-192.

Rogers, S. 2003. Supercooling and ice nucleation in citrus leaves. International Society of Citriculture Congress. (9th 2000: Orlando, Florida), Orlando, Florida USA. 1: 396-400.

Valiente, J. I., and L. G. Albrigo. 2003. *Modeling flowering date of sweet orange trees in central Florida based on historical weather.* International Society of Citriculture Congress. (9th 2000: Orlando, Florida), Orlando, Florida USA 1: 296-299.

Rogers, S. Hail damage: Physical meteorology and crop losses. 1996. Proc. Fla. State Hort. Soc. 109:97-103.

26

Fruit Quality

F.S. Davies and D.P.H. Tucker

INTRODUCTION

Many factors affect fruit quality including climate, scion, rootstock, tree spacing, tree age, and cultural practices such as nutrition and water management, pruning, pest and disease control, and the application of plant growth regulators. However, it is important to manage each component properly. For example, excessive nitrogen decreases fruit size, diminishes peel color, delays color break, and increases peel coarseness, thus greatly reducing fresh fruit quality. It is therefore possible to have too much of a good thing. Generally, peel blemishes caused by diseases or pests are major factors in reducing fresh fruit quality, although their impact on processing fruit quality is much less pronounced and usually not economically important. It is important to understand the balance and relative significance of the factors affecting fruit quality presented in this chapter.

Fruit quality is important in determining the value and profitability of fresh and processed citrus. While quality requirements vary according to market, season, and supply and demand, an overall commitment to quality is essential to maintaining our industry. External fruit quality factors include: fruit size, peel color, fruit shape, and the extent of peel blemishes. Internal fruit quality factors include total soluble solids (TSS), acids and the ratio of TSS to acids, juice content, and the presence of aromatic or bitter compounds such as limonin or naringin. Many factors affect fruit quality, including tree canopy position (Figure 26.1), climate, scion, rootstock, tree spacing, tree age and cultural practices such as nutrition and water management, pruning, pest and disease control, and the application of plant growth regulators. Some of these factors have consistent effects on quality, while the effect of others may vary. While primary fruit quality factors are established in the field, fruit harvesting and postharvest handling practices can also have a significant impact on the quality of fruit or juice in the market, and on its shelf life. In practice, many grapefruit blocks are grown for the fresh market, but often 10 percent to more than 50 percent of the fruit are eliminated from fresh use by inferior quality factors and are utilized for juice.

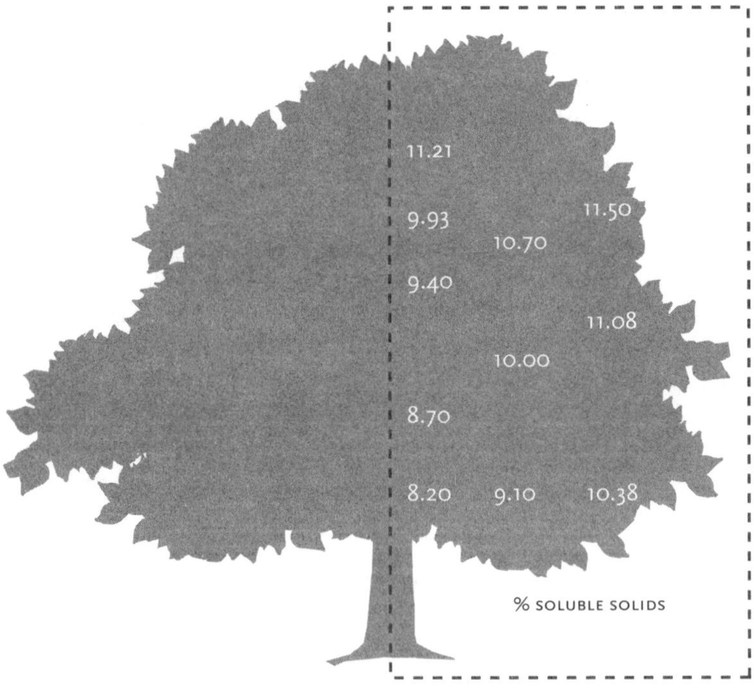

Figure 26.1 When testing for fruit quality, it is important to remember that internal quality varies with canopy position and samples should be collected accordingly.

QUALITY DEVELOPMENT

Gradual seasonal changes in juice content and in the levels of sugar (Brix) and acid determine fruit maturity for citrus; acid content decreases and sugar content increases as the fruit matures (Figure 26.2). Minimum juice content, Brix, and Brix to acid ratio are regulated for marketing most Florida citrus fruits. Citrus does not continue to ripen after harvest, so the fruit must be allowed to reach peak maturity on the tree. Cool fall and winter temperatures enhance the yellow and orange color in the peel of the fruit, which consumers perceive as an indication of fruit quality. The color development can be enhanced or accelerated by "degreening," the process of exposing the fruit to ethylene, causing a breakdown of chlorophyll, but the internal quality of the fruit is not changed. Fresh fruit of marginal internal quality harvested and marketed too early, may provide high short-term economic returns but may discourage consumers and weaken the industry's seasonal market development. Florida's citrus industry cannot expect to remain globally competitive unless it can maintain and enhance its reputation as a source of the very highest quality fresh fruit and juice products.

CLIMATE

Citrus cultivars differ widely in their performance in the various citrus growing regions of the world. These differences are apparent in growth and yield, in the time between bloom and market maturity, and in the time marketable fruit will hold on the trees. Climate is the single most important factor affecting growth, yield, and fruit quality; it essentially dictates where citrus may be profitably grown (Figure 26.3). The most important climatic factors for fruit quality development are temperature, rainfall, soil moisture, relative humidity (RH), and sunlight. Temperature during fruit development is the most dominant of these factors.

In regions with consistently high average temperatures, such as the low tropics, fruit grows rapidly and peel color is poor; oranges and mandarins tend to remain green throughout the year. In addition, total soluble solids and acids are low and internal quality is poor. In contrast, fruits that are naturally high in acids, such as limes and grapefruit, attain good external and internal quality under high temperatures. In subtropical areas with moderately high average temperatures, such as Florida, fruit growth is slower, but peel and juice color are considerably better than in the low tropical regions. Most importantly, total soluble solids and juice content are high, factors that contribute to Florida's preeminence as a major citrus processor. In contrast, peel and juice colors are generally best where average temperatures are moderate and large daily fluctuations occur. Such conditions are typical of Mediterranean-like climates. Juice content is usually low, total soluble solids are intermediate and acids are high. Fruit from these regions are usually sold on the fresh market.

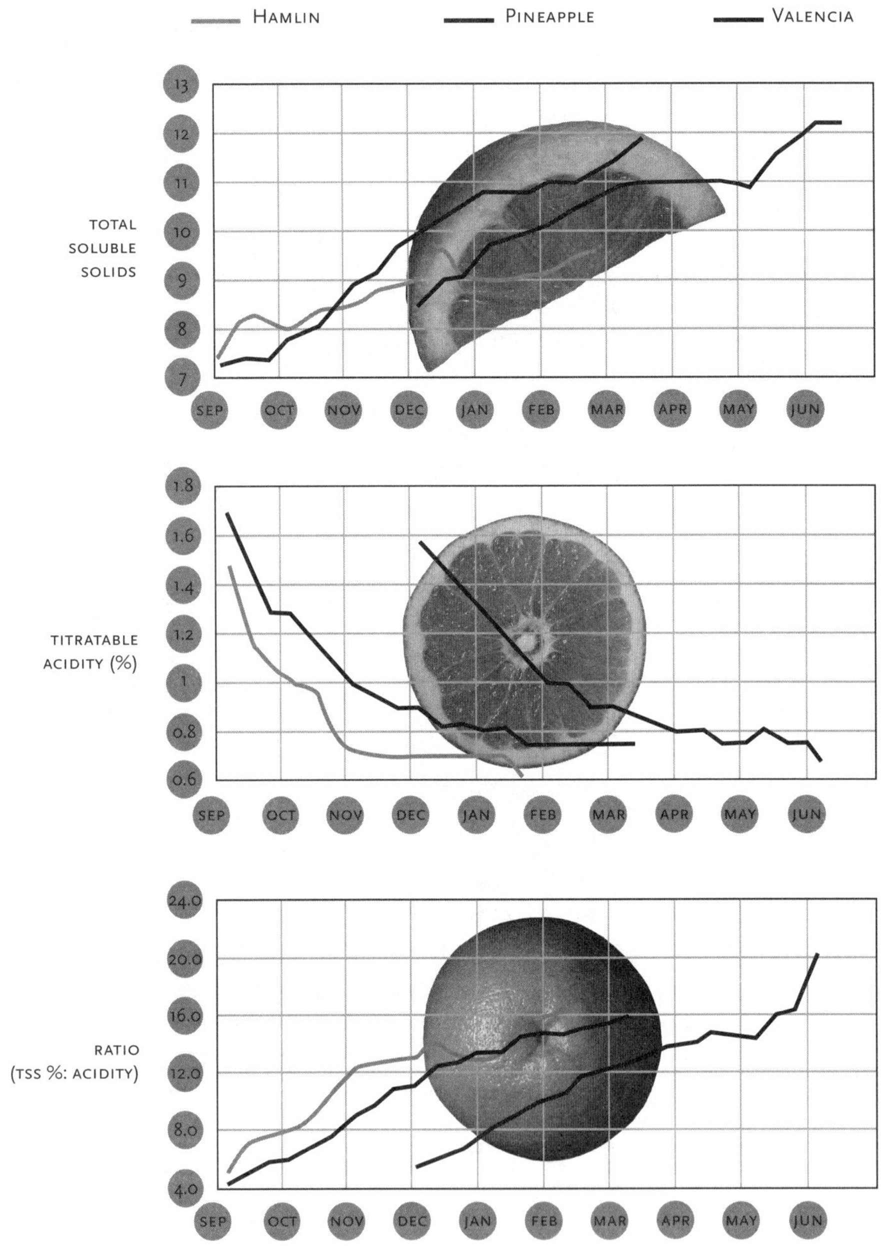

Figure 26.2 Seasonal changes influence the development of total soluble solids, acidity and ratio as shown in these data for three oranges varieties. From P.L. Harding, J.R. Winston, and D.F. Fisher. 1940. Seasonal changes in Florida oranges. USDA Technical Bulletin, no. 753. Washington, D.C.

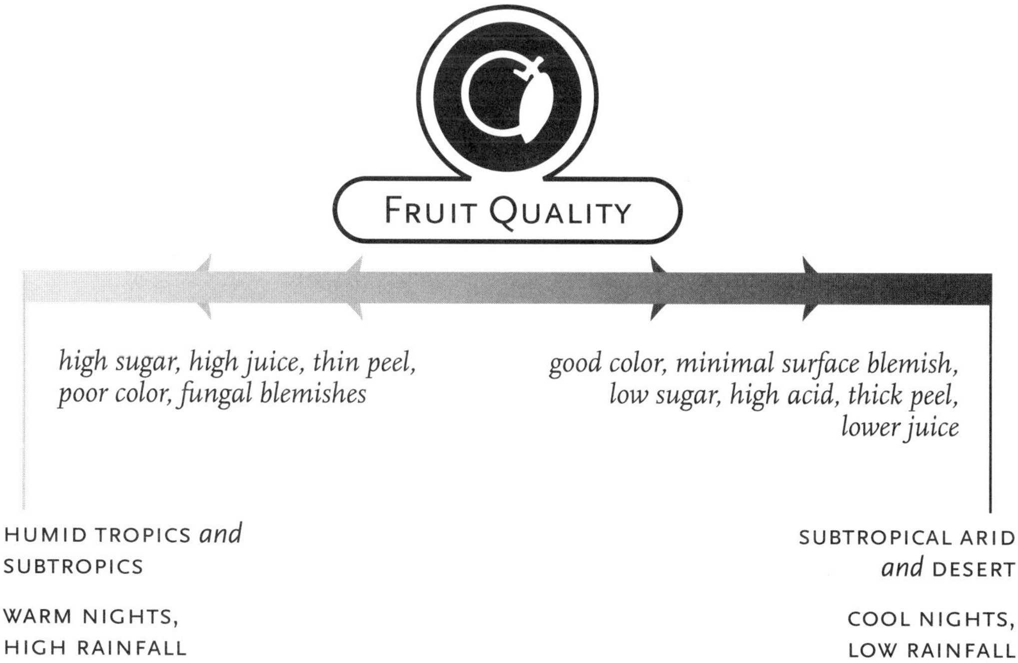

Figure 26.3 Impact of climatic regions on fruit quality.

The amount and distribution of rainfall affects fruit quality both directly and indirectly. High rainfall, distributed uniformly throughout the year, increases fruit size, growth rate, and juice content. In areas with low rainfall and limited irrigation, fruit size is greatly reduced. Excessive rainfall also reduces TSS and acidity, and may produce fruit with a coarse peel. Rainfall indirectly affects the extent of peel blemishes by increasing disease incidence. Not surprisingly, wet growing regions experience much greater disease pressures, and fruit peel is more likely to be blemished. Fresh fruit production in such regions often requires the application of fungicides to control disease and blemish organisms. Semiarid or arid regions have far fewer disease-related blemishes. Even within a particular location or season, disease incidence and associated peel blemishes vary considerably with rainfall distribution and amount.

Average relative humidity is correlated with rainfall, and also directly or indirectly affects fruit quality. High RH reduces transpiration losses and water stress, and is important for increasing fruit size. Water loss due to transpiration and evaporation is generally greater in more arid growing regions. The effects of RH on other fruit quality factors are variable and have not been widely studied High RH is necessary for spore release and growth of many fungi. Thus, RH has direct effect on the incidence of peel blemishes, especially those caused by fungi such as greasy spot or melanose. Typically, areas with high average RH, such as Florida, have a higher incidence of peel blemishes than arid or semiarid regions.

Adequate sunlight is very important for satisfactory peel color development. Increasing sunlight levels also improve fruit size and TSS. Wind reduces fruit quality through desiccation, but more importantly, may cause wind scarring of the peel, which is a major blemish problem in Florida and worldwide. Windbreaks, usually consisting of dense rows of trees surrounding blocks of citrus, reduce the severity of wind damage but do not completely eliminate it. Tall, aggressively growing windbreak species that have been used in Florida shade the adjoining grove and compete for water and nutrients. As a result, windbreaks aren't widely used in Florida. New windbreak tree species, low marketability of blemished fruit, and the value of windbreaks in controlling citrus canker (a bacterial disease) may increase the use of windbreaks in Florida citrus.

Even very small variations in climate may have a significant effect on citrus fruit quality. For example, fruit from Orchid Island in Florida is known for its excellent external appearance, due primarily to the unique microclimate in that area. Obviously, growers cannot control the climate. They can, however, influence the microclimate in the grove through a variety of cultural practices, especially irrigation. In addition, because of the tremendous impact of climate on fruit quality, climate must be a deciding factor in site selection.

Climate largely determines the suitability of a specific cultivar for commercial production in a particular growing region.

SCION AND ROOTSTOCK

Scion selection has a significant and long-term impact on fruit quality. Some cultivars are best suited for fresh fruit production, some for processing fruit production, and some are suitable for both. Variations among cultivars in mature harvest period can also be substantial; for example, the peel and juice color, and TSS of Valencia oranges are superior to that of Hamlin or Parson Brown oranges, and are preferred by both processors and packers. Yet, for processing fruit, yields and pounds solids per acre are also very important, and Hamlin is often superior to Valencia in this regard. Navel oranges are grown primarily for the fresh market because of their seedlessness, easy peeling and inherent large size. They are not well suited for processing because of the presence of limonin, a bitter compound that is largely masked (as a glycoside) in the fruit, but is released after juicing and heating. Fruit quality also differs greatly among tangerine cultivars. For example, Murcott, Sunburst, and Robinson have color, internal quality, and flavor superior to K-Early. Grapefruit cultivars differ primarily in peel and juice color, although TSS is usually higher in Duncan, a seedy cultivar, than in the preferred seedless cultivars. Increasingly, many fresh fruit markets prefer red grapefruit, and whites are grown primarily for a few Asian markets (notably Japan). Some of the deepest red grapefruit cultivars, such as Star Ruby and Flame, require more careful horticultural management than other cultivars.

The rootstock interacts with the cultivar to influence fruit quality. In general, for a specific cultivar, lemon-type rootstocks (rough lemon, *Citrus volkameriana*, *C. macrophylla*) produce large, coarse-peeled, often poorly colored fruit with low TSS. These rootstocks are usually avoided for groves targeting the fresh market; yet the often-superior yields produced on these rootstocks make them desirable for the processing fruit market. Conversely, most cultivars growing on Cleopatra mandarin rootstock have greater TSS and better peel quality than those on lemon types, but the fruit size is usually smaller. Internal and external fruit quality are generally excellent for cultivars grown on sour orange and Swingle citrumelo; sour orange has been the mainstay of the fresh fruit industry in many areas of the world, but is now avoided in Florida because it is susceptible to citrus tristeza virus (CTV). The grower must consider yields and potential markets as well as fruit quality when selecting the rootstock and cultivar.

CULTURAL PRACTICES

The grower can improve fruit quality within the bounds of climate effects through appropriate cultural practices. In particular, nutrition, water management, pruning, pest and disease control, and the use of plant growth regulators are the most important cultural factors affecting fruit quality.

NUTRIENT MANAGEMENT

Among the nutrients essential for citrus, nitrogen has the greatest overall effect on fruit quality. Insufficient nitrogen decreases yields and pounds solids. However, excessive nitrogen decreases fruit size, diminishes peel color, delays color break, and increases peel coarseness, thus greatly reducing fresh fruit quality. High nitrogen also slightly decreases juice content and increases acids, with inconsistent effects on TSS. Potassium levels are important, primarily for fresh fruit quality. Proper potassium nutrition is essential to maintaining adequate fruit size, especially in inherently small-fruited cultivars; however, as potassium levels increase fruit size peel thickness may also increase, and the peel may become coarse. High potassium levels generally decrease juice content and TSS, and increases acid. Phosphorous has less effect on external fruit quality than nitrogen or potassium. Increasing phosphorous levels may decrease juice acidity, but effects on other internal quality factors are inconsistent.

Micronutrients rarely affect fruit quality with the exceptions of boron and copper. Boron deficiency may produce lesions in the fruit, an unusually thick peel, and internal gumming. Severe copper deficiency produces brownish lesions on the peel.

IRRIGATION

The effects of irrigation are comparable to those of rainfall. Proper timing and amounts of irrigation increase fruit size and juice content for most cultivars. In some instances, excessive irrigation increases peel coarseness, reducing TSS and acids present in the fruit, thus decreasing fruit quality. Effects of irrigation on other quality factors are inconsistent.

PRUNING

As mentioned previously, sunlight is important for fruit color development and TSS production. In mature trees, light becomes limiting in the inner canopy. As trees become taller and grow together, shading of the lower canopy occurs. A regular hedging and topping program improves light penetration into the canopy and generally increases fruit size and TSS, and improves peel color. Opening up the tree may also decrease peel blemishes caused by fungi, since fruit dry more rapidly and there is better spray penetration into the canopy. It is thought that skirting (the removal of the lower part of the canopy) may improve fresh fruit quality by reducing peel blemishes caused when fruit come in contact with the soil. A regular hedging and topping program, especially as foliage density increases, is necessary for maintaining good fruit quality.

PEST AND DISEASE CONTROL

Peel blemishes caused by diseases or pests are major factors in reducing fresh fruit quality, although their impact on processing fruit quality is much less pronounced or economically important. Major peel blemishes in Florida are caused by rust mite, melanose, Alternaria, scab, and greasy spot rind blotch. Plant bugs and katydids that chew on the peel also damage fruit. Control of these pests and diseases requires timely spray applications. There is a strong interaction between weather conditions, pest development, and effective duration of pest control measures. Consequently, many spray programs include use of weather data for each block to schedule efficient spray applications.

PLANT GROWTH REGULATORS

Plant growth regulators have been used to improve fresh fruit quality for many years, and their use and effects are covered in Chapter 23.

REFERENCES

Agusti, M., V. Almela, M. Juan, F. Alferez, F.R. Tadeo, and L. Zacarias. 2001. *Histological and physiological characterization of rind breakdown of 'Navelate' sweet orange.* Annals of Botany 88: 415-422.

Albrigo, G. 1986. *Peel morphology and fruit blemishes. Citrus Flowering, fruit set, and development.* http://www.fcprac.ifas.ufl.edu/UF%20IFAS%20Short%20Course%20Proceedings/fruitset.htm University of Florida, Gainesville.

Albrigo, L. G. 1988. Stress factors affecting external and internal quality of citrus fruit, p. 69-77, eds. J. J. Ferguson and W. F. Wardowski. *Factors Affecting Fruit Quality.* University of Florida Press, Gainesville.

Barry, G. H., W. S. Castle, and F. S. Davies. 2003. *Juice quality of 'Valencia' sweet orange among citrus-producing regions in Florida and between canopy positions.* International Society of Citriculture Congress. (9th 2000: Orlando, Florida), Orlando, Florida USA 1: 308-314.

Castle, W. S. 1995. *Rootstock as a fruit quality factor in citrus and deciduous tree crops.* New Zealand Journal of Crop and Horticultural Science 23: 383-394.

Goldschmidt, E. E. 2003. *Reassessment of climatic effects on fruit maturation in Citrus towards the development of a fruit maturation and quality model.* International Society of Citriculture Congress. (9th 2000: Orlando, Florida), Orlando, Florida USA 1: 300-302.

P.L. Harding, J.R. Winston, and D.F. Fisher. 1940. Seasonal changes in Florida oranges. USDA Technical Bulletin, no. 753. Washington, D.C.)

Hilgemann, R. M. 1966. Effect of climate of Florida and Arizona on grapefruit enlargement and quality: Apparent transpiration and internal water stress. Proceedings of the Florida State Horticultural Society 79: 99-106.

McCoy, C. W. 1988. The biology of the citrus rust mite and its effects on fruit quality, p. 54-68, eds. J. J. Ferguson and W. F. Wardowski. *Factors Affecting Fruit Quality*. University of Florida Press, Gainesville.

Miller, W. M., and W. L. Verba. 1996. *Technique for abrasive damage assessment of citrus*. Proceedings of the Florida State Horticultural Society 109: 254-257.

Patil, B. S. 2001. *Location and rootstock affect sheepnosing in grapefruit*. HortScience 36: 710-713.

Quaggio, J. A., H. Cantarella, D. Mattos Jr., and A. Novello. 2003. *Fruit quality of two sweet orange varieties affected by NPK rates in tropical soils*. International Society of Citriculture Congress. (9th 2000: Orlando, Florida), Orlando, Florida USA 1: 581.

27

Fruit Harvesting, Handling, and Transport

G.K. Brown

INTRODUCTION

If the citrus manager can get through all the trials and tribulations of designing, planting, growing, and taking care of his trees, there will hopefully be a profitable crop to be harvested at the end of the season. Today most of this fruit is harvested by hand, but there is a trend to move toward mechanical harvesting methods. Nevertheless, the efficient use of hand labor will be of continued importance as the industry attempts to maintain its fresh market and enhance its specialty citrus business, but the industry needs to reduce its dependence on hand labor for harvesting processing fruit. Many independent operations are required to get the fruit from the grove to the consumer. One of the most basic forms of equipment is a type of grove truck known as a "loader" that is used to deliver empty containers, pick up the filled containers, and move the fruit to a highway trailer. Efforts to improve the efficiency of hand harvesting have largely been unsuccessful, so various mechanical mass-removal methods have been studied. Grove design and physical conditions greatly affect harvest cost; contracting for mechanical harvesting is best arranged on a cost-per-acre basis rather than cost per box of fruit. This chapter reviews the basics of harvesting with a special emphasis on the mechanical harvesting that is expected to become more common in the future.

While total production of Florida citrus is not projected to increase, the supply of hand-pickers is expected to decrease as many workers leave the citrus harvesting workforce. The efficient use of hand labor will be of continued importance as the industry attempts to maintain its fresh orange, grapefruit, and specialty citrus business, but the industry needs to reduce its dependence on hand labor for harvesting processing fruit. Currently, the Florida grower is a relatively high-cost producer of citrus for juice, in part because Florida harvesting costs are two to three times higher than that of international competitors. Efficient harvesting is essential to the citrus industry to maximize recovered yield, minimize harvest-related damage to equipment and grove infrastructure, protect workers and trees, and provide workers with a reasonable income while minimizing harvesting costs.

HAND HARVESTING

When fruit is picked for processing, fruit removal and harvesting speed are the primary concerns. Each fruit is grasped by hand, then quickly twisted and pulled so the stem is bent along the fruit surface and snapped from the fruit at the button or calyx. When done properly, the stem breaks at the surface of the button, or the button is removed with the stem, leaving the fruit rind intact. When done improperly, either the stem breaks a few inches from the fruit surface or an area of rind pulls out of the fruit, leaving what is called a "plugged" rind that exposes the fruit flesh. Both results are undesirable. Processing fruit can be picked while wet, whereas fresh fruit must be picked dry.

When fruit is picked for the fresh market, care must be taken to avoid physical injury to the peel. Injury results in decay or poor appearance when the fruit reaches the retail stores. To avoid peel damage and plugging, navel orange and tangerine stems may be clipped with shears instead of snapped. This is rarely done in Florida. Picking is often delayed on cool damp mornings until the fruit is dry, and/or peel turgor decreases and the likelihood of rupturing oil glands in the peel is reduced.

Citrus for the fresh or processed market is picked using tapered ladders made from wood, fiberglass, or aluminum (Plate 27.1). Picked fruit is typically placed in a fabric picking sack that is supported by straps over the shoulders. Picking starts at the top of the ladder and finishes at the bottom or on the ground. A full sack will hold 70 to 90 pounds of fruit. An industry-wide training video provides general guidelines for effective hand harvesting.

Picked fruit is measured in field boxes. Florida law states that one field box of oranges occupies 4,800 cubic inches of volume and weighs 90 pounds. All containers must be clearly marked showing their capacity in numbers of field boxes of fruit. The usual sizes for bulk “pallet bin” or “tub” containers range from 9 to 11 field boxes.

The productivity of hand-pickers is measured in picked field boxes of fruit per hour of work. The industry average picker productivity is 9 to 10 field boxes per hour for processed harvesting and 8 to 9 boxes per hour for fresh market harvesting. However, adverse conditions can decrease productivity to 6 boxes per hour or less, whereas favorable conditions can increase productivity to 12 or more boxes per hour. During picking, fruit is snapped from the tree at the rate of 20 to 50 per minute, or clipped at the rate of 20 to 40 per minute. Productivity is affected by picker experience, skill, motivation, quality of supervision, grove floor conditions (soil type, wet or dry soil), ground profile (level or ditched), depth of grass, amount of trash on the grove floor, height of the tree, and number and size of fruit on the tree. Both pallet bins and tubs are used as a legal means of measuring the daily productivity of each picker. Pallet bins are almost always used for fresh market fruit. They may be constructed of plywood, metal, or plastic. Plastic bins are now replacing plywood and metal bins because fruit in plastic bins receives less peel abrasion damage during transport, plastic surface can be cleaned more easily, and newer construction makes the bins reasonably durable. All fresh market fruit must be carefully emptied from the picking sack into the pallet bin. Conical plastic tubs are usually used for process fruit, and less care is required when filling the tubs.

In the mid-1990s, pickers received about $0.72 per field box for harvesting oranges. Contractors, who furnished the complete harvesting operation to the grower charged another $0.72 per field box for that service. Thus, the total harvest cost to the grower to pick the fruit and put it in the highway trailer was about $1.44 per field box. The total cost approached $1.75 per field box in 2003.

A crew of up to twenty pickers usually performs hand harvesting. The crew is transported to and from the grove in a crew bus. The bus carries drinking water and first aid supplies and remains at the work site. A complete set of Department of Labor posters may also accompany the bus if they are not already on display within the required distance from the work site. A grove truck, ladder and tub trailer, and portable toilet with handwashing facility also accompany each crew.

On site supervision and work records for each picker are the responsibility of the crew leader, who must be a licensed Florida labor contractor. The crew leader is often the grove truck operator. A field-rider, employed by the grower, processor, or packinghouse serves as liaison with the crew leader. The field-rider provides general work assignments, inspects the quality of the work, schedules the delivery of empty fruit trailers and pickup of full trailers, and verifies that all state and federal regulations are met. Anyone who has management or supervision responsibility for a harvesting crew must be knowledgeable of the many regulations that apply to farm workers and employers and may also need to be a licensed farm labor contractor.

A grove truck (typically known as a "loader" or "goat," with a hydraulic arm for lifting and dumping containers) is used to deliver empty bins or tubs, to pick up the filled containers, and to move the fruit out of the grove to a highway truck. This operation is called road siding. Bins of fresh market fruit are picked up by a hydraulic mechanical arm, stacked on a flatbed truck, transported out of the grove, and restacked onto a flatbed highway truck (Plate 27.2). Tubs of processing fruit are picked up by a slightly different arm assembly; the fruit are dumped in bulk into a 6,000- to 9,000-pound capacity box on a high-lift grove truck, transported out of the grove, and dumped again in bulk into a 45,000- to 55,000-pound capacity bulk highway trailer (Plate 27.3).

MECHANICAL HARVESTING

Commercial growers recognized the eventual necessity of mechanical harvesting of citrus for processing as early as the 1950s. Research projects to increase the efficiency of hand harvesting and to develop methods of mechanical harvesting were funded by the Florida Department of Citrus, the University of Florida, and the USDA-Agricultural Research Service in the late 1950s and early 1960s. The work was conducted at the Lake Alfred Citrus Research and Education Center, and several inventors and equipment manufacturers also joined in the search for new harvesting technologies. The cooperative research projects were ended in the mid-1980s after devastating freezes reduced bearing acreage and research funds.

Although efforts to improve the efficiency of hand harvesting were largely unsuccessful, various mechanical mass-removal methods, in which the fruiting canopy, the major scaffold limbs, or the tree trunk were shaken to remove the fruit, proved very effective. Tree vigor and yield were largely unaffected by the use of these devices, and studies continue to show little negative impact on the trees.

The abundance and affordability of labor for hand harvesting discouraged the commercial adoption of mechanical harvesting technologies in Florida for many years, but mechanical harvesting now offers significant economic advantage. It is expected to become a cost-effective method for harvesting citrus for processing as growers are forced to rely on resident U.S. workers. Mechanical harvesting systems using three to six workers are comparable to two to six standard crews of twenty pickers each. New groves should be designed to accommodate mechanical harvesting equipment, and existing groves will ultimately be modified or replaced to allow efficient mechanical harvesting.

There are several distinct mechanical harvesting systems available or under development. These operate in one of the following ways:

1. Hand pick and drop, mechanical fruit pickup from the ground
2. Mechanical pull and drop, fruit collection on catching surfaces.
3. Area shaking of the fruiting canopy, fruit collection on catching surfaces,
4. Trunk shaking, fruit collection on catching surfaces (Plate 27.4)
5. Continuous mechanical shaking of the fruiting canopy, fruit collection on catching surfaces (Plate 27.5)
6. Continuous airblast shaking of the fruiting canopy, mechanical fruit pickup from the ground

The Table 27.1 summarizes the characteristics and potential savings associated with each of these methods of mechanical harvesting listed above. The estimated harvesting cost per 90-pound field box assumes grower ownership of the harvesting system and represents all machine ownership and operation costs plus all labor costs to harvest the crop. Contractor ownership and operation might add about 30 percent to these estimates. In most cases, a seven-year depreciation period has been assumed, and at least 800 hours of harvesting each year. Costs do not include 5 percent to 10 percent of the crop that may be left in the groves because of imperfect mechanical harvest. When fruit prices are high and hand labor is available, a gleaning operation may increase total grower return. When fruit prices are low and labor is scarce, abandonment may be the best choice. Grove design and conditions greatly affect harvest cost; contracting for mechanical harvesting is best arranged on a cost-per-acre basis rather than cost-per-box of fruit.

Table 27.1 Summary of potential mechanical harvesting system characteristics as listed above.

Method no.	Acres required	Rate FB/hr	Crew size	Estimated savings	Grove Design
1	500	320-480	18-20	10%-20%	Mature, skirts 18 inches high, no deep swales
2	450	90-120	3	15%-30%	Mature, skirts 18 inches high, 22 foot wide rows max
3	400	80-120	4-5	15%-30%	Mature, skirts 18 inches high, no deep swales high yield
4	500	240-480	3	25%-50%	Max trunk dia. < 10 inches, uniform spacing, 2-row beds or level
5	2000	800-1600	6	50%-75%	Canopy width 12 feet max, no deep swales, hedgerow
6	1000	400-800	2-4	25%-50%	Mature, skirts 18 inches high, natural shape, skirts, no deep swales

GROVE DESIGN FOR MECHANICAL HARVESTING

The diverse range of tree ages and sizes, tree spacing, trunk heights, skirt heights, irrigation types and emitter placements, bed systems and swale profiles, headland space, and hedging and topping practices results in the need for several different harvesting systems. Most existing groves can be adapted to at least one of the mechanical harvesting systems shown listed above.

Management of tree size and shape is an important requirement in groves that are harvested mechanically. Tree design begins in the nursery. Most harvesters require a consistent clear skirt height to provide visibility, an unobstructed area for fruit collection, and a well-defined lower harvesting height for the machines. Some harvesters are designed for a maximum height tree as well and will not remove fruit that are produced above that height, which demands regular grove topping. Some harvesters only remove fruit from the outer 5 to 7 feet of the tree canopy, so hedging to limit the canopy width is needed with such equipment. Most mechanical harvesters break deadwood from the tree during each harvest, especially during the first two years of mechanical harvesting. Most of the deadwood can be removed from the fruit by detrashing devices on the harvesters, but some increase in harvest trash will be experienced at the unloading area of the processing plants.

When establishing a new grove, growers should select a compatible harvesting system and grove design. New groves intended for processed fruit production must be planned for highly efficient mechanical harvesting systems if the industry is to remain profitable. In general, this requires the use of rootstock/scion selections that mature quickly into moderate-sized trees shaped to suit the intended harvesting system. Tree trunks should be free of limbs to at least 24 inches above the ground, preferably 30 inches, particularly for trunk shakers. The trees can be pruned to develop scaffold branching in the range of 24 to 36 inches or 30 to 36 inches above the ground. Skirting to clear limbs and foliage from the first 36 to 42 inches of canopy height at the dripline will also be a standard practice in mechanically harvested groves to facilitate catching frame equipment (Plate 27.6). If this is not done, the average harvest rate and fruit removal will be low, tree damage and equipment damage will be high, and potential savings will not be realized. The spacing between rows and trees in the rows should also be uniform within each block of trees.

Other grove layout factors that need to be reconsidered for mechanical harvesting are: headland space (at row ends); row length; irrigation system and emitter location; bed and swale design; and distance to loading locations for the highway trailers.

Mechanical harvesting systems require more headland space than is typically found in existing groves. Clear headlands of at least 25 to 30 feet may be needed, and headlands of 10 to 20 feet are inadequate. Even small increases in the time to maneuver the equipment between rows due to tight headlands can significantly increase the cost of the harvest. In addition, equipment damage and tree damage are more likely when headlands are tight. The headland should also be relatively level from the last tree to the next limiting obstacle, next row or canal. In general, longer rows are desirable for mechanical harvesting. Short rows increase nonproductive turn-around time.

The irrigation system style and the location of the emitters are also important for mechanical harvesting. Solid-set irrigation risers may interfere with some harvest systems. The solid-set equipment is usually rugged enough that it can withstand contact from the harvesters, although it may prevent the catching, sweeping, or mechanical recovery of some fruit. Plastic microjets or drippers cannot withstand similar contact, so they must be located where the harvesting equipment will not damage them. Mechanical harvesters that remove and catch the fruit in a single operation are less likely to damage the plastic irrigation equipment, while mechanical fruit sweep and pickup systems are quite likely to damage these systems. This is especially a concern if skirt heights are 1.5 feet or less, making it difficult for the sweeper operator to see the irrigation system under each tree. In general, microjets should be less than 12 inches above the ground, in line with the row of trunks, and 3 feet or more away from the trunk. Drippers on the ground may be less vulnerable to damage than microjets on risers.

The bed and swale design for groves affects the choice of harvesting systems, the harvest rate, and the portion of the crop recovered by the machine system. In some groves, swales are designed with varying profiles and depths to facilitate water removal. Most mechanical harvesting systems that catch the harvested fruit are able to work in these swales. However, those systems currently available that drop fruit to the ground for later mechanical pickup do not work well in these conditions. Mechanical pickup machines work best on flat ground that has been prepared (mowed, detrashed, smoothed, etc.) just prior to the harvest operation. They can work in swales that maintain a uniform slope from the trunk line to the row center and that do not drop more than about 1.5 feet in elevation. This is of special relevance for a harvest system, as envisioned by many Florida growers, which uses an air harvester for shaking the tree canopy and removing the fruit, followed by mechanical pickup. Mechanical pickup of the crop is essential for such a system, but cannot be accomplished if the swales are deep or irregular. It should be noted that this air-harvest system requires the use of an abscission chemical to loosen the fruit prior to the air-harvest operation. In the 1970s, research trials using abscission chemicals showed that 5 percent to 80 percent of the crop dropped prior to the actual day of harvest depending on weather conditions, which sometimes reduced final juice quality.

Mechanical harvesting systems may be capable of harvesting up to 24 highway trailer loads of fruit per day, when the grove conditions are ideal. The distance from the end of any row to a loading area for highway trailers should not exceed three-fourths of a mile, and the road should permit rapid travel. Excessive round-trip time for one grove loader or goat may result in an inefficient operation requiring a second vehicle.

PREPARATIONS FOR HARVESTING

Preparing for harvesting requires scheduling, coordination, and training well in advance of the harvest. The harvester may be either a subsidiary of the grove operation, a representative of the purchasing processor, or an independent contractor. If the grower owns a mechanical harvesting system and hires the operators, the responsibility to have the equipment in working order and the operators fully trained rests with the grower. If a contractor provides the harvesting service, it is still advisable for the grower to verify that the contractor has dependable equipment, competent repair services, trained operators, and is complying with all federal and state regulations.

The harvesting department must communicate with the processors to ensure they are aware that the fruit is being mechanically harvested. Their cooperation is needed to be sure the harvesting system can work at full capacity each day. It is important that empty trailers be available throughout the harvest period and that processors are able to accept large numbers of trailers from one location. Proper scheduling includes the effects of abscission sprays. Determining whether a fruit catch system or a fruit pickup system should be used for harvesting is essential since fruit that is used to make unpasteurized products should always be harvested using a catch system.

Cultural operations that facilitate mechanical harvesting include mowing, ground preparation (such as disking, ground smoothing, and removal of grove debris for efficient fruit pickup) and appropriate scheduling of pesticide and abscission sprays (so that the number of days to harvest is within the desired range).

SCHEDULING THE HARVEST

Each citrus variety has its own harvest period, which varies somewhat each year. The actual dates of the harvest depend on weather, tree condition, fruit condition, and a variety of market-related factors including yield, fruit value, and fruit delivery allocations. The early- and mid-season orange harvest period usually starts between November 1 and November 21 and continues until February 1 to February 15. Several cultivars (Parson Brown, Hamlin, and Pineapple) mature during this period in roughly the same sequence each year.

The Valencia variety presents special complications for mechanical harvest. There are essentially three phases of Valencia harvest, occurring from late February through early June. The initial Valencia harvest is much like the harvest of other early- and mid-season cultivars, as only the mature fruit are on the tree. However, from early to mid-March, bloom begins on all trees and lasts until early April. Although some concern has been expressed, there is no evidence that continuing to mechanically harvest Valencia during the bloom period reduces yield of the following crop. As the green Valencia fruit develop in size beyond about 0.75-inch diameter, however, they become susceptible to removal during mechanical harvesting operations. Trunk shakers and air harvesters tend to remove more of the green crop than do other systems. After early May, these two methods of harvest must not be used in order to protect the following crop. A selective fruit abscission chemical is needed for Valencia to selectively loosen the mature crop without affecting the immature crop. Such a material would enable late-season Valencias to be harvested in the same way as the earlier fruit, but is not yet available. The most selective harvesting methods available should be employed for harvesting Valencia methods 4 and 6.

In order to realize the potential savings of mechanical harvesting, it is essential that the harvest proceed at full capacity once it begins, interrupted only by extremely adverse weather conditions or mechanical failure. The cost of the harvest continues to increase when equipment and operators are idle. Coordination between the processor, the grower, the transporter, and the harvest provider are essential to keeping each mechanical harvesting system working at maximum efficiency. At present, mechanical harvesting does not include the use of abscission chemicals. The fruit that is mechanically harvested is typically in good condition and holds in highway trailers about as well as hand-harvested fruit does.

When fruit abscission chemicals become available, however, harvest scheduling will depend on the abscission spray schedule and the effects of weather on fruit loosening. Improper harvest scheduling may result in a significant portion of the crop being left in the grove, either on the tree, because it was not loosened in time for the harvest, or on the ground, because it was loosened too soon.

REFERENCES

Burns, J. K., and E. Echeverria. 1988. *Assessment of quality loss during commercial harvesting and postharvest handling of 'Hamlin' oranges.* Proceedings of the Florida State Horticultural Society 101: 76-79.

Brown, G. K. 2002. *Mechanical harvesting systems for the Florida citrus juice industry.* ASAE Paper No. 021108, 18 pp. St. Joseph. Mich.

Grierson, W., and W. C. Wilson. 1982. *Mechanical Harvesting And Citrus Quality: Cannery vs. Fresh Fruit*. University of Florida, Citrus Research and Education Center, Lake Alfred. Report: 1-13.

Kostenyuk, I. A., and J. K. Burns. 2004. *Mechanical wounding and abscission in citrus*. Physiological Plantarum 122: 354-361.

Whitney, J. D. 2003. *Performance of mechanical citrus harvesters in Florida*. International Society of Citriculture Congress. (9th 2000: Orlando, Florida), Orlando, Florida USA 1: 277-279.

Whitney, J. D., T. A. Wheaton, W. S. Castle, and D. P. H. Tucker. 1995. *Tree height, fruit size, and fruit yield affect manual orange harvesting rates*. Proceedings of the Florida State Horticultural Society 108: 112-118.

28

Fresh Citrus Utilization

W.M. Miller

INTRODUCTION

Once a crop has been successfully harvested and moved from the grove to the processing or packing plant, optimal utilization of the citrus crop is paramount. Quality fresh citrus depends on the production and packing of fruit with minimal surface blemishes and high internal quality, careful harvesting and expeditious handling of the fruit to minimize damage and deterioration, proper conditions for storage and transportation of the fruit, and a quality assurance program that includes fruit inspection. In the case of citrus canker, growing and shipping asymptomatic fruit are an important part of the fresh fruit program. The harvested fruit should immediately be transported to the packinghouse or fungicide drench facility, and rapid handling and cooling is especially important in early fall or late spring when field and transport temperatures may be over 95°F. Clearly, production, harvest, transportation, postharvest handling, packing, and domestic and foreign sales must be carefully managed and coordinated to maintain the market for fresh Florida citrus.

A BMP for quality fresh citrus depends on the production and packing of fruit with minimal surface blemishes and high internal quality, careful harvesting and expeditious handling of the fruit to minimize damage and deterioration, proper conditions for storage and transportation of the fruit, and a quality assurance program that includes fruit inspection and supervision of postharvest treatments.

The per capita consumption of fresh citrus in the United States has remained constant for several years, despite well-publicized associated health benefits including lower cholesterol levels and reduced incidence of heart disease. In general, U.S. markets have expanded significantly only for new consumer-convenient items. One important factor that limits the market for fresh citrus is seasonal availability. Florida citrus stores fairly well; the peel provides a protective covering for the edible flesh, and the fruit is nonclimacteric (i.e., they do not continue to ripen after harvest as do apples and pears). However, the fresh citrus industry is currently unable to provide a uniform product throughout the year. Early fruit, marketed at minimum maturity standards, may not appeal to many consumers and may reduce subsequent sales. Similarly, fruit that is marketed late, with evidence of aging and deterioration, also undermines consumer confidence and support. A year-round supply of fresh citrus fruit of reasonable quality could significantly contribute to the expansion of the fresh market for Florida growers. Some market gaps including earlier and later than now, may be filled by new cultivars under development. Widespread use of long-term controlled temperature storage could be used to extend the season, but cost of storage may not be recovered.

Packinghouse management is critical in coordination of sales and harvesting activities (Figure 28.1). With few long-term refrigerated storage facilities in Florida and the cost of storage, most fruit is stored "on-tree." Hence, it is critical for the packer to have the right cultivars of proper sizes delivered to the packinghouse. The fruit may come from independent growers or controlled sources. Controlled sources would include fruit from members of a cooperative or the owner's fruit in a privately run packinghouse.

Quality fresh citrus depends on the production and packing of fruit with minimal surface blemishes and high internal quality, careful harvesting and expeditious handling of the fruit to minimize damage and deterioration, proper conditions for storage and transportation of the fruit, and a quality assurance program that includes fruit inspection.

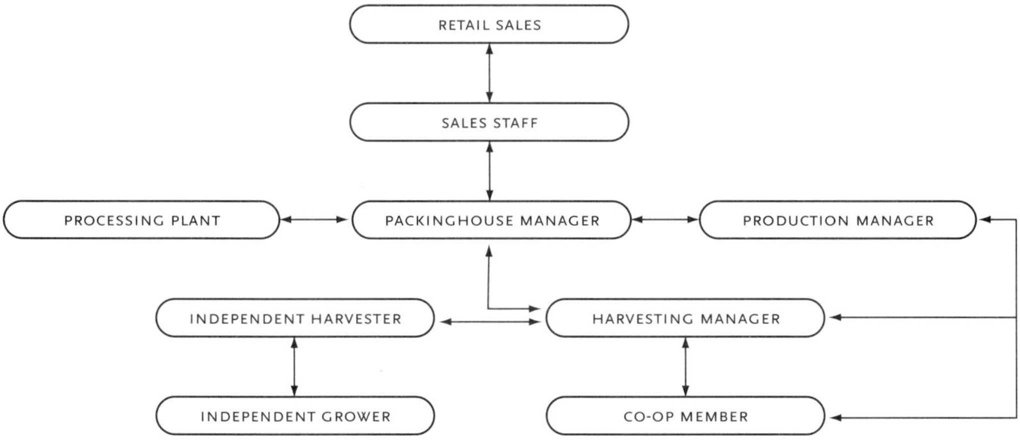

Figure 28.1 Interactions of the packinghouse manager and other industry elements in procuring and selling fresh citrus.

FRESH FRUIT SALES

Fresh Florida citrus is sold through commercial packinghouses and grove stands or gift fruit markets. Gift fruit may be sold directly to the consumer on-site or ordered by mail, phone, or over the Internet. Commercial packinghouse sales account for the great majority of fresh fruit sales, and a typical sales procedure is illustrated below. There has been a general decline of terminal markets, while sales through integrated wholesale and retail channels have increased (Figure 28.2).

There are few refrigerated storage facilities in Florida, so most fruit is stored "on tree," meaning that fruit is harvested when an order is received and is quickly packed and shipped. The packinghouse may purchase fruit from independent growers or from a cooperative, or it may be privately owned by a grower or cooperative and packs the owner's fruit exclusively. A wide range of marketing options is available for the fresh citrus producer. Several of these are discussed in Chapter 32.

PRODUCTION FOR THE FRESH MARKET

Growers who intend to sell their citrus in fresh markets should be aware of the attributes that are considered especially desirable in fresh fruit. Fresh citrus must have an attractive and uniform size and shape, with good color. Internal quality must be high, and the fruit must be essentially free of external blemishes. Consumers prefer fruit that is easily prepared for consumption and that is seedless, or nearly seedless. In addition, the fruit should readily degreen and store and transport well with minimal deterioration. Supply and demand for fresh citrus varies throughout the season, so growers should pay careful attention to harvest dates when selecting cultivars for production. Demand for fresh fruit is traditionally high in December to support the holiday gift and fundraising markets, and supplies tend to be low in late spring and summer.

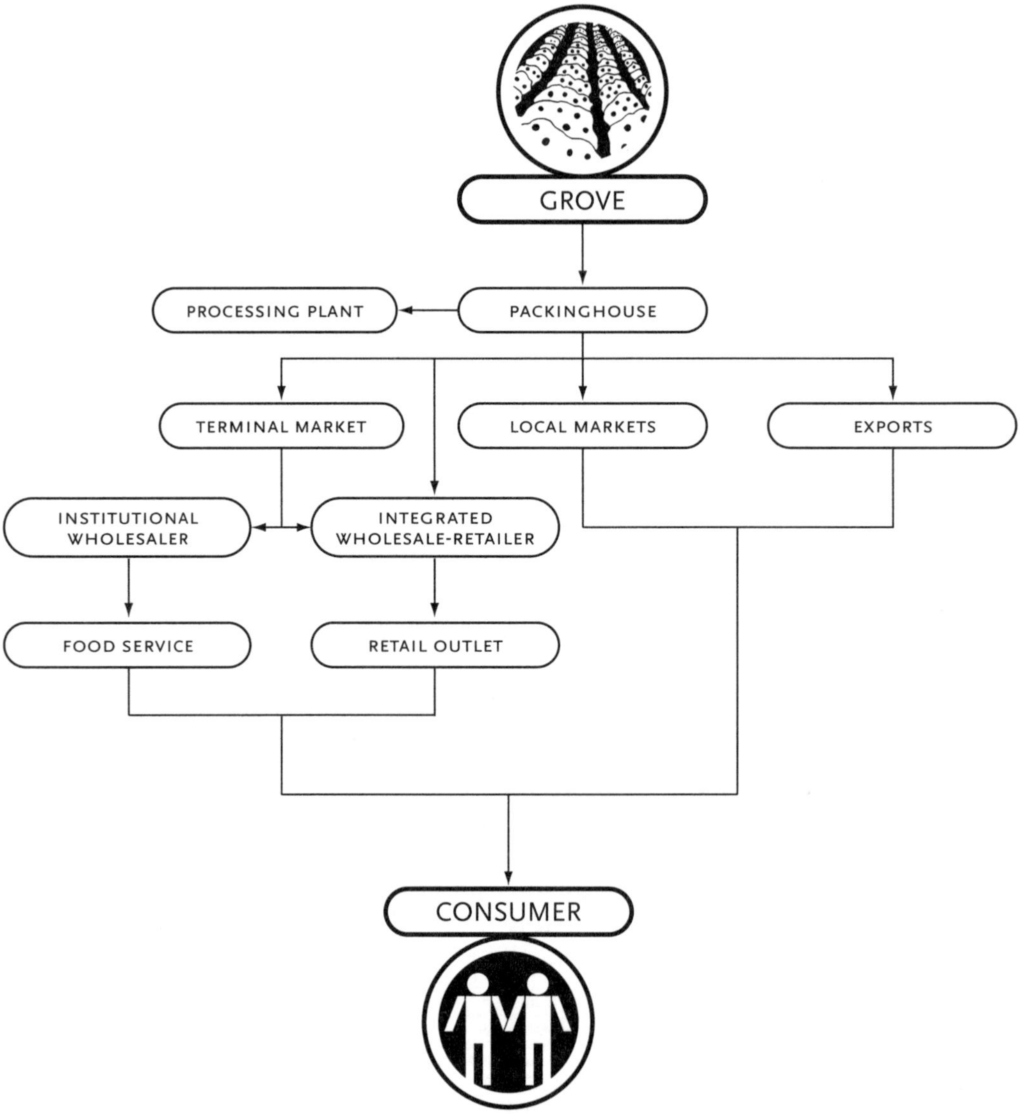

Figure 28.2 Marketing systems for fresh Florida citrus.

Production of high-quality fresh citrus requires intensive grove management to protect the fruit from pests and diseases, especially fungal diseases that may damage the external peel of the fruit. Nutrients and water must also be closely monitored to insure desired fruit size and internal quality. Hedging and topping are practiced to minimize alternate bearing and to facilitate harvesting. Growers evaluate the condition of the fruit periodically in each grove, and fruit that is not deemed appropriate for the fresh market is slated for processing, and its management is altered. This spares the grower unnecessary continuation of grove management expense for practices directed at reducing external blemishes and fresh fruit harvesting.

HARVEST AND TRANSPORT

Careful harvesting is critical for fresh citrus and should be closely supervised. The fruit is hand harvested into a picking bag at the tree and subsequently transferred into a pallet bin. Picking bags have straps that go over pickers shoulders to support the fruit weight and drawstring at the bottom so that pickers merely lean over the bins and allow fruit to empty from the bottom. Bins are usually equivalent to 10 Florida field boxes, or about 900 pounds of oranges or 850 pounds of grapefruit. Pallet bins should not be overfilled or fruit will be damaged in transport, primarily when bins are stacked. Plastic bins are increasingly common because they are lighter, more durable, and easier to keep clean than plywood, solid wood, or metal-framed bins (Plate 28.1). The fruit should never be collected from the ground, since this greatly increases the levels of decay-inducing fungi. Fruit that is damaged by rough handling or plugged during the harvest is vulnerable to a variety of postharvest infections and disorders. The harvested fruit should immediately be transported to the packinghouse or to a fungicide drench facility; rapid handling and cooling is especially important in early fall or late spring when field and transport temperatures may be over 89°F.

POSTHARVEST HANDLING AND STORAGE

Pallet bins are transported to a roadside location in the grove and loaded onto a semitrailer for highway transport to the packinghouse or to an off-site drench facility (Plate 28.2). The fruit may be drenched with the fungicides thiabendazole (TBZ) or imazalil, particularly if it requires degreening or will not be packed immediately. The fungicide suspension also contains chlorine to eliminate fungi that are tolerant to the applied fungicides. After arrival at the packinghouse, fruit bins are unloaded and early season fruit are usually degreened. Degreening removes chlorophyll from the rind of fruit that has met internal maturity standards but still has a green, unripe appearance. The fruit is placed in a specially constructed room at the packinghouse where temperature and relative humidity are controlled (recommended conditions are 86°F,

95 percent RH). Ethylene gas is introduced to maintain a concentration of 5 ppm. The degreening process may require 72 hours or earlier in the season, but only about 12 hours late in the fall (Plate 28.3).

The packing line operations are outlined in Figure 28.3. Individual pallets are emptied onto a conveyor where large field debris and any decayed or damaged fruit are removed by hand after trash eliminators remove leaves and twigs. The fruit is then washed on brush roller beds. High-pressure washers, operating at 300 pounds per square inch (PSI) nozzle pressure or more, have been installed at most commercial packinghouses to clean the fruit more thoroughly. After washing, the fruit is presized and pregraded. The presizers are typically mechanical belt- and roll-type, where unacceptable large and small fruit are diverted to processing as eliminations. Recently, machine-vision sizing and grading have been implemented at some packinghouses. These reduce labor costs and increase fruit output (bins of fruit per hour) on the packing line.

After washing, the fruit is usually pregraded to remove most eliminations before receiving a fungicide and wax treatment that is normally applied to fruit, whether or not it is drenched before entering the packinghouse. The fungicide may be applied as a nonrecovery spray or in the wax. Currently approved fungicides include SOPP (sodium o-phenylphenate), imazalil, and TBZ. Some customers now demand "Chem-Free" fruit with no fungicides applied postharvest. Water waxes, formulated from polyethylene, shellac, or carnauba materials, are applied as a liquid by brushes in the packing line. For proper application of these waxes, the fruit surface is dried with heated air both before and after the waxing process. The predrying facilitates adhesion and uniform coverage, while the postdrying is necessary to vaporize the water and allow the wax to set quickly.

Finally, the fruit is again graded visually (Plate 28.4) or by specialized equipment (Plate 28.5). The fruit may be classified into one of several packable grades or diverted to processing for juice. Packout levels (percentage of fruit that is actually packed for fresh sale) vary significantly and depend on the fruit variety, grove maintenance and care in harvesting, seasonal variability, and the anticipated markets. Fruit is manually packed into 2/5- or 4/5-bushel cartons or put into 3-, 5-, 8-, 10-, 14-, or 20-pound bags for direct retail sales. Fruit sold in bulk to large U.S. grocery chains receives a small PLU (price lookup code) label on each fruit designating origin, variety, and size. The cartons are palletized (arranged on pallets for easier transport) and either moved into cold storage or accumulated in the packinghouse for immediate shipment.

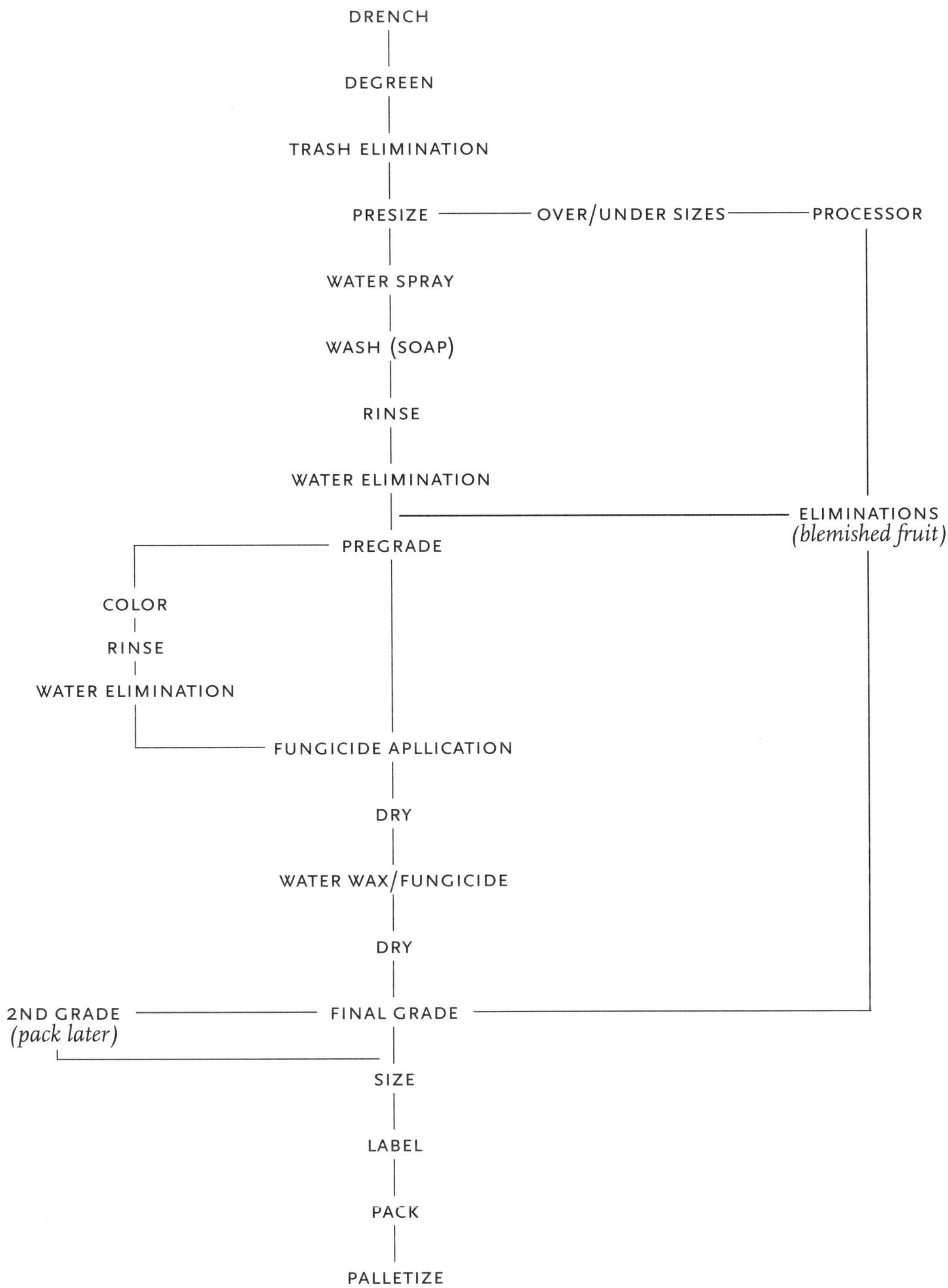

Figure 28.3 Outline of fresh fruit packinghouse operations for Florida citrus.

It is important to maintain sanitary conditions in the packinghouse. Damaged or decayed fruit should be removed immediately after dumping, and pallet bins, degreening rooms, and cold storage rooms should be routinely sanitized.

Fruit that is being exported may need preconditioning in cold storage at high humidity to reduce decay, weight loss, and peel pitting. Recommended storage conditions for grapefruit and oranges are 10°C and 1°C, respectively, with 90 percent RH or higher. Refrigerated storage may also extend fruit availability into the summer months. Valencia orange and grapefruit can be stored for eight weeks or longer to provide both fresh fruit and fresh-squeezed juice. Fruit handled domestically is transported by refrigerated truck; export fruit is loaded onto shipping vessels at port facilities or into shipping containers for delivery to a ship (Plate 28.6).

QUALITY CONTROL

Fresh fruit packers who demonstrate adequate quality control initiatives may be granted a "top quality" rating under the Partners in Quality (PIQ) Program administered by the Division of Fruit and Vegetable Inspection in the Florida Department of Agriculture and Consumer Services (FDACS). The PIQ Program allows citrus packers who routinely meet federal and state environmental and food quality standards to operate without daily inspections, instead relying on their own quality control systems. Packers who assume responsibility for quality control must maintain detailed records to document the effectiveness of their system and must be prepared for unannounced visits by FDACS inspectors. Quality control records are also available to large buyers of fresh citrus who may have established their own quality standards.

COORDINATION AND COMMUNICATION

Production, harvest, transportation, postharvest handling, packing, and domestic and foreign sales must be carefully managed and coordinated to maintain the market for fresh Florida citrus. Timely, effective communication is essential for a successful fresh-fruit marketing program. The production manager, harvest manager, packinghouse manager, and sales manager must work together to bring a high yield of highly desirable fruit to the consumer.

REFERENCES

Affeldt, H. A. 2003. *Current automated citrus sorting technology for detection of defects caused by mechanical harvesting and other factors.* International Society of Citriculture Congress. (9th 2000: Orlando, Florida), Orlando, Florida USA 1: 288.

Hall, D. J. 2003. *Twentieth Century Developments In Handling Florida's fresh Citrus Fruit: An Overview.* Proceedings of the Florida State Horticultural Society 116: 369-374.

Larrigaudiere, C., A. Bonet, and J. Pons. 2003. *Harvest maturity, a key factor in determining the quality of controlled-atmosphere stored mandarins.* International Society of Citriculture Congress. (9th 2000: Orlando, Florida), Orlando, Florida USA 2: 1161-1163.

Miller, W. M., H. Dou, and M. Talbot. 2001. *Evaluation of refrigerated conditions and fruit quality in cooling fresh Florida citrus.* Proceedings of the Florida State Horticultural Society 114: 181-185.

Russo, G., A. M. Rinaldi, M. G. Corona, L. Massignan, and R. Lovino. 2003. *Effect of cold storage on four grapefruit cultivars.* International Society of Citriculture Congress. (9th 2000: Orlando, Florida), Orlando, Florida USA 2: 1169-1170.

29

Citrus Processing

R.M. Goodrich, J.T. Griffiths, and M.E. Parish

INTRODUCTION

The entry of the Coca-Cola Company into the growing and processing of citrus in the 1950s represented a large infusion of capital from outside Florida into the state's citrus processing industry. From that point forward, many changes occurred in the industry that improved the quality and quantity of citrus delivered to consumers around the world. In Florida, all citrus processors cooperate with the State of Florida Fruit and Vegetable Inspection Division and with USDA processed food inspectors. A grower for the processing industry is usually not as concerned with the number of boxes of citrus produced per acre as with the volume of juice and number of pounds of juice solids produced. Further, there are a number of by-products of citrus production. These by-products include citrus peel, molasses, alcohol, washed and/or juicy pulp cells, washed pulp solids, and essential oils derived from the peel of oranges and grapefruit, such as oil and aqueous phase essences, cold-pressed oils, and d-Limonene oil. Unfortunately, growers never see contractual and return accommodations for these by-products. Although difficult to understand at first, there are a number of simple calculations that citrus managers can use to improve their ability to determine the value of the crops they produce. This chapter covers the essentials related to understanding utilization and value of processed citrus crops.

The commercial processing of citrus fruit for juice has become the largest and dominant utilization of the citrus crop in Florida. A fruit processing BMP should preserve the best and most distinctive qualities of the fresh fruit, minimize yield loss through the processing plant, minimize impact on the environment and optimize the end product value to the consumer.

The growing, packing, and processing operations are interrelated, and the organizations responsible for each of these operations are mutually dependent on each other. In fact, some organizations are structured to perform essentially every function from the production of the fruit to the processing of the juice. However, many organizations are responsible only for growing, packing, or processing, and these organizations must maintain cooperative relationships with others.

The development of commercial citrus juice processing has occurred largely since the 1930s. Single-strength pasteurized canned juice was originally the most common product, as it could be stored on the grocer's shelf or at home for months without refrigeration. This product declined in popularity as large home refrigerators and freezers became common in the 1950s and 1960s. Although several Florida processors continue to produce this juice in significant volume, it was ultimately replaced by retail-pack frozen concentrate, and more recently, by ready-to-serve chilled juices. Frozen concentrated orange juice (FCOJ) reached peak sales around 1980. By the mid-1990s, FCOJ, reconstituted juice from concentrate, and pasteurized "not from concentrate" (NFC) juice each represented about a third of the total U.S. consumption. Since that time, FCOJ has declined as a percentage of total consumption and premium NFC has increased. In the late 1990s, unpasteurized chilled juice essentially became available because of health concerns. This juice has claimed a small but important share of the juice market. Orange and grapefruit are the most commonly available juices, but tangerine juice is also available on the market. Several other juice products, including dehydrated orange juice crystals and pasteurized juice in aluminum foil aseptic packs, have been marketed with modest success.

In the early days of the industry, processing plants were often established by local entrepreneurs many of who were also growers. By the mid-1940s, processing plants owned by individual growers or operated by cooperative associations of growers dominated the market, and the industry remained largely Florida based. The entry of the Coca-Cola Company into the growing and processing of citrus in the 1950s represented the first large infusion of capital from outside Florida into the state's citrus processing industry. Currently, more than half of the total of Florida citrus processing capacity is owned by interests outside the state including companies from France, Brazil, and Japan.

In the 1950s and 1960s, most FCOJ was sold in Florida already packaged for retail sales, primarily to buyers for chain stores and major wholesalers. By contrast, by the mid 1990s most Florida FCOJ was transported out of Florida in bulk tankers and processed into retail-sized FCOJ or reconstituted juice at sites outside the state. It is important to note, the company and/or plant that processes the fruit into juice does not necessarily package the final juice product.

QUALITY OF FRUIT FOR PROCESSING

In Florida, all citrus processors must cooperate with the State of Florida Fruit and Vegetable Inspection Division and with United States Department of Agriculture (USDA) processed food inspectors. Generally, state inspectors oversee the raw fruit and have responsibility up to the point of juice extraction. After processing, the USDA inspectors assume responsibility. The State of Florida requires USDA inspection for compliance with U.S. Standards for Grade for both orange and grapefruit juice; however, these inspections are not mandatory in other citrus processing areas of the United States. Continuation of USDA product inspections has recently been questioned by industry representatives and may be altered in the near future, but greater concerns for food safety argue for higher inspection requirements.

Fruit is always inspected when received at a processing plant. Fruit that is received must be whole, sound, mature, and recently harvested. This fruit must be free of physical damage, internal insect infestation, and mold damage. Any fruit arriving at a processing plant found unfit for processing must be destroyed at the expense of the fruit owner under the personal supervision of the inspector.

In addition to state and federal requirements, the processor will have established certain standards for the fruit to ensure the consistent production of high-quality juice. In particular, the citrus variety and the Brix/acid ratio are directly related to the quality of the processed juice. The processor's minimum Brix requirement (USDA standards) will depend on the juice product. Single-strength juice products have slightly lower requirements for Brix than FCOJ. These products are typically packed from slightly lower Brix oranges and from oranges with maximum juice content. Surprisingly, fruit with a lower Brix/acid ratio is more readily enjoyed fresh than as juice.

In some seasons, processors have offered a premium for fruit that was especially low or high in acidity, or which was highly colored. However, market trends determine which juice characteristics are most desirable in any given season. Good color, in particular, usually but does not always increase the value of the juice. Valencias and Pineapple oranges generally have excellent juice color, while the early orange cultivars generally have poor juice color. Only Valencia receives a premium price, partially for color. Rohde Red Valencia has been propagated especially for its superior juice color.

All processing plants perform field test sampling to ensure that minimum maturity standards are met before harvest, especially at the beginning of the season. Representative samples are evaluated prior to receipt of a truckload of fruit. Spot picking of the most mature fruit may be performed to ensure that minimum standards are met. However, this is expensive early in the season. Spot picking is more important and more common in the fresh fruit industry. When multiple-bloom fruit (some years have prolonged bloom with widely varying maturity) is to be processed, processors must be especially careful of fruit quality. The minimum Brix/acid ratio may be met overall, but the juice may be bitter, with a characteristic immature taste unless all of the fruit has matured to meet the minimum regulatory and quality standards. In this case, it is advisable to delay the harvest a few weeks to ensure that the last bloom on the trees has met the standards.

After the fruit is delivered, it should be processed as quickly as possible to minimize deterioration. Fruit loads held in the staging area should not be held longer than a single day. As fruit loads are scheduled for processing, packinghouse eliminations (culls) should be utilized more rapidly than "field-run" fruit, since they have been harvested earlier and may develop decay earlier. It is always difficult to pregrade or initial-grade unwashed fruit or fruit on heavily loaded conveyors. However, every attempt must be made to remove any obviously unacceptable or objectionable fruit to prevent the accelerated decay of other fruit in holding bins. If at all possible, both trailers and holding bins should be swept clean of any leaves, debris, and any broken or split fruit after each load. Such cleaning is mandatory in some parts of Florida for canker control.

SALE OF FRUIT FOR PROCESSING

In the early 1950s, citrus fruit for processing was purchased by the box (85 pounds for grapefruit and 90 pounds for oranges). As the market for FCOJ developed, processors began to pay for pounds of fruit solids (primarily sugars). The pounds of solids per box can be calculated by multiplying the percentage soluble solids (or the Brix) by the number of pounds of juice in 90 pounds of fruit. A grower for the processing industry is usually no longer concerned primarily with the number of boxes of citrus produced per acre, but rather with the volume of juice and number of pounds of fruit solids produced (see sample calculations of pounds solids, yield, and value to the grower). The price, or a method of determining the price, is specified in a contract between the grower and the processor prior to the harvest. These contracts vary widely, differing primarily in the share of risk accepted by the grower and the processor. Some current marketing options are described in Chapter 32.

Pounds-solids = Pounds of juice per box × (Brix ÷ 100)

Example 1:

If a box of oranges contained 45.0 pounds of juice and the Brix was 10.0%, the pounds-solids per box would be

$$45.0 \times \frac{10.0}{100} = 4.50$$

Example 2:

a. Actual yield is 51% juice × 90 lb box = 45.9 lb juice box.

b. If the true Brix is 10°, then

$$45.9 \text{ lb juice} \times \frac{10°\text{Brix}}{100} = 4.590 \text{ lb solids per box}$$

c. If the Brix measured is 10.1°, then

$$45.9 \text{ lb juice} \times \frac{10.1°\text{Brix}}{100} = 4.6359 \text{ lb solids per box}$$

Comparison:

a. The difference in lb solids per box in Examples 1 and 2 would be

$$4.636 - 4.500 = 0.136 \text{ lb}$$

b. If the pounds-solids price is $0.70 per pound, this difference becomes

$$0.136 \times \$0.70 - \$0.0952 \text{ per box}$$

or

$38.08 for a 400 box load

Quantity of juice per box can be expressed either as weight (pounds) or volume (gallons). Although gallons/box is a valuable measure of fruit maturity, as a standard for cannery payments it is inappropriate because it measures water, rather than solids. The yield of juice per standard box is therefore commonly calculated as follows:

$$\text{Pounds (yield) of juice per box} = \frac{\text{weight of juice} \times \text{weight of std. box}}{\text{weight of fruit sample}}$$

Example 3:

a. Weight of fruit sample (oranges): 50.0 lb

b. Weight of juice: 25.0 lb

$$\frac{25.0 \text{ lb juice} \times 90 \text{ lb box}}{50.0 \text{ lb of fruit}} = 45.0 \text{ lb per box}$$

c. To obtain the gallons of juice per box (which is used as a maturity standard) divide the weight of juice per box by the weight per gallon. *Presume that this juice has 12° Brix, and one gallon weighs 8.724 lb. Hence

$$\frac{45.0 \text{ lb juice per box}}{8.724 \text{ lb per gallon}} = 5.16 \text{ gallons per box}$$

d. But cannery contracts often pay for pounds-solids. At 12° Brix, and 45 lb juice/box (see Step c above) then each box of oranges contains

$$45.9 \text{ lb juice} \times \frac{12°\text{Brix}}{100} = 5.5 \text{ lb solids per box}$$

Processing Fruit:

Inspection of a representative sample ensures that only wholesome citrus fruit that meet legal requirements for minimum quality are processed. At the same time, the information provided by the tests is utilized by the cannery to determine whether the fruit is suitable for concentration purposes, how the load should be blended with other oranges as they are processed in order to maintain a uniform product, and finally, the basis upon which payment to the seller shall be made.

Processing fruit may be sold in two ways, pounds-solids or pound of juice per box. Both afford the grower a return commensurate with the internal fruit quality. In the case of pounds-solids, the price is calculated by multiplying the total pound-solids by the price per pounds-solids, for instance:

Example 4:

a. Pounds-solids per box: 5.4 lb

b. Price per pound of solids: $0.90

c. Price per box:

$$5.4 \times 0.90 = \$4.86 \text{ per box}$$

* To find weight per gallon, visit www.fred.ifas.ufl.edu/citrus/pubs/ref/conv.htm#148/

Similarly, the price, based on pounds of juice per box, is computed by multiplying pounds of juice by the price per pound, as shown in the following example:

Example 5:

a. Pounds of juice per box: 45.0 lb
b. Price per pound of juice: $0.10
c. Price per box (pound juice basis):

$$45 \times 0.10 = \$4.50 \text{ per box}$$

Inspection of a representative sample ensures that only wholesome citrus fruit that meet legal requirements for minimum quality are processed. At the same time, the information provided by the tests is utilized by the processor to determine whether the fruit is suitable for concentration purposes, how the load should be blended with other oranges as they are processed in order to maintain a uniform product, and finally, the basis upon which payment to the seller shall be made. Processing fruit may be sold in two ways, pounds-solids or pounds of juice per box. Both afford the grower a return commensurate with the internal fruit quality.

Growers often have a different set of interests to those of the processor. The move from the grove to the plant may follow different avenues ranging from the growers who own the plant or are a part of cooperatives to those who sell their fruit on the tree. The fruit may go to cooperatives or participation pools or be sold on the cash market. For fruit going to the cash market, timing of the sale is important as very early or late harvesting is usually associated with higher cash prices but lower solids yield. In such transactions the financial strength of the buyer is an important consideration. For fruit going to a pool or participation plan, the harvesting date should coincide with maximum pound solids per acre for best returns. Achieving this goal is not as easy as it appears as there are a number of influencing factors to consider including:

1. state of maturity and seasonal variations;
2. possibility of freeze and other weather events;
3. tendency of lbs/solids/box to increase and finally decreases as the season progresses;
4. tendency of droppage to increase with late harvest;
5. tendency of juice, hence weight boxes, to increase and then decrease with time of harvest;
6. changing plant deducts for cullage;

7. potential for affecting adversely next year's crop by later harvest;
8. availability of harvesting facilities and pickers;
9. availability of processing plant quotas.

The NFC processor needs a prolonged and steady flow of fruit meeting not only the legal requirements, but in-house requirements, to minimize the needs for storage and blending. While meeting the above minimums, it is likely that juice percentage can be more important than high Brix, when fruit is purchased on a pound solids basis. The concentrate processor likely has different priorities, such as high Brix, high juice color, and high plant efficiency through processing at a high rate for most of the season. Both types would like a uniform, predictable flow of raw materials free of fluctuations or interruption because of weather, or labor outputs and demands. Sound fruit, free of debris and only hours removed from the tree is preferred also.

Early oranges usually reach maximum pounds solids of about six pounds per box, while Valencias average close to seven pounds per box in most seasons. Fruit drop, and fruit drying or spoilage may reduce the yield late in the season. Brix generally increases with the maturity of the fruit and pounds of juice usually reach a maximum near the middle of the processing period for a given variety. Thus, the highest pounds solids and, consequently, the greatest income for the grower, are usually obtained soon after the mid-point of the season. Late harvesting may also adversely affect the next crop, since tree resources are diverted away for the next crop set. Intermediate fruit handlers and the processor generally decide the time of the harvest, considering these factors as well as the availability of equipment and labor, the processing capacity at the plant, and the likelihood of freezing conditions or other damaging weather.

When the fruit is delivered to the processing site, the trailer, which generally holds around 500 boxes, is identified and weighed (Plates 29.1 & 29.2). A device removes a representative sample of fruit, as the trailer is unloaded fruit are stored in bulk (Plates 29.3 & 29.4). That sample is then sent to the in-house USDA operated test facility where the pounds of juice, the Brix, and the percent acid are calculated on a per box basis. These test house procedures are intended to consistently assure both grower and processor of the accuracy of the solids per box measurement. Figure 29.1 represents a typical inspection certificate. Information on the form includes pounds juice per box, the percent solids (Brix), and the calculated pounds solids per box. Test house technologies and procedures have been the subject of some controversy, since they ultimately determine the value of the fruit.

FORM FV-432
(10-1-87)

MEMORANDUM NO.

UNITED STATES DEPARTMENT OF AGRICULTURE

AGRICULTURAL MARKETING SERVICE

FLORIDA DEPARTMENT OF AGRICULTURE

DIVISION OF FRUIT AND VEGETABLE INSPECTION

INSPECTION CERTIFICATE FOR PROCESSING PLANTS

This certificate is issued pursuant to the Agricultural Marketing Act of 1946, as amended (7 U.S.C., 1621 et seq.), and the Statutes of the State of Florida, and is prima facie evidence in all courts of the United States and Florida. WARNING: Any person who knowingly shall falsely make, issue, alter, forge, or participate in any such actions, is subject to a fine of not more than $1,000 or imprisonment for not more than one year or both.

PROCESSOR	ADDRESS	DATE
Any Processor	Winter Haven, FL	12-08-2004

SUPPLIER	ADDRESS
1018 John Doe Fruit Co.	Pensacola, FL

LOAD IDENTITY	TRAILER NO	RAMP NO.	TYPE FRUIT	DESIGNATED AS
047430	0364	2	Early - Mid	ORG.

LBS. SAMPLE	LBS. JUICE	LBS. JUICE PER BOX	
55.23	32.58	53.091	Juice content determined in accordance with Florida Department of Citrus rule.

% ACID	% SOLIDS (BRIX)	RATIO	LBS. SOLIDS PER BOX
.70	10.69	15.27	5.6754

REMARKS

I, the undersigned, a duly authorized Inspector of the United States Department of Agriculture, do hereby certify that the samples of the herein described products were inspected and the quality and/or condition as shown by said samples were as herein stated.

INSPECTOR J. R. Inspector #090

The information below is for the convenience of the Processor and Supplier and its accuracy is not vouched for by the Inspector.

FIELD TICKET NO			
GROSS LBS.	HAULED BY	FIELD BOXES	BIN NO.
TARE LBS.	PICKED BY	BLK	
NET LBS.	CONTRACT NO.		
GROSS BOXES			
BOXES CULLS			
NET BOXES			
LBS. SOLIDS PER LOAD			
PRICE PER LBS. SOLIDS			
PRICE PER BOX	REMARKS		
TOTAL PAYMENT			

SCALE OPERATOR

FRUIT RECEIVING TICKET NO.

Figure 29.1 "Inspection Certificate for Processing Plants" used by Florida Department of Agriculture to determine sugar and juice content for grower payment. Used with permission.

The USDA test house calculations are reported to the individual grower and to the State of Florida Department of Agriculture, where the data may be used to calculate state averages. The grower receives a report that shows the total number of boxes received by the processor, the volume of culls on the load, and the fruit quality determined at the test house.

It is to the advantage of the grower and the processor to maximize the pounds solids recovered in processing. Pounds solids recoveries have steadily increased with improved processing procedures and with the allowance of pulp wash solids as an in-line additive during production of citrus concentrates. This practice was prohibited until the passage of the Nutrition Labeling and Education Act of 1990, which preempted certain Florida statutes and Florida Department of Citrus (DOC) rules.

PROCESSING

Citrus juice processing requires numerous fruit handling operations, including destemming, washing, juice extraction, blending, and packaging. FCOJ must also be evaporated and frozen, and NFC must be chilled (Figures 29.2 & 29.3). Most juice is also pasteurized to avoid, reduce, or eliminate the possibility of illness caused by microbial pathogens, and to eliminate spoilage microorganisms. This is of special importance for higher pH fruit juices such as grape, apple, blueberry, and various fruit juice blends, but is of some concern in citrus processing as well. NFC juice is often stored aseptically in bulk tanks (Plate 29.4).

UNPASTEURIZED JUICE

Chilled unpasteurized citrus juice (frequently sold as fresh squeezed; Figure 29.4) had become significant in citrus sales because of its superior flavor. However, because of the potential threat to public health from unpasteurized juice, there are substantial differences between production practices required for pasteurized and unpasteurized juice that increase the cost of producing fresh squeezed juice. Fruit processed for unpasteurized juice must be of very high quality and should be produced in groves where no raw animal manure has been used as fertilizer. Every effort must be made to avoid possible contamination of the fruit with pathogenic (disease-causing) microorganisms. This means that the fruit must not contact the ground during the harvest, and the fruit must be protected from contact with insects, birds, or other animals during harvest, transport, storage, and processing.

The washing, sanitizing, and grading of fruit is essential for fresh (unpasteurized) juice production. These operations greatly reduce the number of microorganisms on the surface of the fruit, which might otherwise contaminate the juice during processing. The thorough cleaning of all incoming fruit and elimination of any defective fruit significantly reduces the risk of disease.

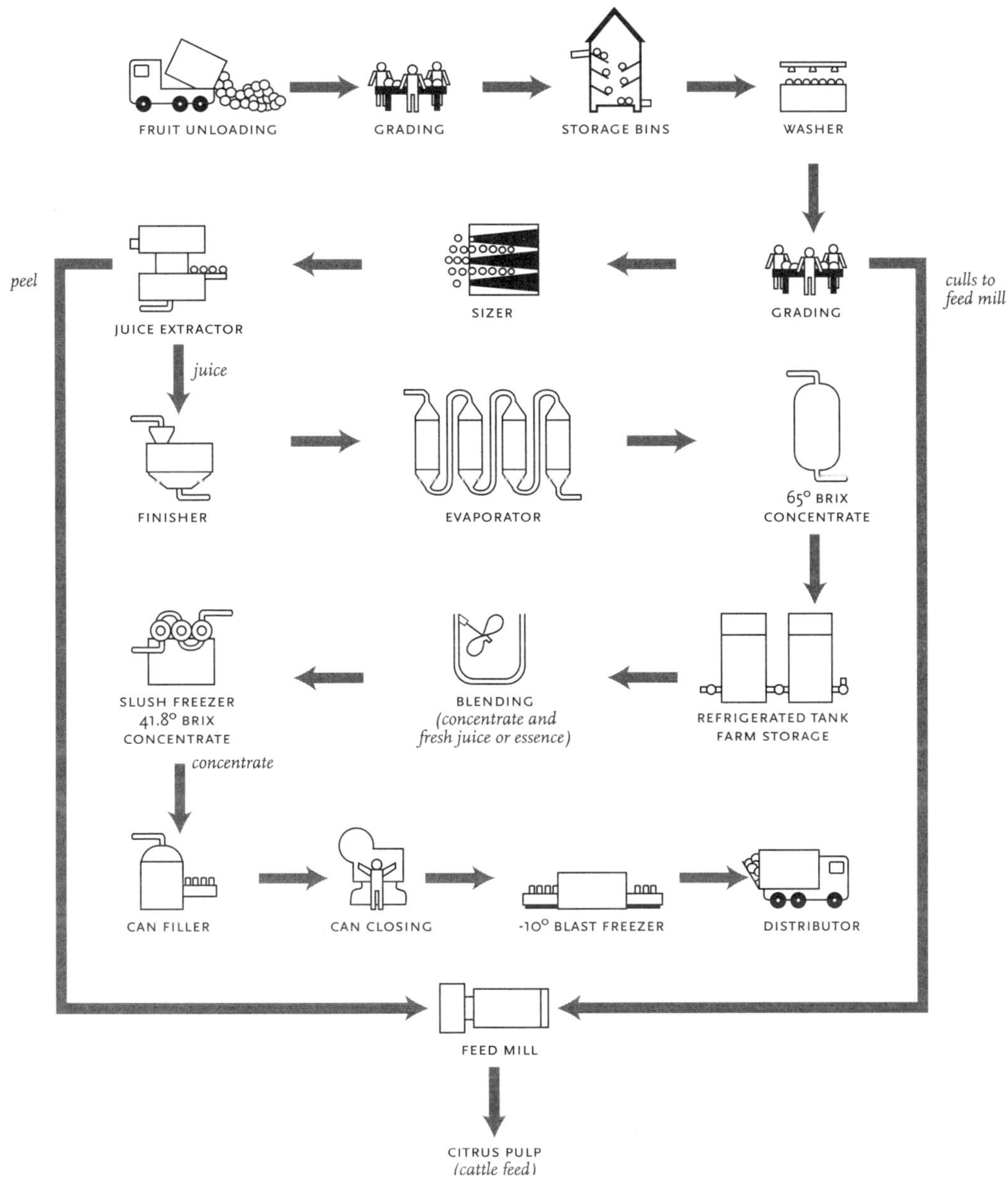

Figure 29.2 Diagrammatic steps in the production of frozen concentrated orange juice (R. F. Matthews, 1994).

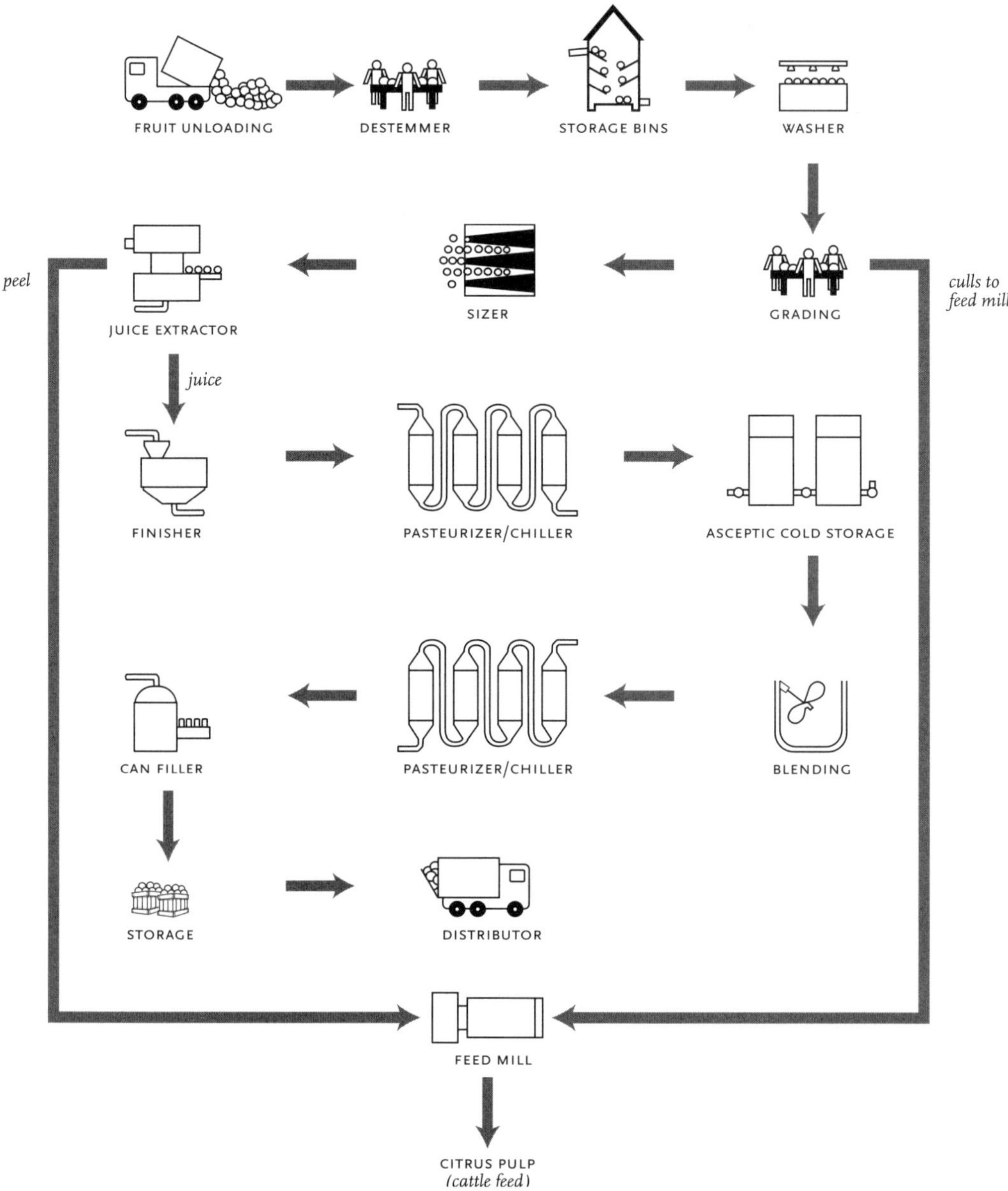

Figure 29.3 Flow chart for production of chilled not-from-concentrate juice.

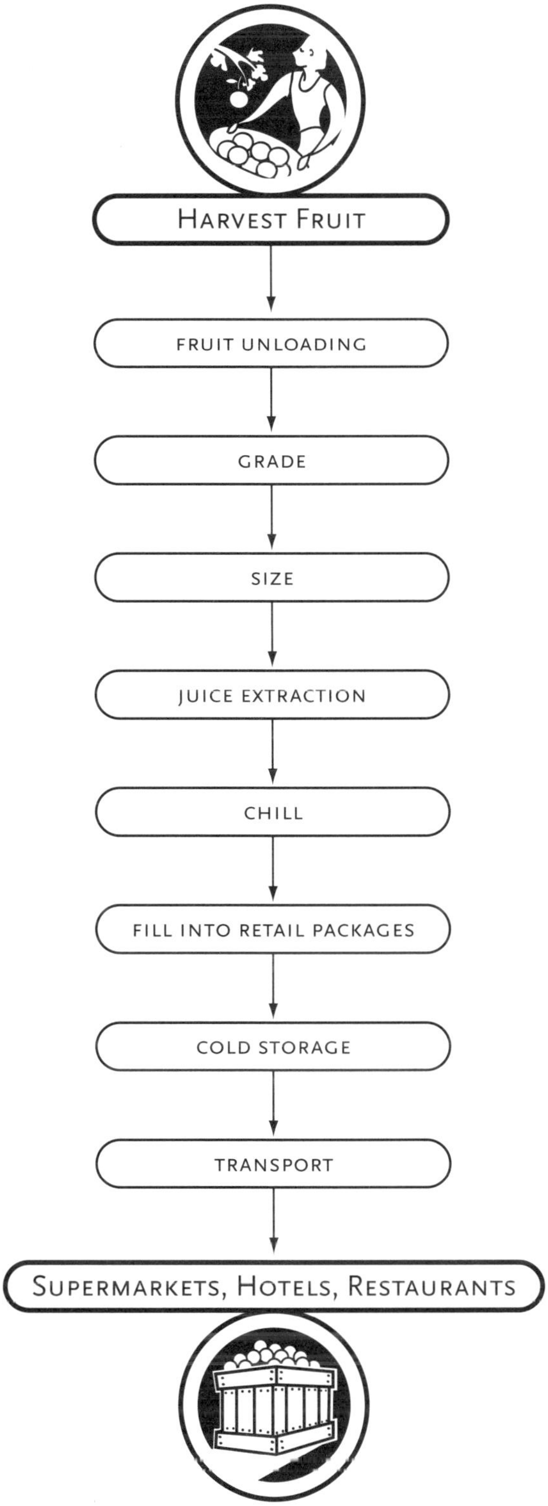

Figure 29.4 Flow chart for production of chilled unpasteurized juice.

After the extraction process, the juice is quickly chilled to temperatures below 35°F (1.7°C) to prevent the growth of microorganisms that can cause spoilage of citrus juice and extends its shelf life. The chilled juice is then packed and placed in cold storage before transport to customers.

All equipment at the processing plant must also be kept clean and sanitary, and pests must be controlled at the site. Scrupulous cleaning and sanitizing protocols are required for production of nnpasteurized Juice. The use of written standard HACCP plans and operating procedures (SOPs), sanitation SOPs (SSOPs), and good manufacturing practices (GMPs) are required. State and federal regulations require most juice processors to develop and implement a Hazard Analysis and Critical Control Points (HACCP) plan to reduce prevent or eliminate the risk of disease.

Salmonella organisms have been found to contaminate various fruit juices, including unpasteurized citrus juices. The U.S. Food and Drug Administration requires a "NOT PASTEURIZED" warning statement in labels on unpasteurized fruit juices that have not received a treatment to reduce the presence of possible pathogens such as Salmonella. Additionally, the U.S. FDA requires citrus juices that are not produced under a HACCP program to have labeling warning of the possible presence of harmful bacteria. These initiatives are limiting the market growth of fresh, unpasteurized juice.

CITRUS PROCESSING BY-PRODUCTS

Various by-products of the juice processing operation may be sold, and the grower may share in the profit from these sales. Because so many citrus by-products are used for human food, either directly or indirectly (as livestock food), it greatly complicates the registration of citrus pesticides. These by-products include citrus peel, molasses, alcohol, washed and/or juicy pulp cells, washed pulp solids, and essential oils extracted from the peel of oranges and grapefruit, such as oil and aqueous phase essences, cold-pressed oils, and d-Limonene oil. Some processors make and sell virtually all of these by-products, others none at all.

Citrus peel, the largest by-product by volume, is often made into animal feed, largely in the form of dried feed pellets. Waste heat evaporators are widely used in citrus feed mills because of their efficiency and low operating costs. They serve as scrubbers for the exhaust vapor flow, thus also aiding pollution control.

Citrus oil phase and aqueous phase essences are obtained from essence recovery trap systems. The oil level helps determine the ratio of essence oil to aroma (aqueous phase) collected, and also affects the composition of these two fractions. The ratio of aqueous to oil phase may vary. A typical ratio for oranges is 10:1, so for every 50 U.S. gallons of aqueous phase collected, 5 U.S. gallons of oil phase essence are obtained. Grapefruit produces less of the oil phase than oranges, and a 20:1 ratio is fairly typical. These oils are used primarily as flavoring compounds, and prices vary tremendously, depending upon world market conditions. Essence oils are usually worth more than the aqueous fractions. Grapefruit oil phase essence has varied from about $5.00 to as much as $125.00 per pound. Orange oil prices have been more stable, generally ranging between about $1.00 per pound and $5.00 per pound. Grapefruit cold pressed oil prices have varied from below $1.00 a pound to $30.00 per pound. Today there are at least five installations in Florida that buy oils and essences, then fractionate and blend these into lines of flavor systems that are sold to processors or other food processing and flavor-using industries.

The incentive to recover d-Limonene as a by-product has increased in recent years because the EPA considers it a pollutant when emitted into the atmosphere. TASTE (Thermally Accelerated Short Time Evaporators) have been equipped with an internal condenser from the vent lines to draw off d-Limonene in the tube bundles, condensing it, and separating it to prevent its discharge with the evaporator barometric water. Over the past 50 years, d-Limonene has sold for as little as $.05 a pound to as much as $1.50 per pound.

CITRUS PROCESSING WASTE PRODUCTS AND POLLUTION

Citrus processing plants, like most other plants, generate materials of little or no commercial value that need to be removed from the processing site. These materials, which can be classified as solid, liquid, or gaseous hazardous wastes, are conveyed from the plant through waste streams. Citrus oils, press liquor for molasses manufacture, and waste heat evaporator "black water" effluent can be serious contaminants of the environment. Contamination of land, lakes, or streams with these materials could cause irreparable damage. Limonene is especially problematic in the waste stream; even very small amounts significantly increase the potential for environmental damage. Various tests, including measures of Dissolved Oxygen (DO), Biological Oxygen Demand (BOD), and Chemical Oxygen Demand (COD), are used to determine the potential environmental damage associated with the waste streams.

The State of Florida Department of Environmental Protection imposes strict regulations on citrus processors through the permit process and pollution control compliance requirements. Florida citrus processors must also comply with federal regulations, especially those developed by the Environmental Protection Agency (EPA), such as the Clean Water Act; the Clean Air Act; the Toxic Substance Control Act; the Safe Drinking Water Acts, and others. For the citrus-processing industry to remain viable, processors must meet or exceed current pollution control requirements and anticipate future regulations.

REFERENCES

Braddock, R. J. and R. M. Goodrich. 2003. Processing technologies to enhance fresh flavor of citrus juice, p. 292-301, eds. K. R. Cadwallader and H. Weenen. *Freshness and Shelf Life of Foods.* Washington, D.C., American Chemical Society, Oxford University Press

Parish, M. E., R. M. Goodrich, J. A. Narciso, and L. M. Friedrich. 2001. *Microflorae on orange fruit surfaces and juice from fruit in processing facilities: Preliminary results.* Proceedings of the Florida State Horticultural Society 114: 174-176.

Roe, E. A., R. A. Gilbert, W. A. Miller, T. K. Das, and E. Lerin. 2003. *Impact of the FlipGate sampler on the Florida citrus processing industry.* International Society of Citriculture Congress. (9th 2000: Orlando, Florida), Orlando, Florida USA 2: 1247-1249.

Rouseff, R., P. Jella, and J. Lin. 2001. *Determination of seasonal and thermal processing changes in grapefruit juice aroma active components using GC-olfactometry.* Special Publication - Royal Society of Chemistry 274: 336-345.

Shaw, P. E., B. S. Buslig, K. L. Goodner, and M. G. Moshonas. 1998. *Monitoring changes in citrus juice quality during processing.* Proceedings of the Florida State Horticultural Society. 111: 280-282.

Taggart, T. A., J. H. Keithly, and S. R. Rizer. 2003. *Early maturing 'Valencia' sweet orange cultivars offer improved juice quality characteristics earlier in the harvesting season.* International Society of Citriculture Congress (9th 2000: Orlando, Florida), Orlando, Florida USA 2: 1255-1257.

30

Energy Utilization

W.M. Miller

INTRODUCTION

The Florida citrus industry is highly energy dependent. Energy efficiency and conservation have been relatively low priorities in the United States, including the Florida citrus industry, but energy efficient operation also reduces the impacts of short-term energy crises on the industry and prepares the Florida citrus industry for the ultimate, inevitable rise in energy costs as traditional energy resources decline. Many ICM practices that are already widely implemented in the Florida citrus industry have produced energy savings as a secondary benefit. This chapter presents the available published information about how energy is utilized and managed in the Florida citrus industry.

The Florida citrus industry is highly energy dependent. Fossil fuels and electricity are used for the operation of field vehicles and stationary operations such as irrigation pumping, light, and power in the grove facilities. Consumable materials, especially fertilizer, also represent a significant energy cost. In addition, production of capital goods, such as tractors and production of agrichemicals, requires substantial energy use. Further energy is needed for the transportation of the raw fruit and the operation of packinghouses and processing plants. Finally, energy is required for refrigerated storage and final delivery to market.

In 1994, Florida orange production consumed 30.3 percent of the direct energy and 18.1 percent of the total primary energy used for all Florida agriculture. Over 11.7 trillion British Thermal Units (Btu) of direct energy were required, primarily in the form of diesel fuel for farm equipment and irrigation systems. Production of fertilizer, especially nitrogen, requires significant energy and has been estimated at approximately 30,000 Btu per pound of ammonia produced. Annual energy requirements for grapefruit and specialty cultivars are comparable to that for oranges. Since cultivars other than oranges constitute approximately 25 percent of Florida's citrus production, the total direct energy requirement for citrus production can be estimated at more than 14.5 trillion Btu (Figure 30.1).

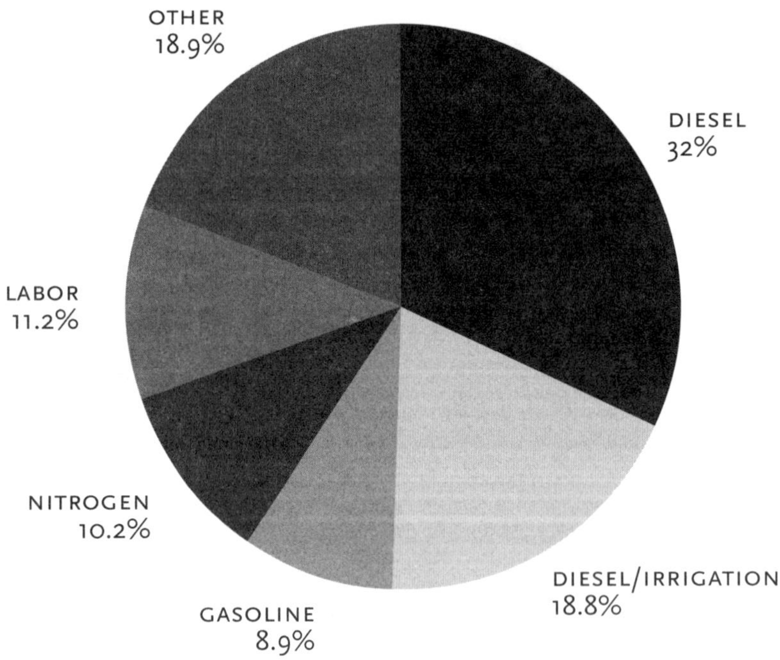

Figure 30.1 Energy utilized for fresh citrus production.

A 1982 study indicated that the energy needed to transport fresh citrus to the packinghouse, the packing operations, and the transport of the packed fruit to market each was greater than that for the field production for fresh fruit. Truck transport to a market located 1,600 km from the packinghouse required the greatest direct energy use at 735 Btu per pound of citrus. In general, combined postharvest operations for fresh citrus require 150 percent to 200 percent of the energy required for field production (Figure 30.2).

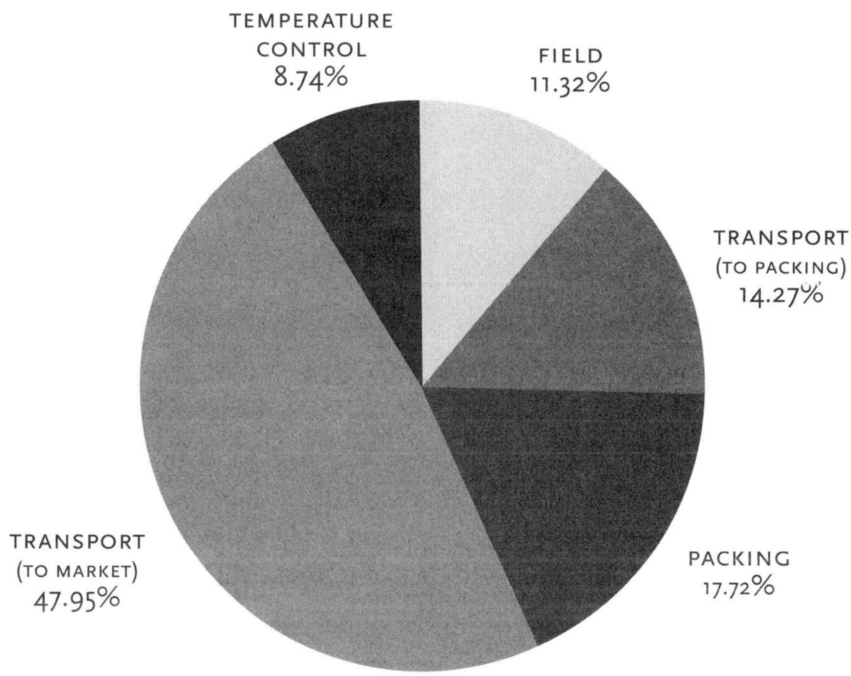

Figure 30.2 Citrus on-farm energy input components (from Fluck, 1994)

In 1980, it was estimated that 23 trillion Btu of energy were used directly at Florida citrus processing plants for juice evaporation and pulp drying in feed mills. A survey of various plants indicated four categories of direct energy use. More than 50 percent of the energy was used in feed mills (Figure 30.3).

Energy costs in the United States are generally low, and the depletion of energy resources is not immediately apparent. Thus, energy efficiency and conservation have been relatively low priorities in the United States, including the Florida citrus industry. However, energy management is an important element of ICM. As with virtually every component of ICM, the benefits of sound energy management are multifaceted.

Energy conservation reduces production costs throughout the industry and reduces industry-related pollution, especially air pollution. Energy-efficient operation also reduces the impacts of short-term energy crises on the industry and prepares the Florida citrus industry for the ultimate, inevitable rise in energy costs as traditional energy resources decline.

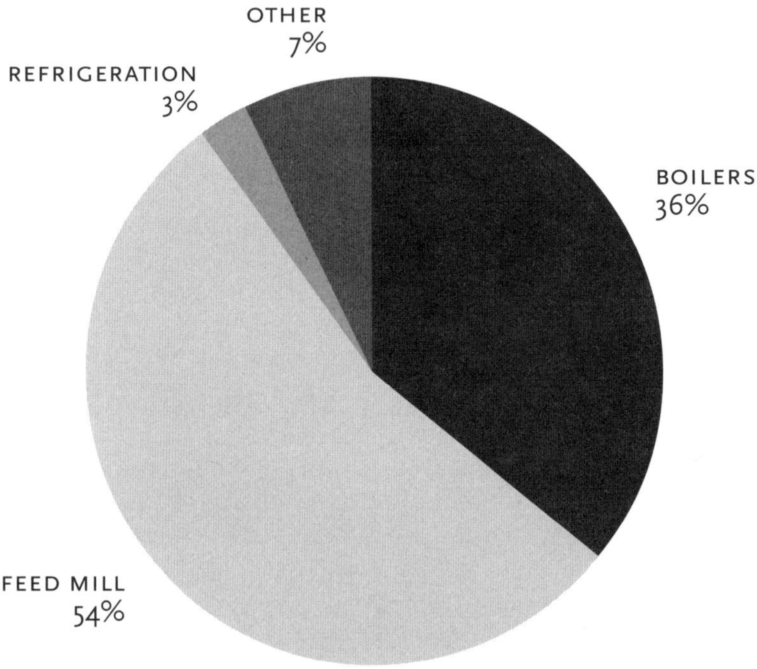

Figure 30.3 Energy utilized in citrus processing.

Many ICM practices that are already widely implemented in the Florida citrus industry have produced energy savings as a secondary benefit. For example, low-volume irrigation and the use of irrigation, rather than grove heaters, for freeze protection reduce energy use. Integrated pest management reduces both pesticide use and the energy required for pesticide application. The use of high-analysis fertilizers reduces weight and, therefore, the energy costs associated with hauling and spreading.

Some specific energy-saving measures have also been implemented, especially in the processing plants and packing houses. For example, insulation has been upgraded in many refrigerated and frozen storage facilities, as well as in feed mill dryers and juice evaporators. Since the 1980 energy-use survey, many processing plants have implemented energy audit and conservation programs, especially in their feed mill operations. Cogeneration of energy is now common at processing plants in Florida. Electricity is produced on site, and the resulting low-pressure steam or hot vapor is then used in the fruit processing operation. The current rise in popularity of single-strength juice reduces the requirement for boiler energy; however, the energy savings may be offset by the increased need for refrigerated storage.

As with every aspect of ICM, the first step in energy management is a careful assessment of current practices. Energy use should be scrutinized at every stage of citrus production and marketing. Efficient energy use offers immediate benefit to the citrus grower and is vital to the viability of the industry.

REFERENCES

Benson, J. 1994. *Energy efficiency opportunities for citrus processing*. Transactions of the Citrus Engineering Conference 40 79-95.

Carter, R. D. 1989. *Heating steam control in a taste evaporator using a microprocessor controller*. International Citrus Congress (6th 1988: Tel Aviv, Israel), Tel Aviv, Israel 4: 1717-1723.

Clark, C. W. 1982. *Energy recovery by production of fuel from citrus wastes*. Food Technology 36: 253-255.

Fluck, R. C., C. D. Baird, and B. S. Panesar. 1992. *Energy requirements for Florida citrus production*. Proceedings of the Florida State Horticultural Society 105: 84-87.

Leo, M. A. 1982. *Energy conservation in citrus processing*. Food Technology 36: 231-233.

Matthews, R. F., and C. S. Chen. 1981. *Energy conservation in citrus processing-freezing orange concentrate*. University of Florida, IFAS, Cooperative Extension Service, Circular 497 8 p.

Miller, W. M. 1985. *Prediction of energy requirements and drying times for surface drying fresh produce*. Applied Engineering in Agriculture 1: 87-90.

Miller, W. M. 1985. *Evaluation of solar demonstration projects for florida citrus packinghouse*. Proceedings of the Florida State Horticultural Society 98: 191-194.

Miller, W. M., W. F. Wardowski, and S. Nagy. 1982. *Trends in energy use in Florida citrus packinghouses.* Food Technology 36: 227-230.

Salyani, M., and M. Farooq. 2003. *Sprayer air energy demand for satisfactory spray coverage in citrus applications.* Proceedings of the Florida State Horticultural Society 116: 298-304.

31

Labor Requirements

F.M. Roka

INTRODUCTION

Florida citrus production is labor intensive and depends on large numbers of workers to plant, grow, harvest, package, and process citrus products for the final consumer. Peak operation of processing plants also occurs during the harvest season. Citrus harvesters are usually seasonal farm workers who are hired on a day-to-day basis as needed, and citrus harvesters typically migrate into a region when the harvest season begins (October) and depart the region when the last Valencia oranges are harvested (May). Most harvesters are not employed directly by the grower, but instead work through labor contractors who are better able to provide seasonal farm workers with steady employment. Many Florida citrus harvesters travel during the summer months to North and South Carolina, New York, and Michigan where they harvest peaches, apples, tobacco, and vegetable crops. The Florida citrus industry depends on imported foreign labor, particularly for the harvest. In an effort to reduce the cost of production and secure a more stable work force, the citrus industry is actively working to develop and adopt mechanical harvesting systems for processing fruit. However, labor shortages will likely continue despite the adoption of mechanical harvesters, unless agricultural employers develop strategies to attract and retain skilled and reliable seasonal workers.

Florida citrus production is labor intensive and depends on large numbers of workers to plant, grow, harvest, package, and process citrus products for the final consumer. Effective training and management of this workforce is essential to the continued success of the industry. These workers may be seasonal (employed primarily during harvest) or permanent year-round employees. The distinction is significant; seasonal farm workers have limited employment and generally do not receive employment benefits. Other than worker's compensation insurance, which covers job-related injuries, seasonal farm workers generally do not receive any health benefits, nor do seasonal farm workers usually receive paid vacation or sick days. In addition, their income is less predictable and may be dramatically reduced by factors beyond their control or influence.

PLANTING

Young trees are planted by hand to establish a new grove. Labor accounts for more than one-fourth of planting costs. It is always the grower's desire and intention to plant a grove only once; however, widespread disease, severe freezes, or changing market conditions periodically require the grower to replant sizeable blocks of acreage. Even under normal conditions, 2 percent to 5 percent of the trees within a grove die, and individual trees are reset annually. The task of tree planting is generally contracted to companies that specialize in this service. Since citrus trees can be planted year round, tree planting crews tend to be staffed by permanent workers.

GROVE MAINTENANCE

Grove care activities occur year round, and workers engaged in these activities are usually permanent employees, either of the grove or of companies that specialize in specific grove maintenance services. Much of the work of grove maintenance is performed with the aid of tractors and other power equipment. As a general rule, one full-time employee is required to work on 150 acres or more of grove. Generally, labor accounts for 10 percent to 15 percent of total grove care costs. However, worker's compensation insurance, training, worker safety, and other government mandated programs are significant additional costs in sustaining a labor workforce.

FRUIT PACKING AND PROCESSING

Packinghouses and juice processing plants prepare the harvested fruit for the consumer, either fresh or as juice. Packinghouses operate concurrently with the harvest season, typically from Labor Day to Memorial Day. As many as 90 percent of the workers in a packinghouse are seasonal workers and are employed as fruit graders, packers, carton makers, pallet stackers, forklift operators, etc. Peak operation of processing plants also occurs during the harvest season. Only about 25 percent to 30 percent of the workers in a processing plant are seasonal employees, meeting the peak processing demand with the aid of automated equipment. The remaining workers are permanent employees, occupied during the summer in blending and packaging the bulk juice.

HARVESTING

Most citrus farm workers are hired as harvesters. Citrus growers may hire and manage their own harvesting crews, but many choose to employ their harvest labor through an independent contractor. It may be difficult for the grower to obtain workers as needed solely for the harvest period. For example, a grower who needs to harvest a forty-acre block of oranges averaging 500 boxes (ninety pounds per box) per acre may require the labor of approximately fifty-five people for one forty-hour week. Since labor contractors service many groves, they are better able to provide seasonal farm workers with steady employment. A labor contractor also supervises the workers and may assume responsibility for some of the accounting and record keeping required for seasonal workers. However, the grower should be aware that the U.S. Department of Labor might judge that the grower is a joint employer with the contractor and shares legal responsibility for their actions.

The technique for collecting fruit has changed very little since Spanish explorers planted the first citrus tree in Florida more than 300 years ago. Harvesters systematically pick individual mature fruits from entire trees into a shoulder strap bag, harvesting while standing on the ground, when possible, or climbing ladders as high as eighteen feet. The equipment used to remove harvested fruit from the grove has evolved from horse drawn wagons to specially designed trucks, or "goats."

Recent Florida labor studies indicate that harvesters can now pick an average of one tub (eight to ten 90-pound field boxes) of oranges for processing per hour. Harvest productivity has increased by more than 20 percent in the past twenty years. The amount that a harvester can pick in an hour or a day depends in large part on his or her strength and agility. There are, however, a number of factors beyond the control of the harvester that affect productivity. For example, a harvester must work more slowly when picking for the fresh market to avoid damaging the fruit's external appearance, and to select for specified size grades. Harvest performance is also related to tree yields; the harvester can pick more rapidly in a high-yield grove. Hedging and topping practices also affect harvest productivity because the harvester must work more slowly when picking from a ladder.

A harvest contractor negotiates a rate, called a "piece rate," or picking price, for each field tub or pallet bin harvested. Piece rates vary according to grove conditions and the availability of harvesters. Between 1989 and 1999, the piece rates for ninety-pound box equivalents ranged from $0.55 to $1.30 per box. A harvester's effective hourly wage depends on both the contracted piece rate and his or her productivity. Federal labor laws require that any worker be paid at least the minimum wage. Thus, a harvester is entitled to at least an equivalent of that wage for every hour they are in a grove. The minimum wage establishes a minimum for hourly and daily earnings, but harvesters have the opportunity to earn more than the minimum if conditions are favorable and the harvester picks quickly. For a recent six years, hourly wages of harvesters averaged $7 and, under superior conditions, a productive harvester earned more than $10 per hour. Most harvesters expect to earn between $50 and $70 for an eight- to ten-hour workday. Citrus harvesters are hired on a day-to-day basis as needed. Inclement weather, a short harvest season, or a surplus of labor can limit the work opportunities for a harvester, and thus reduce income.

THE LABOR FORCE

The social demographic features of seasonal farm workers have changed over time. During the 1970s, almost 60 percent of the citrus harvesters were African-American and only 10 percent of the workers were of Mexican ethnic origin. Nearly 95 percent of citrus harvesters were U.S. citizens and 60 percent lived with families. During the 1980s and 1990s, increasing numbers of foreign-born workers were employed as harvesters. Today, more than 70 percent of citrus harvesters are recent immigrants from Mexico. Most harvesters are not U.S. citizens, and the majority of the workers are single men working in Florida only during the harvest season. Citrus harvesters typically migrate into a region when the harvest season begins in October and depart the region when the last Valencia oranges are harvested in May. Many Florida citrus harvesters travel during the summer months to North and South Carolina, New York, and Michigan, where they harvest peaches, apples, tobacco, and vegetable crops.

LABOR MANAGEMENT

Recruiting and retaining workers in the citrus industry, particularly harvest workers, has become an increasingly difficult challenge for agricultural employers. A booming economy has lured many workers away from the farm to jobs in construction, manufacturing, and the service industry, where hours are often shorter and employment is available on a year-round basis. Growers throughout the state are working to improve employee satisfaction and motivation in hopes of increasing worker retention.

At the very least, agricultural employers must fulfill their responsibilities as outlined in the Migrant and Seasonal Worker Protection Act and related statutes. Growers must verify that labor contractors are properly licensed by the state of Florida and the U.S. Department of Labor and that they have authorization to provide transportation and/or housing if they do so. Growers need to ensure that field sanitation equipment is adequate and acceptable. The grower should also regularly review payroll records to ensure that workers are properly paid. Many new laws and regulations require that workers receive training in the safe operation of equipment and in practices to mitigate their exposure to pesticides and other potentially hazardous materials. Employers who implement effective safety training programs may receive an insurance premium discount (Plate 31.1).

Modern personnel management focuses on effective communication and motivation. Most employers, including citrus growers, find their employees respond to positive reinforcement and recognition. Growers need to emphasize the value of each worker's contribution and demonstrate their concern for the workers' job satisfaction. Workers should be commended, and tangibly rewarded whenever possible, for consistent satisfactory performance. Disciplinary measures may be required at times for serious or recurring infractions. These measures should be clearly communicated to the workers in advance. The severity of the penalty should relate to the severity of the violation, and penalties must be applied promptly and consistently. Many progressive employers have developed some system of employee involvement or team building to increase the workers' loyalty and commitment. These systems work in conjunction with traditional tools of personnel management such as periodic performance appraisals and individual and group incentive pay programs.

Basic employee benefits, such as medical insurance and pension plans, offer a powerful incentive to workers, especially seasonal workers who are not generally awarded such benefits. Although these benefits may be costly for the grower to provide, they help attract and retain the best workers.

Some growers have provided housing on the farm for their permanent employees but not for their seasonal employees. A few Florida growers have recently begun to offer housing to seasonal workers in hopes of securing a more reliable seasonal work force. The experience of these growers suggests that the investment in housing may be readily recovered in reduced employee turnover and higher quality work performance (Plate 31.2).

LABOR SHORTAGES

Many Florida farm workers are not U.S. citizens, and many workers presently in Florida do not possess legal immigrant documentation. This gives rise to a real concern about the long-term availability of sufficient labor and the impact of a labor shortage on the industry.

A foreign guest worker program offers agricultural employers a means of legally utilizing imported labor. Unfortunately, the current guest worker program, known as H2A, is bureaucratically cumbersome, and very few citrus growers have utilized the program. Amendments to the H2A program, which have strong grower support, have been proposed in the U.S. Congress. However, the political future of any guest worker program remains in doubt, largely due to the avid opposition of farm worker advocacy organizations. These groups maintain that the supply of domestic labor is adequate and that perceived labor shortages are the result of unacceptably low wages.

The citrus industry is actively working to develop and adopt mechanical harvesting systems for processing fruit. Several prototypes have been developed and are currently under evaluation, and it is certain that mechanical harvesting will become the norm for processing fruit in the near future. This will drastically reduce the need for seasonal labor and may alleviate the projected labor shortage. However, the need for seasonal workers will not be completely eliminated. The crews needed to operate the mechanical harvesters are likely to be seasonal employees. More importantly, fruit harvested for the fresh market will continue to be picked by hand for the foreseeable future. Thus, labor shortages will likely continue despite the adoption of mechanical harvesters, unless agricultural employers develop strategies to attract and retain skilled and reliable seasonal workers.

REFERENCES

Emerson, R. D., N. Chunkasut, R. Chung, and L. C. Polopolus. 1995. *Prevailing wage summary results Florida oranges and specialty fruits 1994 season.* University of Florida, IFAS, Cooperative Extension Service, Gainesville. Staff Paper 34 pp.

Emerson, R. D., and L. C. Polopolus. 1992. *IRCA and agriculture in citrus.* University of Florida, IFAS, Cooperative Extension Service, Gainesville. Staff Paper 17 pp.

Muraro, R. P., and W. F. Wardowski. 1992. *Estimated average picking, roadsiding and hauling charges for Florida citrus, 1990-91 season.* University of Florida, IFAS, Cooperative Extension Service, Food and Resource Economics Extension Notes EN-28.

Polopolus, L. C., R. P. Muraro, W. F. Wardowski, and S. Moon. 1985. *The effects of the December-January 1983-84 freezes upon the citrus industry in eleven Florida counties: responses from packinghouses, processing plants, property tax assessors and lending institutions.* University of Florida, IFAS, Cooperative Extension Service. Staff Paper 267, 31 pp.

Polopolus, L. C., and R. D. Emerson. 1991. *Immigration and employment policy impacts: The Florida citrus case.* Choices 1991: 16-18.

Roka, F. M., G. K. Brown, and R. P. Muraro. 2003. *Measuring performance of citrus harvesting systems.* International Society of Citriculture Congress. (9th 2000: Orlando, Florida), Orlando, Florida USA 1: 280-284.

Roka, F. M., and S. Longworth. 2001. *Labor requirements in Florida citrus.* University of Florida, IFAS, Cooperative Extension Services. Food and Resource Economics FE 304. 4 pp.

32

Marketing

C.E. Townsend

INTRODUCTION

Citrus marketing refers to the entire process by which the fresh fruit or juice product ultimately reaches a consumer. A simplistic view is that the grower sells the fruit to a packer or processor, who then sells the packed or processed fruit to a retailer (either directly or through a broker), who then sells it to the consumer. The Florida citrus grower has a wide variety of marketing options for both fresh and processed fruit. There are several levels of marketing that occur. This chapter will discuss the many marketing options in detail. Understanding these market options will help citrus managers improve their bottom line.

Citrus marketing (Figure 32.1) refers to the entire process by which the fresh fruit or juice product ultimately reaches a consumer. In its simplest form, this process may consist of a single "grove stand" transaction between the grower and the consumer. Marketing of Florida citrus can begin with scenic views of Florida promoting Florida citrus (Plate 32.1). Typically, a grower sells fruit to a packer or processor, who then sells the packed or processed fruit to a retailer (either directly or through a broker), who then sells to the consumer.

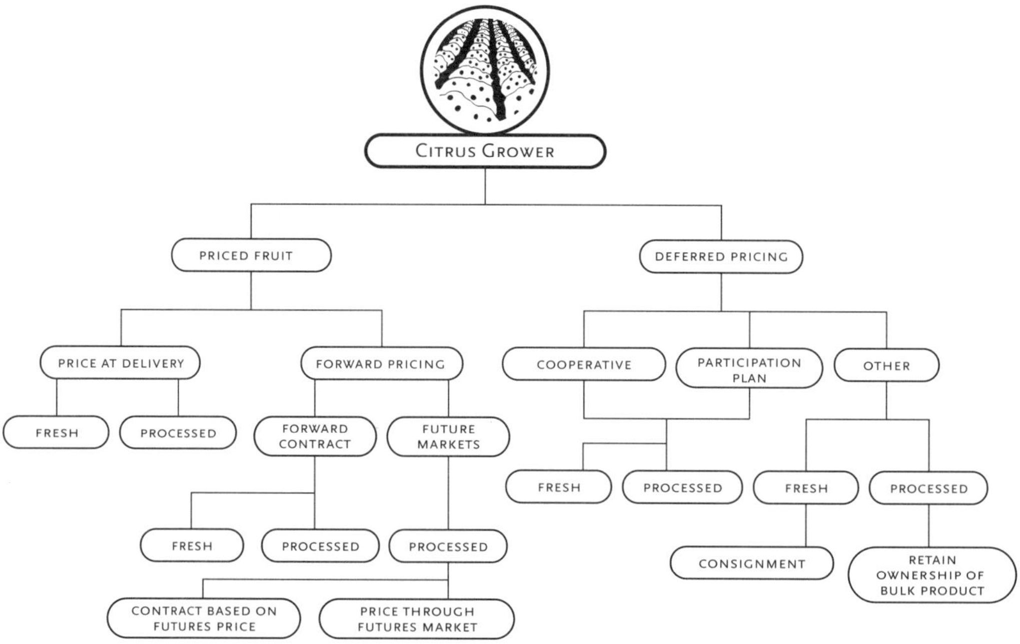

Figure 32.1 Marketing alternatives for Florida citrus growers.

SALE OF FRUIT

The Florida citrus grower has a wide variety of marketing options for both fresh and processed fruit. The grower may retain ownership of the fruit through the packing or processing or may sell the fruit to the packer or processor outright. In each case, the price of the fruit may be computed in one of several ways. The grower must carefully consider the alternatives and balance risk and potential gain.

PRICED FRUIT

The grower may choose to sell the fruit to a packer or processor and relinquish all interest. This is referred to as a "cash sale" for processed fruit, and the grower generally pays for the harvest and transport of the fruit to the processing plant.

For fresh fruit, this is called an "on-tree sale," and the packer or processor typically pays for the harvest and haul. The fruit may be "priced at delivery" ("spot market" priced) or a "forward contract" may be signed, fixing the price in advance. In either case, the fruit may be sold to a processor or packer, a gift fruit or roadside-market vendor, or an intermediate handler ("bird dog"). Fresh fruit is generally purchased by the box (ninety pounds). Processing fruit is generally purchased as pounds of fruit solids, but may be purchased by the box, as gallons of juice, or in "bulk" (total grove production). The grower should consider which method of pricing is most profitable.

If the grower chooses to sell in the spot market, the fruit is not committed to any buyer until harvest. The buyer, based on immediate supply and demand at the time of the harvest, largely determines the price of the fruit. The grower expects to be paid promptly, usually within two weeks. The grower may be able to take advantage of a short-term scarcity of fruit and receive a premium price. However, the grower also accepts the risk of a price drop due to a fruit surplus. Fruit surplus is of special concern in areas where freezing conditions are likely and threatening weather may force many growers to harvest at the same time.

Alternatively, the grower may negotiate a "forward price." The grower agrees in advance to provide fruit to a particular buyer at a specified price, which may or may not include the cost of the harvest and transport. The contract may specify some number of boxes to be delivered or the entire production of the grove and may be for multiple years. In the latter case, the fruit maybe purchased on-tree for fresh use and the packer incurs the cost of picking and hauling. The buyer may provide a cash binder or an advance at the time of contract, leaving the remainder of the purchase price to be paid on delivery. To the grower's advantage, an outlet for the fruit is guaranteed. Of course, the grower may find that the spot price at the time of harvest is higher than the contracted forward price. In some cases, a minimum price may be negotiated with provisions for an increase if the spot market price exceeds some specified amount, or the price may be determined by the futures price at the time of harvest.

Pricing fruit by hedging in the Florida Concentrated Orange Juice (FCOJ) futures market is another forward pricing option. Hedging is limited to oranges grown for processing and can be used most effectively by the grower with uncommitted fruit. Hedging, in its simplest terms, is the establishment of a futures market position opposite to one made in the product market. The orange grower is "long" in the product market since he anticipates harvesting fruit, the value of which is subject to change as prices fluctuate during the season. The grower establishes a price for his product by taking a "short" position on selling contracts in the futures market, promising to make delivery.

Forward pricing through the futures market differs from forward contracting in that hedging implies no original intent to deliver (or accept delivery) on the futures contract. This flexibility is one advantage hedging has over forward contracting, where actual delivery must be made. Hedging allows the trader to control the timing of sales or purchases while affording the flexibility to close out a futures position by offsetting futures trade. Since futures contracts are bought and sold on margin, additional capital is required.

Hedging works because prices in both the cash and the futures markets have a strong tendency to move up and down together. Hedging is the act of taking opposite positions in two related markets. When prices move in one direction, the cash revenue position may sustain a gain or loss, but the futures revenue position will be just the opposite with the result that nothing has happened to the net revenue position, given that no product is lost.

Placing a successful hedge also requires knowledge about the "basis," which is the difference between prices on the cash market and the futures market. The basis may vary due to seasonal trends.

DEFERRED FRUIT PRICING

The grower may join with other growers in a "cooperative" association. In this case, the grower's fruit is pooled with that of the other growers, and the price of the fruit is the average selling price, less the expenses of the cooperative. The grower is guaranteed an outlet for the fruit, but the selling price is not known in advance, and the grower may not receive payment until the pool is "closed" (that is, every grower has harvested, and all of the fruit has been sold). The cooperative may only sell fruit or may be involved in the packing or processing of the fruit as well.

Another common method of deferred pricing is a "participation plan." Participation plans are contractual agreements between citrus growers and handlers in which the grower's price for the fruit depends on the price received by the packer or processor for the final product. Again, the selling price is not known in advance, and the grower does not receive payment until the final sale of the fruit or juice product. Typically, the grower receives a share of the proceeds, less the cost of the processing and selling. Thus, the grower relies on the packer or processor to successfully market his product and bears a substantial share of the risk. Some grower cooperatives pack or process their own fruit and may offer participation plans, but most participation agreements are negotiated with private corporations. Participation plans may be production contracts or limit contracts. They differ from cooperatives in that there are no "retains," which are profits reinvested in the cooperative rather than distributed to grower participants.

SALE OF PACKED OR PROCESSED FRUIT

Although a significant amount of citrus fruit and juice is purchased at fruit stands or roadside markets, the majority is sold to consumers through extensive regional or national chains of retail grocery stores. Retailers study consumer response to determine which products to sell and how to most effectively market those products. Consolidation of the nation's supermarket industry has resulted in downward price pressure and a loss of bargaining position for most fruit packers and processors and, consequently, for most fruit growers. It has also worked to eliminate the market for many cultivars deemed unsuitable in some way for mass retailing. Producers are seeking to regain some market influence through cooperative or collective operations and by controlling the production of new or specialty cultivars. Some growers are conducting their own market research to identify potentially lucrative niche markets for both fresh fruit and juice products. Figures 32.2 and 32.3 indicate the retail cost components per carton of red grapefruit and for single strength equivalent (SSE) gallon of FCOJ. These prices reflect those for one year and will vary with local and global production, effectiveness of marketing programs, and consumer demand for the particular fresh fruit and processed products.

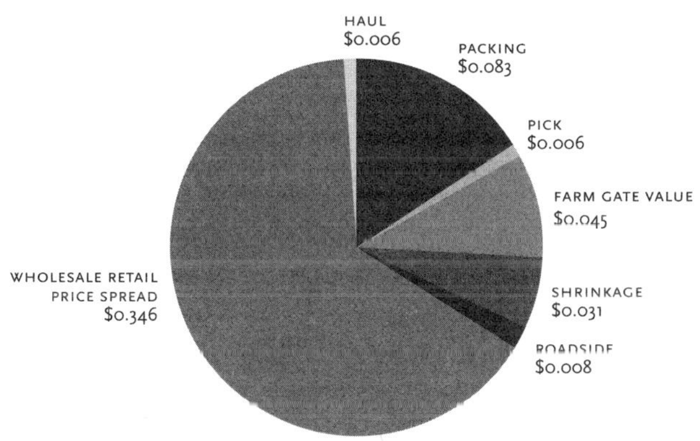

Figure 32.2 Retail FCOJ price breakdown for the 1996-1997 season ($2.98 per single strength gallon average).

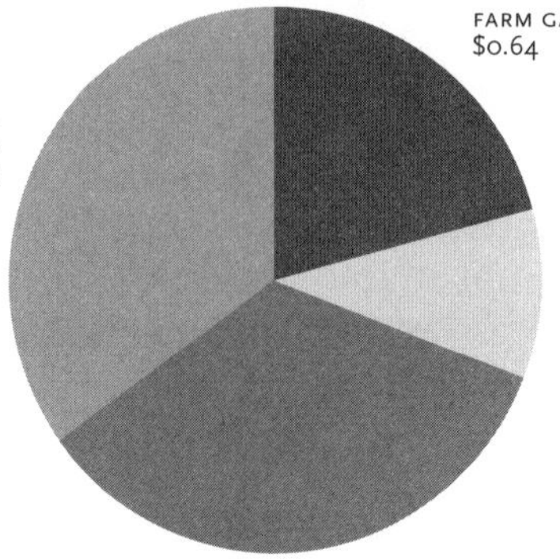

Figure 32.3 Retail fresh red grapefruit price breakdown for the 1996-1997 season ($.525/lb average).

PROCESSED JUICE PRODUCTS

Most processed citrus products are sold to retailers through food brokers. The broker or brokerage company is always the exclusive representative of a particular branded product, but probably represents many different brands and product lines. Brokers work directly with retailers in a variety of ways. The broker discusses marketing and advertising strategies with the retailer, accepts orders from the retailer and arranges delivery, and a representative may make regular store visits to ensure that the product is well stocked and displayed to its best advantage. Most large processors have regional representatives who work directly with their brokers and make sales calls to their most important retail customers. Field personnel of the Florida Department of Citrus (DOC) also work with retailers to increase consumer interest. For example, the DOC currently supports the "Triple Crown" program, which advertises the health and nutritional value of Florida orange juice and grapefruit juice with the endorsement of the American Cancer Society, the March of Dimes, and the American Heart Association.

The juice product may be delivered directly to the retailer or, especially for large retail chains or wholesalers, it may be delivered to a distribution center, where it is stored under appropriate conditions until needed. However, since the broker works with many retailers, the broker may be able to reduce transportation costs to the retailer by combining several orders to fill a truck and making spot deliveries as needed. Concentrated juice is also sold and delivered to local and regional dairies outside of Florida, where it is reconstituted and sold under that dairy's brand name, perhaps through yet another food broker.

World orange juice consumption and trade have shown remarkable growth over the past two decades. After major freezes destroyed many orange trees in Florida in the 1980s, the high prices that followed spurred a major expansion in orange production in both Florida and São Paulo, Brazil. These two regions continue to dominate the world market for orange juice, collectively accounting for approximately 90 percent of world production.

The United States remains the largest market for orange juice in the world, although consumption in the European Union (EU) continues to grow. Canada is also a major destination for orange juice. Consumption growth in Japan has been disappointing. The increasing acceptance of not-from-concentrate orange juice has changed the face of the orange juice market. More than 40 percent of the orange juice consumed in the United States is NFC.

Despite the turmoil stimulated by the negotiation and implementation of North American Free Trade Agreement (NAFTA), there has not been a significant increase in orange juice imports from Mexico into the United States. One reason is that Florida orange juice production has experienced a strong recovery after the freezes of the 1980s. Another reason is little new investment in Mexican citrus, which has left the processing sector in Mexico well behind their counterparts in Florida and São Paulo.

The U.S. orange juice tariff helps support the price of orange juice in the United States and positively affects grower returns. Because the total cost of producing orange juice in São Paulo is lower than the cost in Florida, the tariff protects Florida growers from imports from Brazil. The evolution of NFC consumption in the United States means that Florida has been able to develop a differentiated product, which is costly to ship from Brazil.

If the U.S. tariff on orange juice were to be removed, not all of the tariff savings would accrue to Brazilian exporters. Brazil has a large market in Europe, which, in recent years, has grown to nearly one billion SSE gallons. The tariff imposed by the EU on orange juice imports is 15.2 percent ad valorem. At current prices, the U.S. tariff is comparable to an ad valorem tariff of 40 percent to 50 percent. If the U.S. tariff were eliminated, Brazil would shift exports from Europe to the United States, which would cause the U.S. price to decline and prices in Europe to rise. This process would continue until a new price equilibrium was established. The impact on Florida would be muted because it would be more costly for Brazil to supply the NFC market. It is clear, however, that the U.S. orange juice tariff supports the price of orange juice in the

United States, and returns to Florida growers would decline if the tariff were eliminated.

The continued growth of orange juice consumption around the world is a positive factor for growers in Florida and São Paulo. Rising consumer income in Latin America, China, and Eastern Europe all bode well for the future of the orange juice industry.

PACKED FRESH FRUIT

Fresh fruit is also largely sold to retailers through food brokers, although some packinghouses do market directly. The marketing agency or food broker works closely with retailers to develop special events and promotions. The retailer may need to purchase fruit from several packinghouse sources to support an advertised sale. The produce will be delivered directly to the retail store or to a warehouse where it is stored briefly. Transportation costs are usually included in the contract price. Retailers generally contract for "straight" loads; that is, the trucks are loaded with a single variety of produce. However, especially if a food broker conducts the sale, it is possible for a retailer to order a mixed load. The food broker would be responsible for coordinating the purchase and delivery. To maintain quality, trucks delivering fresh produce usually make a single delivery stop.

Attractive presentation of the product is the produce manager's responsibility. Many retailers are enlarging their produce departments or relocating them adjacent to the bakery or deli (many now also include salad bars). Produce is displayed in attractive fixtures, and samples or preparation demonstrations may be offered. Although this increases retail prices, market trends indicate that a pleasant shopping experience is essential to maintaining a customer base (Plates 32.2 & 32.3).

INTERNATIONAL SALES

The economic impact of global competition on the Florida citrus industry is twofold. Florida producers must compete with citrus producers from growing regions around the world, and these producers may operate with lower cost production and may be less restricted by government regulations in their production and marketing practices. In addition, traditional spring and summer fruits are now imported and available to U.S. consumers during the fall and winter months and compete strongly with fresh citrus.

There is, however, a positive aspect to global marketing. The opening of foreign markets, especially in Asia and Europe, and the introduction of the USDA's export promotion program in the 1980s offered Florida citrus growers an enormous new outlet. Exports of fresh grapefruit quickly grew to more than 60 percent of the fresh crop. Japan is currently the largest export market for fresh Florida grapefruit; the per capita consumption of Florida grapefruit is greater in Japan than in the U.S. Orange juice and grapefruit juice exports currently account for 11 percent and 17 percent of production, respectively.

Most exported fruit is sold to an intermediary foreign import agency, although some large retail chains import directly. In Japan, the large importers (Mitsubishi and Sumitomo) and major wholesalers are responsible for virtually all imports of fresh fruit; direct import is more common in Europe. There is also a fruit auction market in Rotterdam where retailers can purchase imported fruit in the spot market and compare prices. Retailers who purchase from importers or wholesalers expect that the delivered fruit has been inspected and repacked. Thus, the importer typically assumes all the cost and risk of the shipment and storage of the fruit.

The Florida Department of Citrus maintains field offices overseas, with contracted local personnel. These contractors represent the Florida citrus industry and supervise generic advertising, merchandising, and public relations. The FDOC works to increase distribution of Florida citrus products and encourage premium pricing. They do not negotiate for citrus exporters and are not involved in sales in any way.

REFERENCES

Crawford, C., R. Tomlinson, and A. Wysocki. 2002. *Haines City Citrus Growers Association: Remaining competitive and profitable in a changing market.* University of Florida, IFAS, Cooperative Extension Services. Teaching and Learning Paper TLP 02-15. 13 pp.

Davis, K., T. Minton, and S. Thornsbury. 2001. *A first step in utilizing Florida citrus data.* University of Florida, IFAS, Cooperative Extension Services. Staff Paper Series SP 01-12. 34 pp.

Dou, H. 2002. *Citrus production in China and market potential for Florida citrus.* Fifty-third Citrus Processors' and Subtropical Technology Conference, Lake Alfred, FL. Proceedings p. 25-26.

Peterson, B. T. 2003. Global produce marketing today and tomorrow. International Society of Citriculture Congress (9th 2000: Orlando, Florida), Orlando, Florida USA 2: 1104.

Sparks, M. W., and M. A. Ismail. 2003. *Use of health and nutritional information in marketing Florida citrus products.* International Society of Citriculture Congress (9th 2000: Orlando, Florida), Orlando, Florida USA 2: 1068-1069.

Thomas, A. 2000. *Worldwide procurement and marketing of citrus fruit in England.* International Society of Citriculture Congress. (9th 2000: Orlando, Florida), Orlando, Florida USA 2: 1104.

33

Forecasting Citrus Production

Florida Agricultural Statistics Service and L.G. Albrigo

INTRODUCTION

The crop forecast plays a major role in establishing the value of the Florida citrus crop. For any agricultural crop, early information concerning production is extremely valuable in guiding production practices, pricing and marketing decisions, and in planning future crop direction. This is certainly true of citrus, where on-tree storage of fruit and fairly wide harvest windows for most cultivars give growers considerable flexibility in planning their harvest schedule (especially where harvest labor shortages are also an important consideration). The U.S. Department of Agriculture (USDA) first made forecasts of Florida citrus production in 1918. These forecasts were based on the survey opinions of crop observers and statisticians. The need for more accurate and comprehensive forecasts became evident as Florida's citrus production increased and as Florida entered world markets.

Currently, a commercial tree census is conducted in Florida every two years by the Florida Agricultural Statistics Service (FASS). The census uses aerial photographs of about 14,000 square miles of the Florida peninsula covering virtually every citrus-growing area. Photos are taken at 15,000 feet on black and white panchromatic film. Each developed photo records a three-mile swath on the ground and is developed at a scale of 1:30,000. The same flight lines are followed for each census, so year-to-year comparisons are meaningful. The first such photos were taken in late 1965 and used for the January 1966 census. The aerial information is coordinated with ground truthing to provide a complete record, by county, of the ages and total numbers of trees, and the total acreage planted to each type and variety. Researchers projecting fruit production, which helps growers in planning future plantings, use census information. In addition, the census is used in conjunction with survey data collected from a sample of groves to develop the current season forecast, which is invaluable in pricing and marketing production. The four basic parameters in the forecast are: (1) the number of bearing trees, (2) the number of fruit per tree, (3) average fruit size, and (4) the number of fruit dropped. Production indicator in boxes equals the number of bearing trees multiplied by the fruit per tree multiplied by the percent remaining at harvest divided by pieces of fruit per box.

These data are collected from a representative sample of groves and used to forecast total production (Plate 33.1). The number of bearing trees in each sample grove is determined, and the number of fruit per tree is estimated from a count of the fruit on selected limbs, called the limb count (Plate 33.2). Fruit size is measured monthly, beginning in August and continuing until harvest, on two trees in each of about 1,800 sample groves. Fruit loss is estimated from the loss on selected branches. These sample branches are tagged during the initial limb count, and their fruit is recounted monthly. Cumulative fruit loss for the season and historical data from previous seasons are used to project fruit loss until harvest time.

Another feature of the Florida citrus forecasting program is the projected yield of frozen concentrated orange juice (FCOJ), expressed in gallons of concentrate per box of oranges. This projection is especially important since the majority of Florida oranges are processed for juice.

The historical records maintained by FASS have also been correlated with climatic records to examine apparent relations and provide insight into the effect of climate on various aspects of fruit development. For example, large differences in soluble solids (pounds solids per box) from year to year have been noted in Florida. It is known that climate affects the production of soluble solids; Florida's climate generally results in better solids production than either a Mediterranean subtropical, or tropical climate. The question was raised whether year-to-year variations in Florida's climate accounted for the observed variation in soluble solids production. To answer this question, a multiple regression study was undertaken to relate pounds solids and its components (Brix and percent juice) to climatic variables.

Data for Valencia oranges were obtained from the FASS for the years 1968 through 1994 (twenty-six years). These data were evaluated in a multiple regression program against climatic data from the four (until 1984) or five (after 1984) geographic production regions of citrus in Florida as established by FASS. Weather data used in the regression study were average temperatures and accumulated rainfall in two-month increments, starting each year with the December to January period preceding bloom.

A significant correlation was found between increased pounds solids and warmer winter temperatures both before bloom and during final maturation. It appears likely that higher winter photosynthetic rates produce more sugars that are eventually stored in the fruit. Additional variables that were positively correlated with pounds solids production were higher temperatures in February and March and increased winter rainfall. Increases in rainfall during April and May corresponded with decreased pounds solids.

Further evaluation of the data indicated that both winter temperatures and pounds solids increased or decreased in cycles of three to five years. It was determined that these cycles were the effect of El Niño on the winter climate in the southeastern United States. The winter climate of Brazil may be similarly affected, preceding the winter in Florida by six months. However, Brazilian production is also influenced by periods of drought. These winter temperature effects on pounds solids can result in variations of more than one pound of solids per box. These variations could result in a difference of more than two million pounds of solids in Florida between years of high and low yield.

Fruit acidity is also observed to vary significantly from year to year. A similar regression study was conducted correlating fruit acidity levels over several years with temperatures and rainfall throughout fruit development. Higher bloom time temperatures were correlated with lower acidity at harvest in early oranges and grapefruit. Higher temperatures in June and July were also related to reduced acidity at harvest for late oranges and grapefruit. These may be found to be critical periods for acid synthesis or development of the enzyme systems that eventually produce or facilitate transport of the acids.

REFERENCES

Florida Agricultural Statistics Service. Citrus Summary 2002-2003. http://www.nass.usda.gov/fl/rtoc0ci.htm

Terry, R. R. 1998. *Citrus tree numbers—have we reached the top?* Proceedings of the Florida State Horticultural Society 111: 104-105.

34

Grove Maintenance: Renovation, Replant, and Abandonment Options

D.P.H. Tucker

INTRODUCTION

The average annual tree loss in the Florida citrus industry, excluding catastrophic weather events, is 3 percent to 4 percent. Annual percentage losses could increase as citrus canker, greening, and severe tristeza strains become more widespread. Such losses result from a number of factors, including various diseases and disorders such as CTV, blight and foot rot, and insects such as root weevils, as well as encroaching urban development. Localized drainage problems in unusually wet years can also contribute to tree mortality. Losses can vary markedly among individual groves, from 0 percent to 10 percent or more, having substantially different impacts on grove revenues and equity. In addition, the profitability of a grove may decline gradually with age, or a planted variety may lose favor in the market. If the variety has current or future market potential and specific problems have been identified in the grove, rejuvenation may be the grower's best choice. Sometimes the choices are difficult—rejuvenate or push the grove and start over. This chapter presents information regarding the factors to consider when working with declining or abandoned citrus groves.

The average annual tree loss in the Florida citrus industry, excluding catastrophic weather events, is 3 percent to 4 percent. Such losses result from a number of factors, including various diseases and disorders such as CTV, blight and foot rot, and insects such as root weevils, as well as encroaching urban development. Localized drainage problems in unusually wet years can also contribute to tree mortality. Losses can vary markedly among individual groves, from 0 percent to 10 percent or more, having substantially different impacts on grove revenues and equity. In addition, the profitability of a grove may decline gradually with age, or a planted variety may lose favor in the market.

A citrus grove represents a highly capitalized, long-term investment. As trees are lost, or returns decline for any reason, the grower must evaluate the economic feasibility of various grove maintenance options. In general, a grower must choose to do one of the following:

Maintain the grove at the current level.

Maintain the grove at a reduced level.

Renovate the grove.

Rejuvenate the grove.

Replant the grove.

Abandon the grove.

These are fundamental management decisions, and they must be carefully considered. The primary factors that influence the grower's choice are the health and productivity of the grove, the present and future marketability of the variety, and the costs associated with each of the options.

If the trees are healthy and productive and the variety can be profitably marketed, the grower will probably choose to continue normal production practices in the grove, addressing specific problems as they are identified. In this case, the grower may need to consider a program of resetting to replace lost trees. The costs of tree removal, resetting, and other expenses associated with mixed-aged groves can add 10 percent to 15 percent to the annual expenditures of bearing groves. It is important, therefore, to evaluate the economic impact of regular resetting. Grove maintenance strategies based on resetting trees every one, two, or three years have been modeled and compared. In general, if the production of healthy trees in a grove is profitable, then annual resetting appears to be profitable.

If the grove is healthy and productive but the variety is not currently profitable, the grower may choose to reduce the level of grove care to maintain tree vigor but sacrifice productivity.

In this case, the grower maintains the grove in anticipation of the future market potential of the variety and should continue to address any problems that seriously threaten the health of the trees. Again, the grower will probably need to reset lost trees periodically so the grove will be prepared to return to full productivity should the variety become more marketable.

If the grove is healthy and productive but the variety has no current or future market potential, the grower may choose to renovate the grove by topworking the trees to a marketable variety (Plate 34.1). The new variety must be compatible with the scion and the established rootstock. It is important to note that the grafting success rate may be low, and, if the graft is successful, acceptable fruit quality and yield may not be attained for several years. The newly topworked trees will also require more intensive care, especially protection from pests, cold weather, and sunburn of the exposed trunk. Some growers report success with topworking, with less cost and earlier production than expected from replanting, while others have had less favorable results.

Alternatively, if the variety has current or future market potential and specific problems have been identified in the grove, grove rejuvenation may be the grower's best choice. The trees may be pruned to remove dead wood and reduce the canopy, or trees may be removed to reduce crowding. Mature trees can also be moved (Plate 34.2). Grove drainage and irrigation may need to be improved, and a remedial nutrition program may be indicated. Soilborne pests or diseases may need to be controlled, or infected trees may need to be removed. If the grove is unhealthy or unproductive and specific problems have not been identified, the grower should seek the advice of plant specialists with the University of Florida's Institute of Food and Agricultural Sciences (IFAS) and/or private consultants.

Tree removal and replanting should be seriously considered when a grove exhibits a rapid decline in health or productivity or when profitability falls to an unacceptable level. It is essential that limiting factors such as poor drainage at the site be corrected before replanting; the survival, growth, and ultimate productivity of the new trees will be endangered or compromised otherwise. Some correction may be possible after planting (Plates 34.3 & 34.4). The site should be evaluated for soil pH, toxic metal accumulation, soilborne pests and diseases, adequate drainage, and infertile grove areas likely to remain unproductive. The site should be thoroughly cleared of root debris, and severed tree stumps should be treated immediately to prevent vigorous root sprouts among the young trees. The grower should invest in registered planting stock from a reputable nursery when establishing any new planting.

The final and most drastic option the grower may consider is economic abandonment of the grove. The grower may choose to cease all grove maintenance if it appears that any further investment is unwarranted. The economic and legal implications associated with abandonment should be carefully researched. The grower may have continued responsibility for some level of pest, disease, and weed control to protect trees on adjacent property. The grower may also be responsible for fire hazards at the abandoned grove and for disrupted water flow due to drainage and irrigation systems in disuse. Potential loss of agricultural tax status should also be considered. The grower should consult a qualified consultant regarding the advisability of retaining ownership of an abandoned grove.

REFERENCES

Castle, W. S., and J. C. Baldwin. 1996. *Interplant competition and rootstock affect the growth and yield of 'Hamlin' orange trees used for replanting*. Proceedings of the Florida State Horticultural Society 109: 115-117.

Davies, F. S., M. Tignor, and L. A. Mathers. 1996. *Irrigation and fertilization of young interset 'Hamlin' orange trees in Florida*. Proceedings of the Florida State Horticultural Society 109: 77-80.

Derrick, K. S., and L. W. Timmer. 2000. *Citrus blight and other diseases of recalcitrant etiology*. Annual Review of Phytopathology 38: 181-205.

Ferguson, J. J. 2003. *Florida citrus freeze protection methods*. International Society of Citriculture Congress. (9th 2000: Orlando, Florida), Orlando, Florida USA 1: 393-395.

Ford, S. A., and R. P. Muraro. 1990. *An economic analysis of replanting existing groves vs buying new land for citrus production*. Proceedings of the Florida State Horticultural Society 103: 44-46.

Giuffrida, A., F. Intrigliolo, and G. Reforgiato Recupero. 2003. *High density citrus plantings under replant conditions*. International Society of Citriculture Congress. (9th 2000: Orlando, Florida), Orlando, Florida USA 1: 494-499.

Muraro, R. P., and W. S. Castle. 1996. *Resetting and replanting options in Florida citrus groves and the financial consequences.* Proceedings of the Florida State Horticultural Society 109: 128-131.

Stover, E., and W. Castle. 2002. *Citrus rootstock usage, characteristics, and selection in the Florida Indian River region.* HortTechnology. 12: 143-147.

35

Crop Damage: Documenting and Recovering Economic Losses

J. S. Rogers

INTRODUCTION

Tree or crop damage may occur as the result of diseases or pests, unsuitable grove conditions, inappropriate cultivation practices, or extreme weather conditions. In addition, severe damage may result in significant economic losses, which the grower may seek to recover from the responsible party or from their insurance company. Crop damage claims are increasingly resolved through legal channels. It is important for the grower to understand how to identify the cause of the damage and how to document the damage to support a possible claim for recovery of losses. It requires careful and thorough collection and organization of all relevant information.

Tree or crop damage may occur as the result of diseases or pests, unsuitable grove conditions, inappropriate cultivation practices, or extreme weather conditions. In addition, severe damage may result in significant economic losses, which the grower may seek to recover from the responsible party or from their insurance company. Crop damage claims are increasingly resolved through legal channels. It is important for the grower to understand how to identify the cause of the damage and how to document the damage to support a possible claim for recovery of losses. This requires careful and thorough collection and organization of all relevant information (Table 35.1 & Figure 35.1).

Table 35.1 A list of suggested records to help diagnose crop diseases and disorders

Weather	Unexplained symptoms
Chemical applications	Wind (speed and directions)
Crop surveys	Active ingredient per acre
Rainfall	Crop development stage
Treated crop	Temperature (max and min)
Insect populations	Application rate (gal/acre)
Date and time	Photographic documentation
Acres treated	Storms and unusual weather
Progress of diseases	Application date and time
Relative humidity	Date and time of inspections
Chemical name	

Generally, there are three types of agricultural crop damage: (1) preharvest damage, (2) harvest damage, and (3) postharvest damage. Indications of preharvest damage are apparent in the field before the fruit is collected. Fruit may be damaged during the harvest, but this damage may not be evident until the fruit is inspected at the packinghouse or processing plant. Finally, damage may occur after the harvest, during handling, storage, transport, or in the marketplace. The grower is concerned primarily with preharvest crop damage.

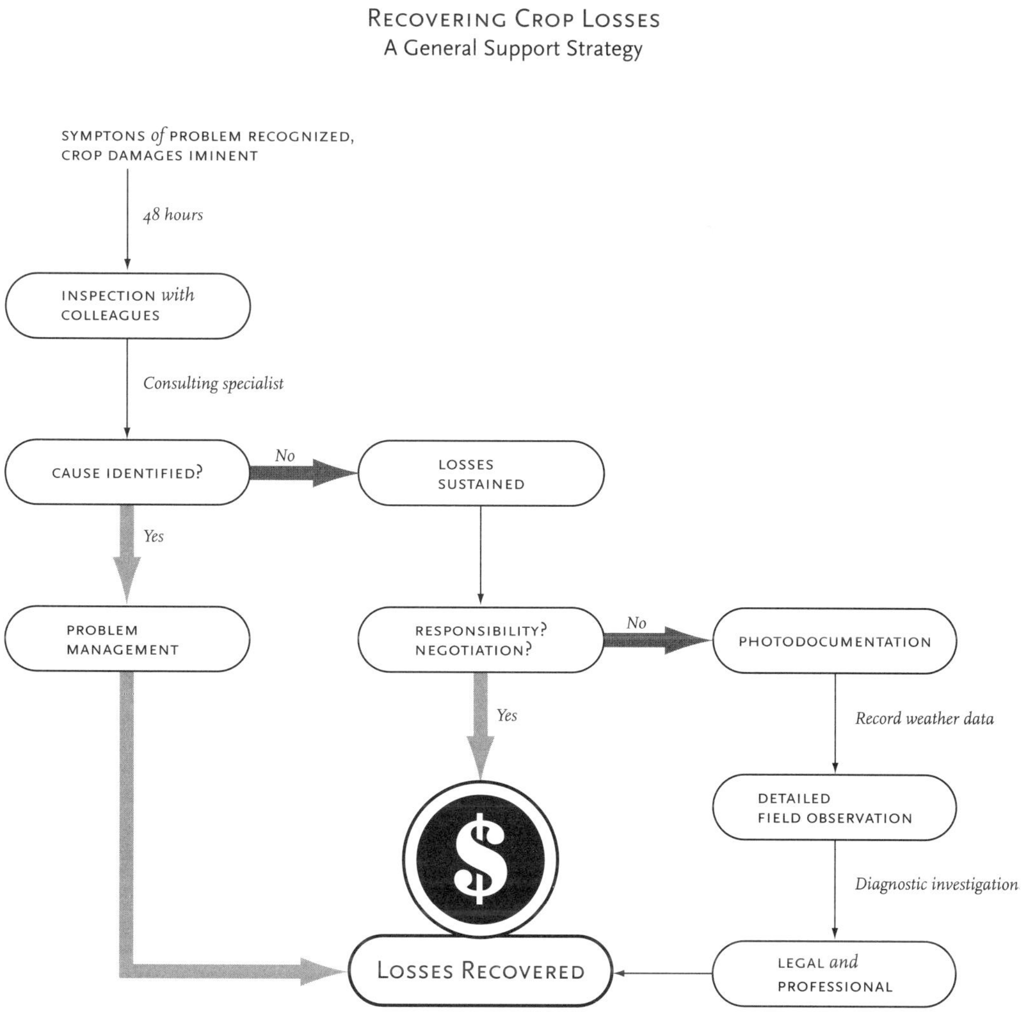

Figure 35.1 A decision support model for recovering crop losses resulting from damage due to plant diseases or disorders. See text for a more detailed description of the model.

Weather and field activity records should be carefully maintained throughout each growing season. Most growers routinely maintain these records for their own information and analysis. In the event of crop damage, these records may also be crucial to plant biologists seeking to diagnose a problem. The development of many pests and diseases is dependent on weather conditions, so weather records may indicate or eliminate possible causes. Weather recording devices that monitor temperature, rainfall, humidity, and wind are available at a reasonable cost, and every grower should invest in a device that is appropriate for their individual operation. Grove maintenance, especially agrichemical applications, may also be related to crop damage, either as a direct cause or by creating a favorable environment for pests or disease. Careful records should be maintained listing the materials applied, application rates, methods of application, and dates and times of applications. Whenever possible, the lot numbers and manufacture dates of the products should also be noted.

Investigation and documentation of a potential problem should begin immediately upon discovery of tree or crop damage. The grower's first effort to identify and resolve the problem should be consultation with colleagues, including production managers and county extension agents. Many growers also employ an independent crop consultant, who may advise them concerning the nature and severity of the problem and offer recommendations. It may be appropriate to contact the greenhouse or nursery that provided planting material or the product representatives of any agrichemicals that have been applied. Initial consultation is most effective if all the interested parties can meet in the field to observe and discuss the damage. The grower's first concern is to identify the cause of the problem and prevent further damage. If significant losses are expected and there is clear responsibility for the damage, it may be possible to negotiate compensation at this point. However, it may be impossible to reach consensus concerning the cause or causes of the crop damage. In this case, the grower may be advised to initiate a formal investigation.

The development of symptoms should be carefully recorded. The grower should carefully observe and document the appearance and spread of symptoms. This epidemiological information is valuable to plant biologists as they consider possible causes of the damage. Note whether the symptoms appeared in a large number of plants simultaneously or whether they appeared to spread from one area to another; and if so, at what rate? Also note the sequence of symptoms as they appeared in affected plants and the rate of plant decline. The grower should be aware that this information might ultimately be reviewed in a legal dispute, so the records must be clear and accurate.

The damage should be thoroughly documented on film. Photographs may be presented for analysis by specialists who are not able to visit the field for observation or as evidence to support a claim for damages. Record in a journal the date, time, and place of any meetings and where and when the photographs were taken. A diagnostic form is an excellent tool for documenting plant disease symptoms. The photographic record should include panoramic photographs of the affected fields, taken with a good quality camera or a video recorder.

These photos should reflect any general information that may indicate the nature of the problem. For example:

Do the symptoms extend across or down the rows?

Do the symptoms extend only down certain rows, exclusive of others?

Is the problem confined to specific regions of the field, such as low-lying areas?

In addition, photos of individual affected plants are needed. It is probably impractical for the grower to photograph each affected plant, but a sufficient number of plant photos should be included to clearly represent their condition. Whenever possible, plants in progressive stages of decline should be photographed. Pictures of normal, healthy plants should also be included for comparison. Again, the photos should document distinctive characteristics of the symptoms, such as:

Are the symptoms confined to one part of the plant?

Are the leaves or fruit wilted, smaller, or larger than normal?

What is the condition of the roots?

How has crop quality or quantity been compromised?

Finally, close-up photographs of the affected plant tissues should be included. These photos will indicate specifically how roots, leaves, and fruit are affected. For example:

If present, are lesions circular or irregular?

Is foliage curled or twisted upward or downward?

Do symptoms extend across leaf veins, or are they confined between the veins?

This photographic documentation may be vital to accurate diagnosis of the problem and fair recovery of damages. Some growers anticipate the need for this record and prepare photographs of healthy trees and fruit for future reference.

If a diagnostic investigation is indicated, the grower should seek professional advice. The grower might require the assistance of a consulting plant pathologist or other trained crop advisor (specialist). If there is disagreement as to the cause of potentially substantial losses, it may also be prudent for the grower to engage an attorney who has trial experience in agricultural law. The attorney may prefer to hire the plant scientist for the grower, as this can allow for the results of that work to remain confidential.

Consulting plant pathologists are specialists in the identification, diagnosis, and control of plant diseases and disorders. They are skilled in the process of scientific problem solving and can help identify and evaluate the various factors contributing to plant and crop damage. This consultant may determine that the specific skills of another specialist, such as a horticulturist, nematologist, or biochemist, are required. An independent plant consultant may be retained, or the grower may rely on a scientist who is publicly employed by a government agency or a university. University researchers tend to be specialists, while extension agents are more likely to be generalists. These publicly employed scientists provide their services free of charge to the grower, but their time and resources may be very limited. If the investigation is likely to be prolonged, an independent crop consultant may be needed. One efficient approach is for an independent consultant to orchestrate the investigation, contacting university specialists or extension agents on behalf of the grower as needed.

The crop consultant might need specimens of affected plant tissues to test for the presence of microbial plant parasites (such as viruses, bacteria, fungi, nematodes, or insects). These specimens should be collected promptly after symptoms appear to reduce the effect of confounding factors such as secondary infecting agents invading diseased plant tissues. Protocols for sanitation and documentation should be followed to protect samples from cross-contamination and ensure proper identification, so the consultant usually collects these specimens. Soil and water samples may also be tested for the presence of toxins and to develop a nutritional profile of the afflicted plants. The consultant will require the photographic documentation and will also analyze the grower's records. Finally, some important diagnostic information may be obtained from field experiments conducted on the site. The consultant should be qualified to design and analyze these experiments to yield relevant information and reliable statistics.

It is essential for the grower to be completely honest with the plant consultant and with the attorney. Incomplete or incorrect information may prevent the scientist from correctly diagnosing and resolving the problem, causing it to worsen or recur. Unreliable information also undermines the grower's credibility and jeopardizes any claim for compensation.

Careful documentation is central to identifying the cause of crop damage and the possible recovery of associated economic losses. The grower should not rely solely on memory to reconstruct the myriad of events and conditions that could be relevant to the investigation, but should maintain records throughout the season. The grower should also resist the temptation to manage an unusual problem without determining the underlying cause. Extraordinary measures, especially indiscriminant agrichemical applications, may veil the symptoms and complicate diagnosis and may, in some cases, exacerbate the problem. The key to resolving the problem and recovering losses is thorough investigation, documentation, organization, and analysis of the facts.

REFERENCES

Adams, D. C., R. L. Kilmer, C. B. Moss, and A. Schmitz. 2003. *Valuing catastrophic losses for perennial agricultural crops: Citrus as a model.* Proceedings of the Florida State Horticultural Society 116: 242-248.

Karl, T. R., and D. R. Easterling. 1999. *Climate extremes: Selected review and future research directions.* Climatic Change 42: 309-325.

Spreen, T. H., M. L. Zansler, and R. P. Murraro. 2003. *The Costs And Value Loss Associated With Florida Citrus Groves Exposed To Citrus Canker.* Proceedings of the Florida State Horticultural Society 116: 289-294.

36

Civic Action and the Industry

J.S. Rogers

INTRODUCTION

The citrus industry is highly regulated and closely monitored by many government agencies, and scrutinized by consumer and activist groups and the media. Diverse state and federal legislative issues and initiatives affect the industry, including laws and regulations relating to labor, food safety, consumer safety, environmental protection, and commerce. It is vitally important that growers and industry representatives be politically informed and proactive, to ensure that their concerns are addressed, their position is understood, and the potential impact of new legislation is fully appreciated. Growers must actively participate in local, state, and federal government and seek public and media support to strengthen their political voice. One goal of ICM is sustainability of the citrus industry, and appropriate civic action is an important step toward achieving this goal.

KNOW THE ISSUES

The first and perhaps most difficult step in civic action is to become informed regarding the issues. The industry has historically delayed aggressive lobbying until unfavorable legislation was imminent. Growers need to continually promote their position and work to redirect legislation during the discussion, development, and drafting phases. Information is available in newspapers, government bulletins and newsletters, Internet sites, and through personal communication with legislators.

KNOW THE LEGISLATORS

It is very important to build relationships with legislators and their staff members. Invite them to address or attend meetings of industry organizations or grower associations. Learn their affiliations and voting records; learn which of them are most influential in specific areas and which are most likely to support agriculture in general, and the citrus industry in particular. Communicate with your legislators regularly by telephone, letter, fax, or e-mail. Keep them informed regarding your concerns, but do not attempt to mislead them. A legislator who is embarrassed as a result of supporting your industry is likely to withhold future support.

Finally, consider the political position of the legislator. Ideally, he or she strives to represent the best interests of their entire constituency. They also seek to maintain popularity in their voting district. Thus, it is to the advantage of the industry to avoid unpopular positions. Instead, growers should work to find a broad basis of support for the industry position or to promote a compromise position with wider appeal.

KNOW THE INTERNET

The Internet has become the single most important vehicle for communication and information access, it can and should be a powerful tool for civic action as well. Growers can use the Internet to keep abreast of developing legislation and to actively lobby their legislators. Most importantly, the Internet can be used to promote the industry position on virtually any issue. Currently, citrus Internet sites primarily provide marketing information, but may also educate consumers about citrus production "from the grove to the glass." The industry needs to develop sites that can be accessed by the public and the media for information concerning the citrus industry and issues of concern to the industry. The University of Florida's Institute of Food and Agricultural Sciences (IFAS) periodically presents seminars on how citrus growers can use the powerful Internet resources.

COMMUNICATION AND THE MEDIA

Growers must continually remind the public that efficient agriculture is essential to global peace and prosperity. United States farmers also must communicate that the food they grow today is among the safest and healthiest produced in the entire history of the world. Growers need to cultivate a positive media image and take advantage of opportunities to communicate with the public by radio, television, or in print.

Many growers and industry representatives are uncomfortable with public appearances and interviews. However, there are several specific suggestions that will improve the grower's effectiveness and confidence:

1. Prepare for the interview. Never agree to an interview without sufficient time to prepare a statement. Focus on a few important points and present them very briefly. Speak simply, and avoid jargon or specialized language. It may be beneficial to write out and practice your remarks and to ask for assistance or candid comments from colleagues prior to the interview.
2. Check your personal appearance. Remember that the general public may harbor outdated stereotypes regarding farmers and agriculture. It is important to cultivate an informed, intelligent, and professional image. Your dress and conduct should create a favorable impression. Again, ask a colleague for a candid opinion.
3. Be frank and cooperative. Maintain eye contact with the reporter or camera. Respond to the reporter's questions thoughtfully and completely, but do not venture outside your area of expertise. If you are unable to answer with confidence, politely defer the question to the appropriate authority. Do not speculate.
4. Be frank and cooperative. Always tell the truth. Attempts to dissemble or deceive are almost certain to fail and reflect poorly on the entire industry.

The Agricultural Institute can provide media training for citrus growers and businesses. It is an organization composed of media and communications professionals from all segments of the agriculture industry who can design a media training program tailored for specific business organizations.

PARTICIPATE IN PUBLIC FORUMS

Government councils frequently hold hearings where members of the public are invited to express their views on controversial issues. Growers and industry representatives must take advantage of these opportunities to speak out on issues of significance to the agricultural community. Many growers are very reluctant to speak publicly, but it is important for their views to be expressed, especially as they may represent a very small minority in the community.

If grower concerns are not expressed, other views may capture public support, even when earlier discussions could have created a consensus that incorporated grower needs. Individuals who choose to attend and address these hearings generally have strong opinions, and the discussions may become heated. This should not discourage participation, but the grower should refrain from becoming angry or emotional. To best represent the industry, the grower should be courteous and professional throughout the hearing. Following are a few guidelines for participating in a public forum:

1. Resolve to express your opinion. Every individual has a right, even a responsibility, to participate in government. Further, you represent a vital industry in Florida and in the United States. Your view should be heard by the council and by the audience.
2. Prepare to speak. Listen to the views expressed by others, and observe the reaction of the council and the audience. Organize your thoughts, and consider what you can add to the discussion. You may wish to disagree with an opinion previously expressed, endorse an opinion previously expressed, contribute additional information, or raise a question.
3. Address the council. First, state your name and occupation and tell the council and the audience where you are from and why the issue is of significance to you. Express your opinion or ask your question clearly and briefly. Finally, thank the council and the audience for their time and their courtesy.

Public hearings are often poorly attended, and participation may be very low. These hearings present a valuable opportunity for growers to influence the decisions that affect the industry.

SUPPORT YOUR SUPPORTERS

Less than 2 percent of the American population is directly involved in the production of agricultural crops; however, the continued well being of the entire population depends on the agricultural community. That 2 percent constitutes a small minority and is often perceived as a special interest. Fortunately, many individuals and organizations are committed to the farming sector and its goals. Agriculture professionals who serve in public office and trade organizations and industry representatives who are active in legislation and lobbying represent the citrus industry in government and to the public. Growers need to support these individuals and organizations by contributing their time, money, and expertise.

WLIANR

The University of Florida sponsors a program called the Wedgeworth Leadership Institute for Agriculture and Natural Resources (WLIANR). One objective of this program is to develop and refine the leadership capabilities of emerging agricultural leaders (between the ages of twenty-five and forty-five) who will ultimately influence local, state, and federal policy related to agriculture and the protection and utilization of natural resources. Participants are chosen by nomination, interview, and committee selection. The program consists of two years of study and travel throughout the state, the country, and the world. During this time, the thirty participants develop a thorough understanding of pertinent political, social, economic, and cultural issues facing decision makers at the community, metropolitan, state, national, and international levels. Emphasis is placed on the impact of these issues to the future of Florida's agriculture and natural resources. Numerous seminars are designed to allow participants the opportunity to interact directly with recognized leaders in private business, government, state and federal agencies, and the academic community.

A healthy public and political dialogue will increase public awareness and understanding of the special needs and concerns of the citrus industry and may help to protect the industry from needlessly restrictive or impractical regulations. However, the industry must also demonstrate responsibility and foresight, especially in the publicly and politically sensitive areas of environmental, consumer, and worker protection. The industry must accept reasonable, practical, well-founded regulations. This is the only way growers can expect their objections to unreasonable, unfounded regulations to be seriously considered. One goal of ICM is sustainability of the citrus industry, and appropriate civic action is an important step toward achieving this goal.

RESOURCES

Florida Legislature Senate, Committee on Agriculture, and Florida Citrus Commission, Department of Citrus. 1987. *A Review of Section 601.04(3), Florida statutes, relating to the Florida Citrus Commission advisory committees or councils in the Department of Citrus*. The Committee, Tallahassee. http://www.flsenate.gov/statutes/index.cfm?App_mode=Display_Statute&URL=Ch0601/titl0601.htm

Lipsey, R. L. 1979. *Pesticide regulations affecting citrus production*. International Citrus Congress (2nd 1977: Orlando, Florida), Orlando, Florida USA 2: 435-437.

Powe, C. E., and M. R. Langham. 1980. *Toward a Policy Testing Model for the Florida Orange Industry*. University of Florida, IFAS, Technical Bulletin 815.

37

Regulatory Agencies and Environmental Assessment

T.J. Smith, J.S. Rogers, and D.P.H. Tucker

INTRODUCTION

The Florida citrus industry is a highly regulated enterprise. Owners, growers, packers, and processors must interact with many federal, state, regional, and local regulatory agencies. They may be required to obtain permits, submit records, or demonstrate compliance with regulations relating to a wide variety of issues. They are expected to be aware of the rules and requirements that apply to the citrus business and to contact the regulatory agencies when appropriate. This chapter includes a table that lists major regulatory agencies but their respective authority should not be considered exhaustive. Further information can be obtained through the local agricultural extension office, University of Florida educational outlets, county, state, and federal agencies, and various industry organizations and consultants.

The current climate of environmental protection has added a new layer of complexity to property transactions. Evaluation of the environmental liability associated with a property is important to the buyer, lender, and seller. As a result, an Environmental Site Assessment (ESA) is routinely prepared for most commercial, industrial, and agricultural property transactions. It is important for the buyer and the lender to investigate the history and environmental condition of the purchase property. Since passage of the Federal Comprehensive Environmental Response, Compensation, and Recovery Act (or Superfund), current property owners may be responsible for the cost of environmental cleanup of contaminated sites, even if they are not responsible for the contamination. A properly conducted ESA may be necessary to support the so-called "innocent purchaser" defense against such liability. Property owners must also be aware of their potential liability for environmental impact to neighboring properties. The costs associated with bringing a property into regulatory compliance, and obtaining permits for intended activities at the site, must also be considered by the buyer and the lender prior to the purchase, as these may significantly increase the cost of the property. Cleanup may also substantially delay or increase the cost of site development. Environmental liability is also a concern of the property seller. The seller is required to disclose environmental conditions that affect the value of the property. Otherwise, the seller may be responsible for costs associated with the cleanup and for damages claimed by the buyer.

THE PHASE I ESA

The initial Environmental Site Assessment is called a "Phase I ESA." A Phase I ESA is a reasonably comprehensive evaluation or assessment of a property for environmental contamination associated with current or past site operations. The goal of a Phase I ESA is to identify and assess any "Recognized Environmental Conditions" that may be associated with the site. For most assessments, "Recognized Environmental Conditions" relate to the actual or potential release of hazardous materials or petroleum products to the soil, groundwater, or surface water. Minimal releases that present no significant impact to the environment or to public health generally are not included in the definition of these conditions. Evaluations for asbestos, lead-based paints, and radon are not included in most Phase I ESAs. These and other items of special concern at a particular site should be included as additional elements of the assessment.

The Phase I ESA should be performed by a qualified environmental professional and should include a review of pertinent records, site reconnaissance, interviews with involved parties, and a final report.

RECORDS AND INFORMATION REVIEW

A review of reasonably accessible data regarding the site and its history of use should be conducted as part of the Phase I ESA. Topographic maps, soils maps, current and historical aerial photographs, city directories, utility records, fire insurance maps, and geological reports are typical sources that may be consulted. A title search spanning at least fifty years should be completed, along with research into any environmental liens on the property. State and federal records should be examined for indications of, or references to, potential environmental problems on the subject property or adjacent properties. Specific research into regulatory compliance files for the subject property or adjacent sites may also be required. Table 37.1 lists the databases that are typically reviewed during a Phase I ESA in Florida.

Table 37.1 Regulatory agencies relevant to Florida citrus production.

Florida Department of Agriculture and Consumer Services (FDACS)
Fertilizer grades and standards
Pesticide usage and registration
Worker protection standards
Surface and groundwater protecting BMPs
Fruit grades and standards
Budwood registration
Food Safety—pesticide residues, microbiological analysis
Plant and exotic pest quarantine, exclusion, eradication
Florida Department of Citrus
Marketing orders
Department of Environmental Protection (DEP)
Storage tanks—petroleum oils, spray oils
Surface and groundwater—quality, quantity
Wetlands
Land use
Effluent water sources
Point source discharges—wastewater treatment facilities
Drinking water wells
Hazardous waste
Solid waste
Application of wastewater residuals
Florida Department of Health or County Department of Health
Farmworker housing
Florida field sanitation

Department of Community Affairs (DCA)

Comprehensive land-use planning
Community planning
Right-to-know and storage of hazardous materials

County Government

Zoning
Building permits
Storage tanks
Health inspections
Environmental protection
Worker housing
Food safety
Well head protection
Wetland protection
Establishment of land-use restrictions

United States Department of Agriculture (USDA)

Animal Plant Health Inspection Service (APHIS)

Plant and pest quarantine, exclusion, eradication

Agricultural Marketing Service (AMS)

Marketing orders, quality grades, and standards
Food safety

Environmental Protection Agency (EPA)

Pesticide registration and regulation–labels, Worker Protection Standards (WPS), Food Quality Protection Act (FQPA), etc.
Water quality
Hazardous materials
Site contamination

Food and Drug Administration (FDA)

Pesticide residues
Food contamination
End product labeling

U.S. Department of Labor (DOL)

General industry standards (packinghouse, processing facility)

Occupation Safety and Health Administration (OSHA)

OSHA field sanitation

Wage and Hour

Farm labor contractors registration
Fair Labor Standards Act (child labor, minimum wage, payroll records)
MSPA enforcement (migrant transportation requirements, terms/conditions of employment, housing)

U.S. Department of Justice Legal Workers enforcement *Immigration Naturalization Service (INS)* I-9 process
Florida Department of Labor (FDOL) Farmworker Contractors Act Unemployment compensation Workers Compensation Child Labor Florida Hazardous Communications and Right-to-Know
Florida Department of Transportation (FDOT) Transportation of migrant farmworkers Vehicle regulation
Water Management Districts (WMD) Surface and groundwater quality and quantity Consumptive use permits and drainage discharge permits Wetlands Land use preservation Public water supplies Nitrate and other groundwater and surface water quality BMPs Well construction permits

SITE RECONNAISSANCE

A trained environmental professional should perform a walking inspection of the property to determine the general physical characteristics of the site and to detect any evidence of possible prior or ongoing contamination. The inspector will note and examine such items as areas of soil staining or stressed vegetation; chemical storage areas and evidence of chemical use at the site; the presence and condition of underground or aboveground storage tanks; evidence of debris dumping, burning, or burial; inadequate disposal procedures and facilities; vehicle storage and maintenance facilities, including aircraft hangars and landing strips; water supply wells; and any discharges through drains or pipes. A reconnaissance of adjacent and nearby sites should also be completed to determine the potential for contamination from off-site sources. The inspector generally uses a checklist to organize the inspection. A detailed sample checklist of potential items to be reviewed along with relevant agencies to be considered during the site reconnaissance is provided in Table 37.1.

INTERVIEWS

Interviews should be conducted with people who are familiar with the current and past activities at the site. These may include current and former owners of the purchase property or neighboring properties, site managers, employees, tenants, and local government and regulatory officials.

REPORT

Finally, a Phase I Environmental Assessment Report, which summarizes the performance and results of the assessment, should be prepared. The report should include a description of the scope of the assessment and the methods employed in the investigation, the findings of the investigation, any supporting photographs or documentation, and the environmental professional's opinion regarding the condition of the site. The report will also include a recommendation for any further research considered necessary. In most situations the report will conclude with one of two possible findings: a listing of Recognized Environmental Conditions associated with the subject property or a statement indicating that no Recognized Environmental Conditions were found.

THE PHASE II ESA

If the Phase I ESA indicates a potential for environmental contamination of the site, a Phase II Environmental Site Assessment may be conducted. The principal objective of the Phase II ESA is to determine whether a release of hazardous materials or petroleum products has occurred. This determination is most often based on the analysis of a limited number of soil and water samples collected from the site. Efforts to fully characterize the contamination in terms of area affected, quantities present, potential for migration of the hazardous material, and overall site impact are beyond the scope of the Phase II ESA.

A Phase II ESA should consist of the development and implementation of an appropriate sampling plan, including the number and type of samples to be collected, the method of collection, and the analytical methods used to assess samples. The plan should specify any safety precautions necessary and protocols for quality assurance. The objective of the proposed sampling plan relative to the potential contamination should be clearly stated and fully understood.

After the collection and analysis of the samples and the interpretation of the results, the Phase II ESA should conclude with a determination of the presence or absence of significant contamination at the site. The report may recommend additional site assessment, that no further action should be taken, or notification of the appropriate regulatory agencies.

ENVIRONMENTAL CONCERNS

Due to the variety of activities that occur at agricultural sites, a wide range of environmental concerns should be addressed during an ESA of these properties. Site contamination may have occurred as a result of improper waste storage and disposal or as a consequence of normal agricultural operations. The inspector should pay special attention to pesticide, herbicide, and fertilizer use and storage practices, previous use of now-banned pesticides, use of petroleum products in equipment operation and maintenance, storage of petroleum products in underground or aboveground tanks, waste disposal facilities and practices, and the possible presence of fertilizer and pesticide residues in storm runoff.

AGRICULTURAL CHEMICALS

It may be necessary to research crop history and management practices to fully evaluate current and past use of agricultural chemicals for potential soil or groundwater contamination. Information on crops and chemical use may be obtained from regional USDA and state extension offices and local colleges or universities. Interviews with current and past landowners in addition to employees will provide further information and may confirm specific chemical use and application practices. Table 37.1 is a partial list of some pesticides and chemicals of concern associated with individual crop types.

Past use at the site of now-banned agrichemicals, especially the pesticides EDB and DDT, is particularly significant. EDB is a soil fumigant commonly used in citrus groves during the 1960s until its statewide ban in 1983. EDB has been found to persist for as long as 19 years in soil and groundwater where it has been used as a soil fumigant. In Central Florida, the Southwest Florida Water Management District (SWFWMD) has prepared EDB-use maps for evaluating water supply permits and regulating water well construction standards. These maps may be consulted in the ESA if EDB use is a potential concern. DDT was also a commonly used pesticide and has been banned because of its danger to wildlife and its persistence in the environment. Due to the widespread historical use of DDT, it is an important concern at agricultural sites with operations dating to the 1950s and 1960s.

Chemical mixing and loading sites have recently been recognized as potential sources of contamination. These sites are used regularly for mixing and loading chemicals into spraying equipment. Because of their long-term use, occasional spills can ultimately produce large areas of contaminated soil and groundwater. Improper disposal of rinse waters may also produce contamination of surface water or groundwater. The

Florida Department of Environmental Protection (FDEP) has developed construction and operational guidelines for chemical mix and load sites that specify roofed containment structures with impermeable ground barriers, secondary containment systems, and safe chemical handling procedures, spill management procedures, and periodic inspection and maintenance programs. Although chemical mix and load sites are not currently subject to regulation, these guidelines may be incorporated into FDEP rules in the future.

Excess agrichemicals may also be present as waste in storage sheds and outbuildings as a result of a change in crop type and associated chemical use, expiration of the material shelf life, collection and storage of empty chemical containers, or as a result of use bans or restrictions for specific products. Since special requirements may apply to the disposal of this waste, excess agricultural chemicals can accumulate in significant volumes. Storage of these wastes may present a considerable threat if released into the environment.

PETROLEUM PRODUCTS

Petroleum products, including gasoline, diesel fuel, lubricants, and solvents are often present at agricultural sites for the operation and maintenance of farm vehicles, equipment, and machinery. These products are generally stored on site for use, and waste products are also usually accumulated on site for recycling or disposal.

Many farm operations have on-site petroleum storage tanks, either underground or aboveground. Accidental discharges or spills from underground storage tanks (USTs) and aboveground storage tanks (ASTs) are important environmental concerns because of the potential for soil and groundwater contamination. During the assessment, it is vital to fully evaluate any recent or historical spills from UST or AST systems. The inspector should also verify that any USTs or ASTs are in compliance with current regulatory rules. USTs and ASTs must be situated a minimum of 500 feet from any water supply system. Current federal and state rules required most UST systems to be upgraded or replaced to meet the following standards by the end of 1998:

Spill Protection: USTs must be equipped with catchment basins to contain spills during fuel deliveries.

Overfill Protection: USTs must be protected against overfills with an automatic shutoff system, overfill alarm, or ball float valves.

Corrosion Protection: New USTs and delivery lines must be constructed of noncorrodible

> materials such as fiberglass. Steel tanks must have corrosion-resistant coatings and cathodic protection. Existing uncoated steel tanks must be equipped with a cathodic protection system, a noncorrodible interior liner, or a combination of the two.

ASTs are subject to similar requirements regarding spill containment, material dispensing systems, secondary containment (often a concrete dike to contain materials following tank leakage), and exterior coatings. Tank systems that do not meet these requirements are subject to closure.

LIVESTOCK WASTE

Although manure and other livestock wastes are not directly toxic or hazardous, storage of these materials may result in environmental contamination. Storm runoff from areas where these animal wastes are stored can produce excessive nitrogen levels in nearby surface water bodies and local groundwater.

GENERAL WASTE DISPOSAL

Prior to the 1970s, it was common for farm operations to dispose of general waste in on-site burn pits or landfill areas. These old farm landfills or burn pits may contain a variety of debris, including spent chemical containers, equipment and parts, tires, batteries, and various household wastes. As a result, a significant contamination potential to soil and groundwater may be associated with these areas.

NONPOINT CONTAMINATION

Contamination can also occur from nonpoint sources such as fertilizer and pesticide residues in storm water runoff or petroleum and solvent contamination from equipment wash areas. These nonpoint sources of contamination may be especially difficult to evaluate in the ESA.

Documented use of Best Management Practices (BMPs) is an important indication that an agricultural facility is operating in a safe and responsible manner to reduce its impact on the environment.

Clearly, the Florida citrus industry is a highly regulated enterprise. Owners, growers, packers, and processors must interact with many federal, state, regional, and local regulatory agencies. They may be required to obtain permits, submit records, or demonstrate compliance with regulations relating to a wide variety of issues. They are expected to be aware of the rules and requirements that apply to the citrus business and to contact the regulatory agencies when appropriate. Table 37.1 lists the major regulatory agencies and their respective authority, but it should not be considered exhaustive. Further information or clarification can be obtained through the local agricultural extension office, University of Florida educational outlets, county, state, and federal agencies, and various industry organization consultants.

REFERENCES

Florida Legislature Senate, Committee on Agriculture, and Florida Citrus Commission, Department of Citrus. 1987. *A Review of Section 601.04(3), Florida statutes, relating to the Florida Citrus Commission advisory committees or councils in the Department of Citrus.* The Committee, Tallahassee. http://www.flsenate.gov/statutes/index.cfm?App_mode=Display_Statute&URL=Ch0601/titl0601.htm

Lipsey, R. L. 1979. *Pesticide regulations affecting citrus production.* International Citrus Congress (2nd 1977: Orlando, Florida), Orlando, Florida USA 2: 435-437.

Powe, C. E., and M. R. Langham. 1980. *Toward a Policy Testing Model for the Florida Orange Industry.* University of Florida, IFAS, Technical Bulletin 815.

38

Purchasing and Leasing Grove Property

S.H. Futch

INTRODUCTION

Investments in citrus production, or any agricultural enterprise, are inherently risky. The profitability of the operation depends on several major factors that are beyond the influence of the owner or grower. A capable manager can reduce, but never eliminate, the impact of unfavorable weather or pests and diseases. Their influence on the total yield and the value of the crop can be profound. Further, the value of the crop depends on global supply and demand and often-unexplainable market preferences. A healthy, well-managed grove has generally represented a sound investment, although the annual profit may vary tremendously. The annual revenue generated by a citrus grove depends on the annual yield and the market value of the fruit. Specialty cultivars (e.g., tangerines, tangelos, and tangors) may also be valuable, but their production expenses may be even higher than other fresh marketed citrus. It is important to understand that citrus is marketed globally, and the specific value of any crop is somewhat unpredictable even right up to the point of sale. The management of a grove property can ultimately affect the bottom line. There are a number of options related to whether or not grove land is purchased or leased, and this chapter reviews some of these options.

Investments in citrus production, or any agricultural enterprise, are inherently risky. The profitability of the operation depends on several major factors that are beyond the influence of the owner or grower. A capable manager can reduce, but never eliminate, the impact of unfavorable weather or pests and diseases. Their influence on the total yield and the value of the crop can be profound. Further, the value of the crop depends on global supply and demand and often-unexplainable market preferences. A healthy, well-managed grove has generally represented a sound investment, although the annual profit may vary tremendously.

There are many factors that can be rated as to their importance in evaluating a citrus grove (Table 38.1). The primary factors to consider in determining the investment value of a citrus grove are: (1) the variety of citrus; (2) the annual yield in boxes, pounds solids, or gallons of juice per acre; (3) the age and general health of the grove; and, (4) the value of the real estate, including any tax advantages the purchase may offer. The annual revenue generated by a citrus grove depends on the annual yield and the market value of the fruit. Annual profit is total revenue decreased by the sum of all costs associated with production and harvesting.

The market value of the crop varies dramatically between cultivars and can fluctuate widely from year to year and even within a given season, especially in the event of an environmental catastrophe such as a freeze. Late cultivars of processing oranges tend to be more valuable than early- or mid-season cultivars. Fruit that is grown for the fresh market will usually, but not always, command a higher price than processing fruit. However, production expenses are also usually, but not always, higher for the fresh market. There does appear to be a geographical component to the general cost of citrus production, with groves on the east coast of Florida requiring greater inputs than groves located more inland.

Specialty cultivars, such as tangerines, tangelos, and tangors may also be valuable, but their production expenses may be even higher than other fresh-marketed citrus. It is important to understand that citrus is marketed globally, and the specific value of any crop is somewhat unpredictable even right up to the point of sale.

Citrus grove yields may be expressed as boxes per acre (for fresh citrus) or gallons of juice or pounds solids per acre (for processed citrus). Most processing citrus contracts are based on the pounds of total soluble solids (primarily sugar). Under current production regimes in Florida, citrus trees usually reach maximum yield between the eighth and twelfth year of production, with slightly increasing yields each year until about age twenty.

Table 38.1 Some factors to consider during grove evaluation.

Category	Importance rating*
Drainage	10
Price of nearby groves	10
Yields	10
Solids	10
Land use and zoning	10
Scion/Rootstock	9
Uniformity of scion	8
Elevation	8
Insect and disease problems	8
Historical temperatures	7
Spacing	6
Water availability	6
Soil variation	5
Grove accessibility	5
Tree height	4
Suitability for harvesting	4
Potential for exotic diseases	4
Uniformity of size	3
Soil metal toxicities	3

*Factor ratings (on a scale of 1 to 10, 10 being of highest importance) may vary according to individual conducting assessment.

Production varies from region to region and grove to grove, with the current average production per acre for all orange cultivars approaching about 420 boxes per acre. Annual productivity can be greatly affected by weather, pests, and diseases.

Historical yield data is a fairly accurate indicator of the production that can be expected in the future. Significant changes in production are unlikely, unless the grove is very young or unless major improvements are undertaken in the grove. The effects of most improvements to an existing grove are long-term and do not usually improve production immediately. A mature grove with very low production generally does not represent a good investment unless the property is valuable for some other purpose.

The age and general condition of the grove are both important to consider, including the rootstock, soil type, climate, and incidence of pests and diseases. These factors affect the grove's life expectancy and potential yield. For example, sour orange rootstock was once widely planted, but the introduction of CTV has made it likely that trees on this rootstock will soon be unproductive. Groves that need extensive tree replacement, top working, or significant infrastructure renovation should be offered at a substantial discount to reflect the cost associated with making these improvements. These improvement programs are often referred to as "rehabilitation." The drainage system should be inspected with particular care, as it may be difficult to alter. A potential buyer should also confirm that permits necessary for renovation could be readily obtained. A private consultant or extension agent may be able to advise the buyer concerning the feasibility of rehabilitation of the grove.

Finally, it should be noted that the property might have short- or long-term value for other purposes, such as real estate development. Many older groves are situated in the path of urban or suburban growth and may be more valuable for commercial or residential development than for growing citrus. Alternatively, the property may be desirable for ecological conservation purposes. In addition, tax advantages may be associated with the purchase of the grove. These considerations should be discussed with a financial consultant before a purchase is finalized.

REFERENCES

Justesen, L., J. E. Reynolds, T. Taylor, and R. Degner. 2000. *Perspectives in land valuation: a case study on citrus land valuation for prudential agricultural investments.* University of Florida, IFAS, Food and Resource Economics Teaching and Learning Paper TLP 00-25, 35 pp.

Tojo, M., P. J. van Blokland, R. P. Muraro, and J. E. Reynolds. 2002. *Investing in citrus production: Buying a new or established grove?* University of Florida, IFAS, Food and Resource Economics Department, Gainesville, FL, 18 pp.

39

Economics

R.P. Muraro

INTRODUCTION

Citrus production cultural practices vary depending upon the market for which the fruit is being produced, i.e., fresh market or processed juice market. Fresh market fruit usually requires a different pest management and fertilization program in an attempt to attain desired fruit characteristics. Likewise, harvesting costs are usually higher for a fresh market citrus fruit. Also, the transportation costs from the citrus grove to the fresh packinghouse are often more costly on a cost-per-unit basis (e.g., cost per box), since total volume of fruit in pallet bins on the transport truck is often less than the bulk citrus fruit transported to a citrus juice-processing facility. This chapter presents information related to evaluating the market for citrus products and interpreting the economics of a citrus production program, including cultural and management components that influence costs and returns.

Citrus production cultural practices vary depending upon the market for which the fruit is being produced, i.e., fresh market or processed juice market. Fresh market fruit usually requires a different pest management and fertilization program in an attempt to attain desired fruit characteristics. The fresh fruit production program costs more because it is designed to produce fruit with the desired external qualities preferred by fresh fruit buyers. However, citrus fruit produced for the processed market requires a less intensive cultural program, since the external fruit quality is of lesser importance.

Likewise, harvesting costs are usually higher for a fresh market citrus fruit. For fresh market citrus, fruit may require stem clipping to prevent damage to the external rind or spot picking for selection of the desired size that is marketed. Also, the transportation costs from the citrus grove to the fresh packinghouse is often more costly on a cost-per-unit basis (e.g., cost per box), since total volume of fruit in pallet bins on the transport truck is often less than the bulk citrus fruit transported to a citrus juice processing facility.

The allocation of the final retail price paid by the consumer with respect to the major cost components of fresh and processed citrus products also differ. Fresh citrus is subjected to possible additional losses due to weight loss (called shrinkage) during the handling of a perishable commodity through the marketing channels. Processed juice products, once packaged, require very little special handling, except for the refrigeration necessary to preserve the juice quality.

Figures 39.1a–c and 39.2a–c present two sets of pie charts that demonstrate the economic components from the grove to the consumer for citrus fruit grown for both the processed juice market and the fresh market. The two cultivars chosen as examples are processed market juice oranges and fresh market grapefruit. The percentages shown in each set of pie charts represent the typical distribution for each variety and market.

The University of Florida prepares annual reports concerning the economics of citrus production in various Florida regions, including an analysis of crop production costs, cost of the crop at the packinghouse or processing plant (referred to as "delivered-in" cost), and final retail cost of the fresh or processed product. A typical distribution of each of these costs, based on historical budget information extracted from these reports, is presented in Figures 39.1a–c for frozen concentrated orange juice (FCOJ) and Figures 39.2a–c for fresh market grapefruit.

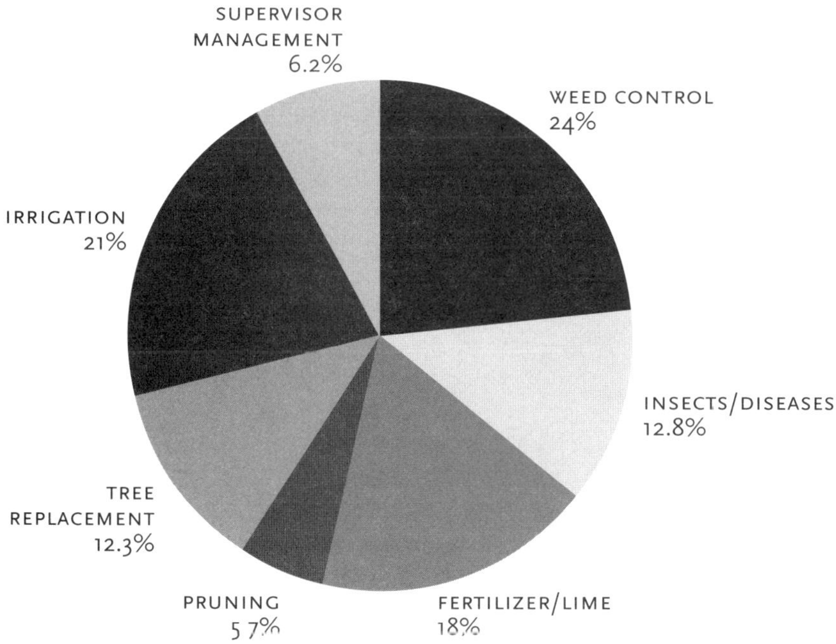

Figure 39.1a Economic components, from grove to consumer, for frozen concentrated orange juice, 1997-98. Total production costs by percentage of each cultural component (Source: Robert Barber, FCM).

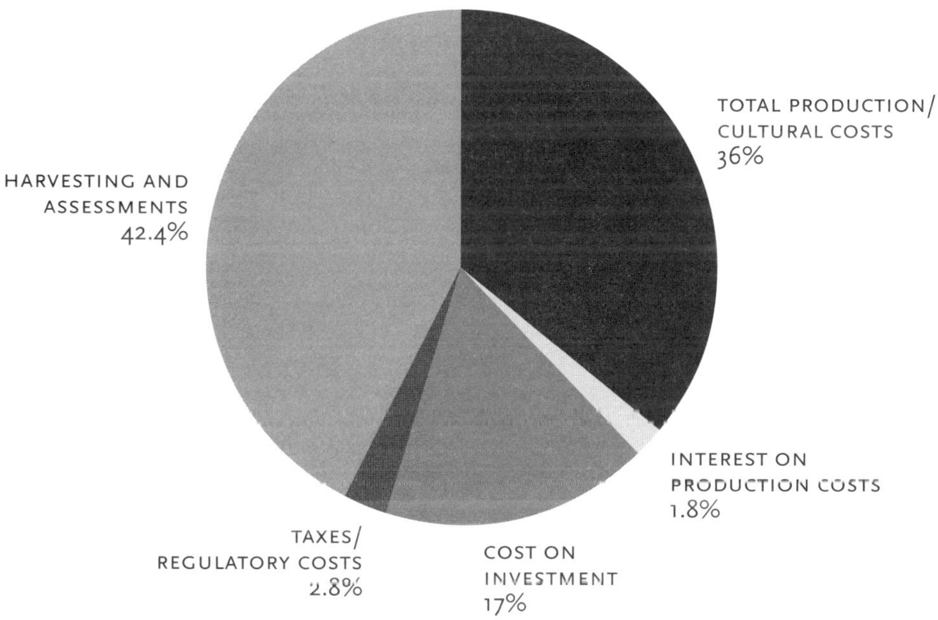

Figure 39.1b Economic components, from grove to consumer, for frozen concentrated orange juice, 1997-98. Total delivered-in costs by percent of each cost component. (Source: Robert Barber, FCM).

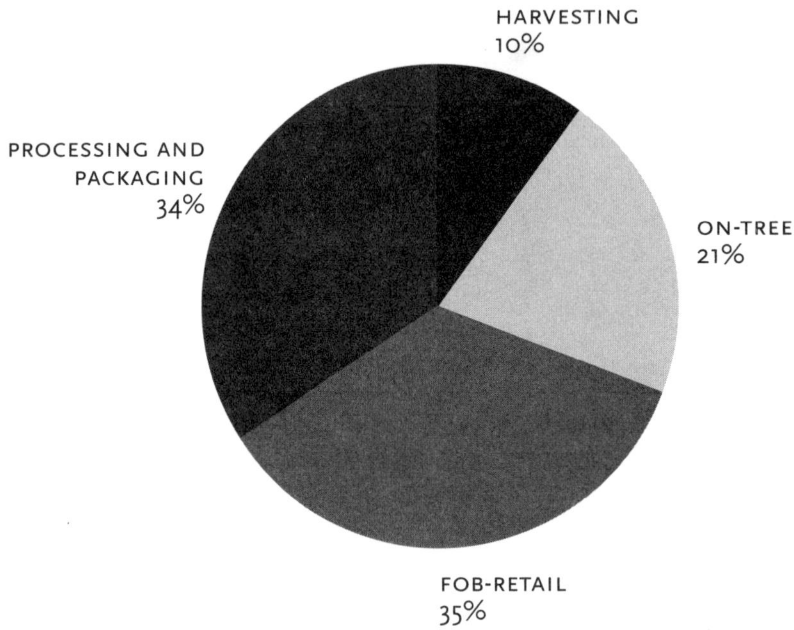

Figure 39.1c Economic components, from grove to consumer, for frozen concentrated orange juice, 1997-98. FCOJ retail price by percent of components (Source: Robert Barber, FCM).

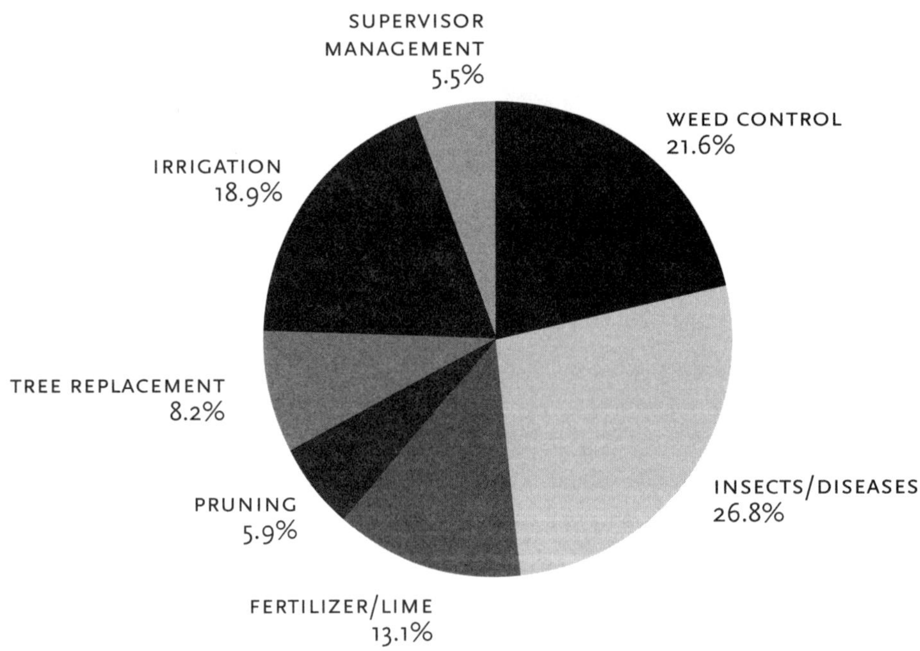

Figure 39.2a Economic components, from grove to consumer, for fresh market grapefruit, 1997-98. Total production costs by percentage of each cultural component (Source: Robert Barber, FCM).

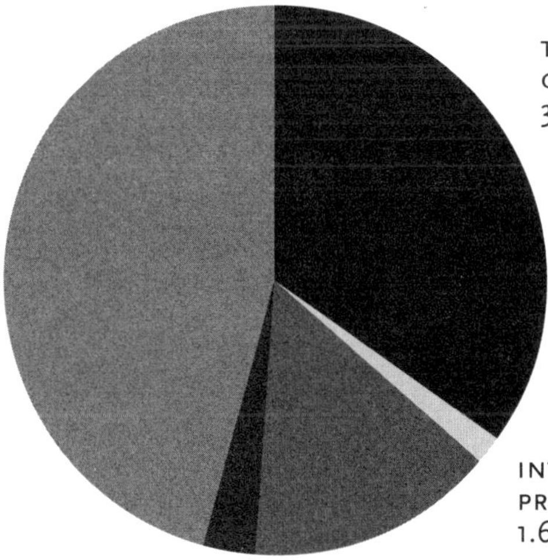

Figure 39.2b Economic components, from grove to consumer, for fresh market grapefruit, 1997-98. Total delivered-in costs by percent of each cost component (Source: Robert Barber, FCM).

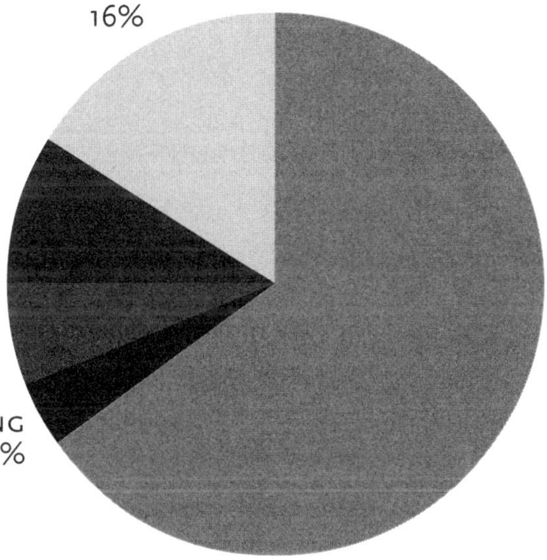

Figure 39.2c Economic components, from grove to consumer, for fresh market grapefruit, 1997-98. Indian River red grapefruit retail price by percent of components (Source: Robert Barber, FCM).

PRODUCTION COST

The major expenses for citrus production are associated with weed management, pest management, nutrition management, irrigation management, tree size management, tree replacement, and general grove management. Standard cultural practices for citrus production depend, in part, on whether the fruit will be marketed fresh or processed for juice. Note that the following discussion of production costs is specific to the production of processing oranges. Fresh market fruit usually requires more rigorous pest management and fertilization programs in order to achieve and maintain the external characteristics preferred by consumers and are generally more costly to produce than fruit for processing.

Weed management accounts for about 25 percent of annual production costs for processing oranges. This includes both chemical and mechanical control of undesirable vegetation under the trees and in the row middles. This vegetation can compete with the citrus trees for water, nutrients, space, and light and can hinder grove access for maintenance. In most cases, the growth of undesirable weeds is chemically controlled, while mechanical and chemical mowing are used to control weed growth in the row middles. It may be desirable to prevent the growth of weeds under the trees, but the objective of weed management in row middles is to maintain the weeds at a height that does not interfere with grove access. It is not feasible or desirable to eliminate all weed growth in the row middles for any extended period of time.

Pest management accounts for about 21 percent of the annual production budget for processing oranges. Costs can be much higher if Diaprepes root weevil is present. Pest management includes the application of various chemical sprays to control diseases, insects, and especially mites. The management program is not intended to eradicate these pests, but rather to limit their economic impact. A much higher level of pest pressure is acceptable in fruit that is intended for processing than in fruit that is to be marketed fresh. Fresh fruit must be protected from external blemishes caused by pests or diseases. Fruit intended for the fresh market typically require four or more sprays, depending on the citrus variety, while processed fruit may only be sprayed once or twice. The more intensive spray program for fresh fruit may double the annual cost of pest management.

The application of materials to the soil or sprayed on the foliage to correct or prevent nutritional deficiencies accounts for approximately 19 percent of the annual budget costs. These nutritional materials may be applied to the soil alone or in combination with herbicides and may be applied through the irrigation system. Many growers employ a fixed low-volume irrigation system beneath the tree canopy to maintain an adequate water supply to the grove and to facilitate application of fertilizers and herbicides. The cost of operating such an irrigation system accounts for about 16 percent of the total citrus production budget. This includes the cost of energy, maintenance, and depreciation of the equipment.

Hedging and topping to control tree size are regular grove maintenance operations performed on a fixed schedule. The grower strives to maintain an ideal tree shape for maximum productivity and an ideal tree size to facilitate grove maintenance and harvesting. These operations account for about 5 percent of the annual budget.

Typical annual tree loss is 2 percent to 4 percent of the planting, and replacement of these trees accounts for about 8 percent of the total budget. Extreme pest or disease pressure, or unusually severe weather, can result in much greater tree loss, and replacement costs can easily exceed 8 percent under these unfavorable conditions. If excessive tree loss is experienced or anticipated, or extensive grove improvements are needed, the grower may be advised to consider replanting the entire grove.

General grove management expenses account for approximately 6 percent of the annual budget. These are basic business expenses, including the salaries of supervisors and other permanent employees.

DELIVERED-IN COST

The growers' existing investment in the fruit when "delivered-in" to the packinghouse or processing plant consists of five major expense categories: harvesting and assessments, production cost, interest on production cost, taxes and regulatory costs, and the cost of the overall grove investment. Harvesting costs and fruit yield assessments (such as marketing order taxes) that are regulated by Florida state statutes account for about 42 percent of the cost of the fruit when delivered for processing or packing. Total production costs account for about 36 percent of the delivered-in cost. These are the costs associated with the various aspects of grove management, as discussed previously. An additional 2 percent of the total delivered-in cost is devoted to the interest on production costs. Investment costs will range widely from grove to grove depending on the cost of the property and its current value. Average investment costs are estimated to account for about 17 percent of the total cost.

Taxes and regulatory costs associated with the grove property account for about 3 percent of the total cost. This is an average cost that includes local property taxes as well as taxes associated with special taxing districts that have been established in some citrus-growing regions.

FINAL RETAIL COST

The final retail value of the processed orange product consists of four major expense categories: the on-tree value of the crop, the retail profit, the cost of harvesting, and expenses associated with processing and packing. The retail profit, referred to as the FOB spread, accounts for the largest share of the final cost and is estimated to be 35 percent of the retail value. Processing and packing costs account for 34 percent of the retail cost; on-tree value accounts for 21 percent; and harvesting accounts for 10 percent of the retail value.

REFERENCES

Hodges, A., E. Philippakos, D. Mulkey, T. Spreen, and R. Muraro. 2001. *Economic impact of Florida's citrus industry, 1999-2000*. Economic Information Report EI 01-02, 14 pp.

Morris, A. 2003. *Economic impact of losing the citrus tariff.* Florida Grower 96: 52-53.

Muraro, R. 2001. *2001-2002 comparative citrus budgets.* University of Florida, Economic Information Report, FE 350. http://edis.ifas.ufl.edu/pdffiles/FE/FE35000.pdf

Van Blokland, P. J. 2003. *Figuring out Futures.* Florida Grower and Rancher 96: 26-27.

40

Risk Management

J.A. Stricker

INTRODUCTION

In an increasingly competitive world, management is becoming more important to business success. Citrus production, like all agricultural enterprises, has many risks. Risk is the possibility that an outcome or event will not meet expectations. Highly variable income from year to year is one indication of risk. A few examples of specific risks include: fruit yield less than expected, price not meeting expectations, or tree loss from insect or disease injury. Most growers are aware of risks but are not aware of the concept of risk management as a part of overall business management. On the positive side, risk involves the chance of gain as well as loss. Negative consequences are referred to as downside risk; positive consequences as upside risk. Risk management is the systematic application of management policies, procedures, and practices to the tasks of identifying, analyzing, assessing, treating, and monitoring risk. In a statewide survey of citrus managers, 90 percent of managers reported that their greatest source of risk was unstable fruit prices, followed by loss of trees/reduced yield from insects and disease (85 percent) and loss of fruit from adverse weather (84 percent). Steps in risk management include: (1) setting goals for the business, (2) knowing your personal risk tolerance, (3) knowing the risk-bearing capacity of the business, (4) identifying sources of risk, (5) estimating the probability of an adverse event, and (6) using appropriate risk management tools. Risk management tools fall into three categories: self-protection (preventive maintenance, diversification, spreading sales); self-insurance (cash reserves, credit reserve capacity, knowledge of regulations); and purchased insurance (casualty insurance, crop insurance, liability insurance, market hedges or options). This chapter covers the basics of managing risks in a citrus operation, including how and why more financially secure growers may even want to use some risk management tools such as growing different cultivars of fruit, maintaining property insurance, and using forward contracts or options to expand profit opportunities while controlling risks.

In a statewide survey of citrus managers, 90 percent of managers reported that their greatest source of risk was unstable fruit prices, followed by loss of trees/reduced yield from insects and disease (85 percent) and loss of fruit from adverse weather (84 percent). Increased cost of inputs such as fertilizer, pest control, machinery, and labor was considered to be an important risk by 80 percent of managers, and 78 percent said that foreign competition was important. Eighty-four percent of managers agreed that there is more risk in citrus production today than ten years ago.

Citrus growers may reduce risks or shift some of their risk to others, at a price. The price may include the premium on an insurance policy, taking the time to remain updated on government regulations, or the added cost of operating groves in different geographical areas. Steps in risk management include: (1) setting goals for the business, (2) knowing your personal risk tolerance, (3) knowing the risk-bearing capacity of the business, (4) identifying sources of risk, (5) estimating the probability of an adverse event, and (6) using appropriate risk management tools.

SETTING GOALS

Setting business goals, as well as personal and family goals, is critical for risk management decisions. Setting goals requires time and thought. A number of goals may be identified, some short term and others long term. In addition, most people have family, business, and personal goals. Some goals are in competition with other goals, while others may be complementary. Goal setting is a dynamic process. Once a goal is reached, new goals may become important. Goal priority may change over time. Also, a person may have unrealistic goals that, if not revised, may be a source of frustration and stress. Examples of some goals might include: increasing the size of the business by purchasing or planting additional grove land; increasing net income by a certain percent by increasing production or reducing costs; or setting aside a sum of money for the kids' education or for retirement. Goals should be written, reasonable, and measurable. Goals should be attainable in the manager's lifetime and should be shared with others involved in the business. Others involved in the business should also share their goals with the manager.

> Goals
>
> Are my goals written, reasonable, and measurable?
>
> Are my goals attainable in my lifetime?
>
> Have I shared my goals with everyone involved in the business and have they shared their goals with me?

RISK TOLERANCE

Individuals vary greatly in their tolerance for risk. They may be classified into four categories: "risk avoiders," "risk takers," "adventurers," and "calculators."

Avoiders are the most cautious risk takers. They do, however, take some risks. They expect the worse to happen and don't take the slightest risk that be can avoided. Unfortunately, they lose by missing opportunities to profit. Farmers who are avoiders typically don't prosper but just manage to survive.

Risk takers are plungers who close their eyes to risk, ignore facts, and go ahead. Risk takers usually don't get involved in farming. They commonly fail because they refuse to take precautions.

Adventurers are people who enjoy taking risks. Risks are challenging and exciting to them. Some adventurers enjoy the excitement of taking risks, but keep the stakes at reasonable levels. Many growers may fall into this category with regard to their marketing activities. As long as financial survival is not threatened, they may enjoy the adventure of playing the market. Many speculators are included in this group.

Calculators understand that they must take some chances to get ahead, but recognize that there is some risk in every situation. Before making a decision or taking action, they collect information and figure the odds. They try to be realistic, recognize the risks, and try to reduce risks to acceptable levels. Eighty-nine percent of citrus growers identified themselves in this category.

RISK ASSESSMENT

Have I identified my risk tolerance?

Have I communicated my tolerance for risk to the professionals who provide me with risk management services?

Which risks can keep me from attaining my goals?

Which risks am I comfortable retaining and managing with my own resources? Which risks will I shift to others? Which will I avoid?

When was my last insurance check-up for health, life, casualty, property, disability, long-term care, and crop insurance?

RISK BEARING CAPACITY

Risk-bearing capacity is directly related to the solvency and liquidity of the business. The stronger the financial position of the business, the better the business is able to handle risk. Risk-bearing capacity is also related to cash flow. Cash flow needs are obligations for cash costs, taxes, loan repayment, and family living expenses that must be met each year. The higher these obligations as a percent of cash flow, the less able the business is to assume risk. Other factors that impact the risk-bearing capacity of the business include: the age of the owner/manager, the form of business organization (corporation, partnership, or sole proprietorship), the size and diversity of the business, location, and the amount of income from other sources.

SOURCES OF RISK

Risk may be broken down into five categories: production, market, financial, legal, and human resource (Table 40.1). While this classification is convenient, there are many interrelationships among the various categories.

Production risk is the potential loss in crop yield from a wide variety of sources including: freezes, hurricanes, tornados, hail, flood, drought, insects, diseases, genetics, and quality of inputs. Fire and theft are also sources of production risk.

Market risk deals with both purchased inputs and sale of farm products. It involves price fluctuations due to weather, foreign competition, government action, changing consumer preferences, and over- or underproduction of farm commodities. Inflation in the general economy can drive up the cost of inputs without a corresponding increase in value of agricultural products.

Financial risk includes the cost and availability of debt capital, the ability to meet cash flow needs in a timely manner, and the ability to maintain and grow equity. Financial risk can involve casualty losses, falling land prices, technological advances that make current investments obsolete, and the potential loss of the farm to pay estate taxes. The ultimate financial risk is the risk of bankruptcy.

Legal risks commonly associated with agriculture include: (1) appropriate business structure, (2) tax and estate planning, (3) contractual arrangements, (4) tort liability, and (5) statutory compliance including environmental and labor issues.

Human resource risk and human resources are both a source of risk and an important part of the strategy for dealing with risk. The core of dealing with human resource risk and the potential for solving human resource problems is managing people. The inability to find enough qualified workers is becoming serious in the Florida citrus industry. In addition, human resource calamities such as divorce, chronic illness, and accidental death can hamper the most carefully made risk management decisions. Risk management should anticipate the likelihood of human resource calamities. Carelessness by the grower or farm workers in using machinery can also cause significant losses.

Table 40.1 Examples of risks and risk management tools and strategies in a company producing citrus.

Production Risk
Disperse risk geographically.
Maintain resource reserves.
Purchase crop insurance.
Maintain irrigation system for drought & freeze protection.
Grow more than one variety of fruit.
Use more than one rootstock.
Grow both fresh and processed fruit.
Keep up to date on pest control strategies.
Financial Risk
Maintain a credit reserve.
Hold a safe solvency position.
Incorporate or use limited partnership to limit risk.
Keep and use good financial records.
Develop an estate plan.
Maintain property insurance.
Maintain liability insurance.
Develop and use cash flow projections.

Market Risk
Determine breakeven price for fruit.
Sell by forward contract.
Use hedging & options.
Sell through coops or pools.
Maintain selling flexibility.
Spread sales over time.
Forward price production inputs.

Human Resource Risk
Have good personnel management program.
Keep informed of labor regulations.
Plan for human resource calamities.
Plan for back-up management.
Maintain health insurance program.
Plan for transfer of assets to next generation.

Legal Risk
Have appropriate business structure.
Practice tax and estate planning.
Meet contractual obligations.
Maintain good pesticide records.
Maintain good labor records.
Keep informed of new regulations.

Not all adverse events are equally likely. The probability of hurricane damage to the grove will be different than the probability that fruit price will meet expectations or the probability of getting sick and unable to manage the business. Identifying sources of risk and estimating the probability that the event will occur helps the manager decide which risk management strategies or tools are appropriate for his or her individual situation.

PROBABILITY OF AN ADVERSE EVENT

Probability is a number that measures the likelihood or chance that a particular event will occur. The number can be stated as a percent. For example, the probability that the sun will come up tomorrow is 100 percent, while the probability that there will be a damaging freeze this winter may be 10 percent. Probabilities are based on the manager's personal beliefs about the chance that a future event will occur. In estimating these probabilities, managers should examine their own experience, available data, and consult whomever time and money allows.

RISK MANAGEMENT TOOLS

A number of risk management tools are available to help minimize the adverse consequences of unanticipated negative events. Risk management tools fall into three categories: self-protection (preventive maintenance, diversification, spreading sales), self insurance (cash reserves, credit reserve capacity, knowledge of regulations), and purchased insurance (casualty insurance, crop insurance, liability insurance, market hedges or options). The decision to use one or more risk management tools depends on each individual manager's personal and business situation. It's important to be aware of risks and consciously evaluate potential impacts of different risks on business success.

CONCLUSION

Growers differ greatly in the level of risk they can tolerate, depending on overall debt-to-equity situation and degree of diversification. One large loss or two or three successive smaller losses could bankrupt a highly specialized grower with a high ratio of debts to equity. A grower in this situation would want to make full use of risk management tools such as forward sales, option contracts, crop insurance, hazard insurance, irrigation, etc., to assure some minimum level of return. On the other hand, a grower with substantial financial reserves or other sources of income may choose to carry greater risks in hopes of achieving higher income in the long run. However, even the most financially secure grower may want to use some risk management tools such as growing different cultivars of fruit, maintaining property insurance, and using forward contracts or options to expand profit opportunities while controlling risks.

REFERENCES

Ira, M. M. 2003. *Process safety management and risk management program (PSM/RMP) Audits—Are you prepared?* Transactions of the 2003 Citrus Engineering Conference, Lake Alfred, FL (March 20, 2003). 49: 25-64.

Spreen, Thomas H. and Ron Muraro. 1999. *The World Market for Citrus Products and Risk Management for Florida Citrus Growers.* Department of Food and Resource Economics, Florida Cooperative Extension Service, IFAS, University of Florida, Gainesville, FL FE 195. 11 pp. http://edis.ifas.ufl.edu/pdffiles/FE/FE19500.pdf

Stricker, J. A., T. D. Hewitt, and R. P. Muraro. 1999. *Risk management for Florida citrus growers.* Citrus Industry. 80: 22-24.

VanSickle, J. J. 2000. *Goal setting for risk management.* Citrus and Vegetable Magazine. 64: 47.

41

Incorporating Wildlife Needs in Citrus Culture

M.B. Main

INTRODUCTION

During the 1980s and 1990s, acreage devoted to citrus production in southwest Florida increased from approximately 14,125 ha (hectacre) to 62,777 ha in 1989. Approximately 70 percent of the area converted to citrus during this period formerly existed as native range. Resource managers have voiced concern about the impact of this increase in citrus production on native wildlife, including thirty-one species listed by the state of Florida or the federal government as endangered, threatened, or as species of special concern. Wildlife surveys conducted in response to these concerns documented 151 (47 percent) of a potential 324 native species of amphibians, reptiles, birds, and mammals within the complex of grove beds, canals, agricultural reservoirs, and isolated wetlands that constitute a typical citrus operation in southwest Florida. Most of these species were recorded within the agricultural reservoirs, which often provide both wetland and upland habitat. For the citrus industry to thrive in a climate of increasing environmental concern, the industry must demonstrate that the needs of wildlife and agriculture are not incompatible. Maintaining suitable patches of habitat, such as reservoirs and natural areas, within the larger citrus complex will increase wildlife habitat and the distribution of wildlife, and promote movement by large mammals across the agricultural landscape.

Agricultural development and urban sprawl are modifying the landscape of Florida, especially southwest Florida. Because relatively little is being done to conserve wildlife habitat in rapidly urbanizing areas, agricultural operations may become increasingly important as areas providing habitat for wildlife and as buffers against urban sprawl. Extensive conversion of natural areas to agriculture also may pose consequences for many species of wildlife. For example, during the 1980s and 1990s, acreage devoted to citrus production in southwest Florida increased from approximately 14,125 ha (hectare) to 62,777 ha in 1989. Approximately 70 percent of the area converted to citrus during this period formerly existed as native range. Resource managers have voiced concern about the impact of this increase in citrus production on native wildlife, including 31 species listed by the state of Florida or the federal government as endangered, threatened, or as species of special concern. Wildlife surveys conducted in response to these concerns documented 151 (47 percent) of a potential 324 native species of amphibians, reptiles, birds, and mammals within the complex of grove beds, canals, agricultural reservoirs, and isolated wetlands that constitute a typical citrus operation in southwest Florida. For the citrus industry to thrive in a climate of increasing environmental concern, the industry must demonstrate that the needs of wildlife and agriculture are not incompatible. To the contrary, through careful grove design and management, the citrus industry can emerge as an important friend of the environment, both by directly providing habitat for wildlife and by acting as a buffer between urban development and natural areas.

The term ecosystem refers to the plants and animals and the ecological processes that are representative of a particular area. Agriculture is a major force in landscape change, and should be considered as an integral and influential component of modern ecosystems. Poorly planned agricultural development or unsuitable farming practices can degrade environments so that they no longer function as healthy ecosystems. In areas of South Florida, as in many areas of the United States, agriculture and other human activities have greatly reduced native habitats and have modified natural ecological processes. Restoration of ecosystems that have been disrupted is difficult and expensive. For example, restoration of the Everglades ecosystem will cost billions of dollars and will be complex and difficult to maintain. The challenge is to promote agricultural practices that sustain and protect important ecosystem functions, provide for the continued existence of native species, and address societal needs and concerns without undermining the industry or eliminating reasonable profit margins (Figure 41.1). The ideal citrus manager should strive to operate where these three areas overlap.

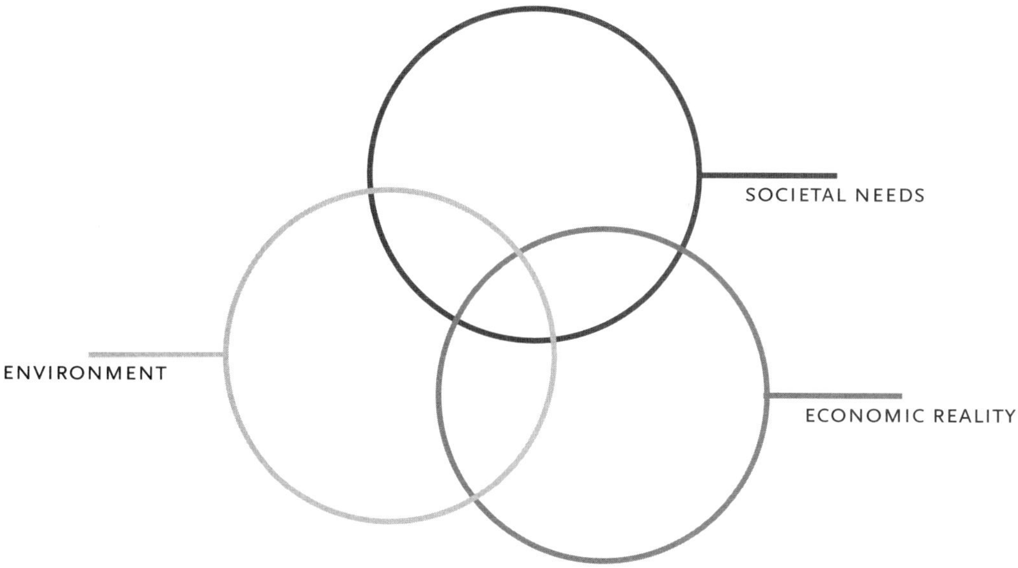

Figure 41.1 Societal needs, economic reality, and the environment are inextricably linked and should all be considered under plans for sustainable agriculture.

Habitat requirements differ among species, and species differ among areas. For this reason, it is best to take a habitat-based approach that utilizes general principles of conservation biology to benefit the greatest number of species when making recommendations for incorporating native habitat into the design of new citrus operations. One principle of habitat conservation is identification and accommodation of "umbrella" species, which are ecologically important species with large habitat requirements; the Florida panther (*Puma concolor*) and the Florida black bear (*Ursus americanus*) are examples. Providing habitat for these species ensures that many other species are also accommodated. Additionally, special consideration should be given to the habitat needs of species that are at risk (threatened, endangered, or of special concern) or that are of extreme ecological importance, such as the gopher tortoise (*Gopherus polyphemus*), whose burrows provide homes for many other species.

The requirement for the large range required by umbrella species can usually be met with a combination of agricultural and native habitats, provided that animals can move freely between areas of preferred habitat. This is typically achieved in one of two ways. One approach is to establish wildlife corridors, which are typically maintained as relatively narrow strips of suitable habitat that physically connects two larger natural areas. Alternatively, suitable patches of habitat may be established close enough together so animals are able to travel across the landscape between them. How close must these habitat patches be in order to be effective and how large? Unfortunately, these are not easy questions to answer, and the answers will differ for different species. Some species, such as tree frogs, require nearly continuous areas of wetland and upland habitat to move, but can survive in relatively small habitat patches. Wider-ranging species, such as the white-tailed deer (*Odocoileus virginianus*) and black bear, are able to travel between habitat patches that are farther apart. For example, radiotelemetry data for the Florida panther suggest that panthers will travel between habitat patches 100 to 200 miles apart, and data for western cougars suggest that these large cats may travel farther between suitable habitat patches (Figure 41.2).

In addition to incorporating habitat patches and larger habitat areas as components of citrus operations, some management of natural areas and created wetlands is needed. Plants and wildlife have evolved and adapted in response to natural pressures and processes in their native habitat. Human manipulation of the environment, including agriculture, disturbs or disrupts many of these natural processes. Two of the most obvious and significant aspects of the south Florida environment that are influenced by agriculture are the quality and quantity of the water supply and the occurrence of fire. Agricultural reservoirs and natural wetlands within the agricultural complex improve water quality by removing nutrients from agricultural runoff, provide habitat for many species of fish and amphibians, and serve as important feeding areas for ducks, wading birds, and other wetland dependent species. Periodic fire is important for maintaining herbaceous wetland communities and for promoting growth and seed production of native plants that provide food and cover for wildlife. Growers also need to control the invasion of nonnative plants, which crowd out native species, alter habitats, and are believed to reduce suitability of habitat for native wildlife.

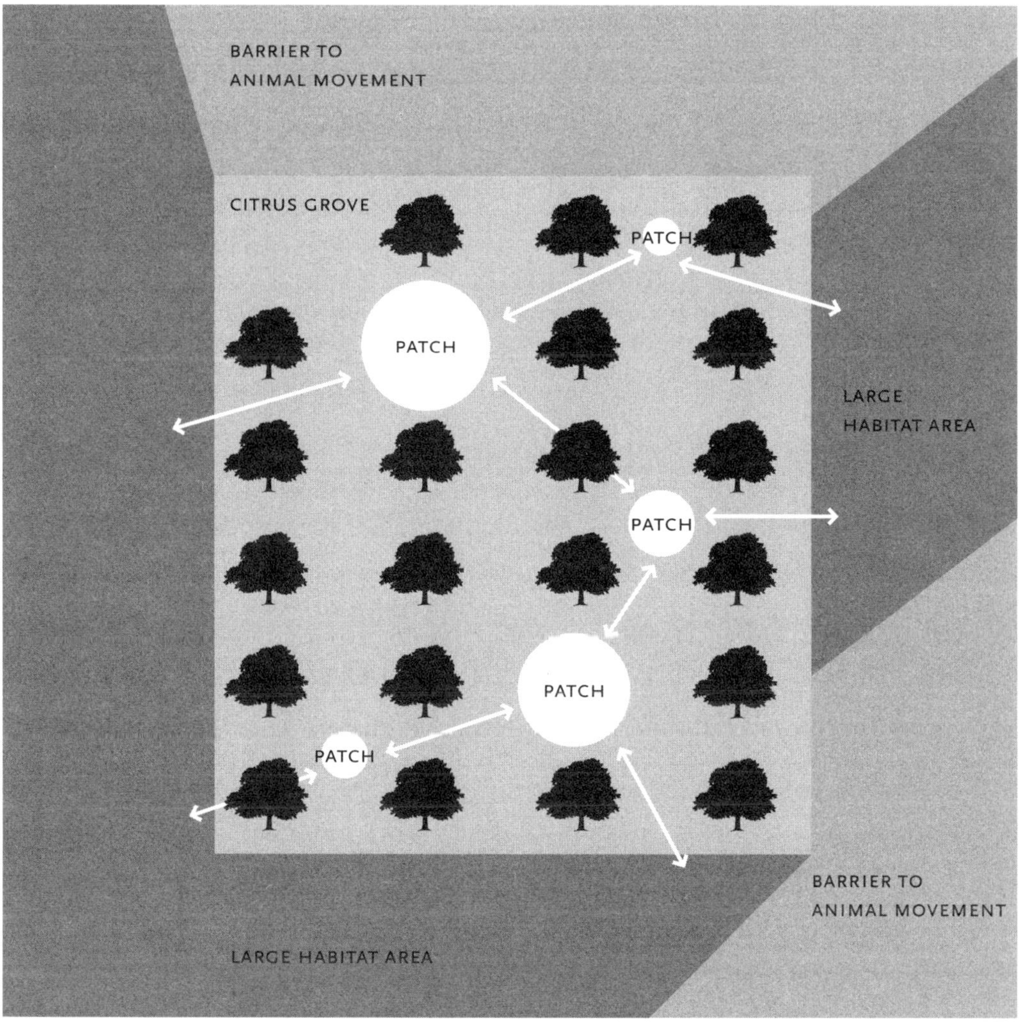

Figure 41.2 Maintaining areas of undisturbed woodland patches within citrus groves provides avenues for wildlife movement.

Increasing environmental concerns regarding declining wildlife and wildlife habitat should send a message to the urban and agricultural communities that the needs of wildlife should not be ignored. The needs of wildlife can be and must be integrated into modern citrus crop management and should be considered a component of modern agricultural costs. Consequently, it is in the best interest of agriculture for growers to take a leadership role in conservation by being responsive, flexible, and foresighted regarding the needs of wildlife.

REFERENCES

Main, M. 2003. *Citrus operations as components of ecosystems: Planning for needs of wildlife.* International Society of Citriculture Congress. (9th 2000: Orlando, Florida), Orlando, Florida USA 1: 320-322.

Neff, E. 1997. *On grove tour, the wildlife is real.* Citrus Industry 78: 26-27.

Pearlstine, L. G., L. A. Brandt, W. M. Kitchens, and F. J. Mazzotti. 1995. *Impacts of citrus development on habitats of southwest Florida.* Conservation Biology 9: 1020-1032.

42

Sustainability of Commercial Citrus Production

D.P.H. Tucker and M.R. Ziegler

INTRODUCTION

Developments in global competition, costly and restrictive regulations, invasive pests and diseases and declining grower profits have prompted concern regarding the long-term sustainability of commercial citrus production in Florida. Individual growers may be uncertain of the future of their citrus operations, and the long-term position of Florida citrus in the global market may appear threatened. However, Florida citrus is a fundamental aspect of the state's economy. Florida has long been identified with its citrus; lovely, fragrant citrus groves and delicious fresh fruit and juice delight residents and attract visitors. Working together with the public, Florida's citrus industry can and should remain viable into the foreseeable future, providing jobs and contributing to the economy of the state.

Sustainable citriculture is a system of production that can evolve indefinitely toward greater human utility, more producer profit, more efficient use of resources, and a balance with the environment that is favorable to humans and most other species.

There are many factors that influence the sustainability of Florida citrus. To become sustainable, farming systems, including citrus production, must share certain attributes. They must (1) allow the producer to receive sufficient share of the retail price of products to provide a reasonable livelihood while supporting viable industries and communities associated with citrus production; (2) optimize production output per unit of capital, labor, land, and energy inputs; (3) provide a quantity and quality of product sufficient to meet present and future demands; (4) conserve and protect natural ecosystem resource bases including water, soils, genetic diversity; (5) be governed by fair and reasonable regulation that does not impose undue financial and administrative burdens; and (6) develop a positive image and harmonious relationships with neighboring businesses and communities.

To remain sustainable, citrus production must meet certain minimal requirements. Essentially, sustainability means that these systems must protect the resources upon which they rely. Sustainable systems must also support the workers on whose labor they depend and operate within the bounds of the society in which they reside. To survive as a business option, sustainability must also attract investors and satisfy consumers. Florida citrus producers, individually and as a group, continually reevaluate traditional growing and marketing practices to abandon those not consistent with the goal of sustainability. Growers are increasingly aware of the environmental impact of their citrus operations and are actively adopting cultural practices that are less consumptive of natural resources. ICM is more than IPM and is central to this transition to sustainability. Most growers support ongoing ICM research and are receptive to new or modified production methods. In addition, the citrus industry has become more efficient and more competitive, adopting a more modern, business-oriented approach to sales and marketing. Growers are embracing new technologies, including precision agriculture and global networking.

IPM AND ICM

IPM is the crop protection system which best meets the requirements of sustainable development and sustainable agriculture. IPM is a component of Integrated Crop Management (ICM), which is developed as a farming system to meet the requirements of long-term sustainability. It is a whole-farm strategy, which involves managing crops profitably, with respect for the environment, in ways that suit local soil, climatic and economic conditions. It safeguards the farm's natural assets in the long term. It includes practices that avoid waste, enhance energy efficiency and minimize pollution. ICM is not a rigidly defined form of crop production but is a dynamic system, which adapts and makes sensible use of the latest research, technology, advice and experience. (Figure 42.1)

The Florida citrus industry and its research community are currently addressing several specific goals to improve and protect the long-term prospects for Florida citrus. The productive life of a grove may be extended to twenty years or more with the development of new rootstock/scion combinations, improved cultural practices, and the control of various conditions associated with tree decline. Mechanical harvesting systems and compatible grove designs are currently under development that will reduce Florida's vulnerability to labor shortages and improve the efficiency and cost effectiveness of citrus production. In addition, new fruit cultivars and products are under development to satisfy specific consumer demands or preferences, thus improving Florida's domestic and global market position.

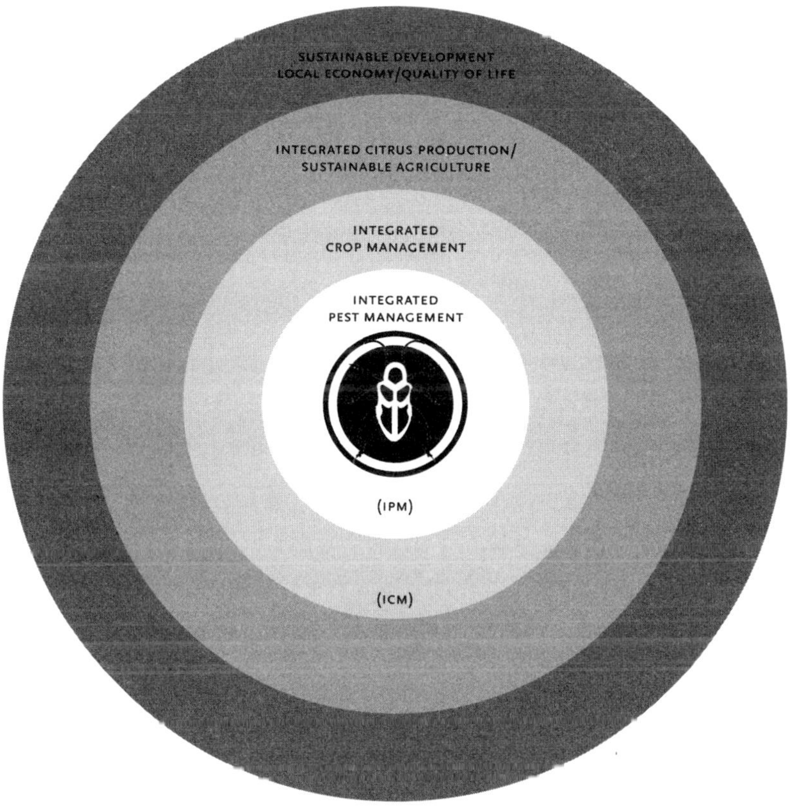

Figure 42.1 IPM and ICM in the context of sustainable development.

Unfortunately, several key requirements for sustainability are not within the control of the growers or their industry. Florida citrus production is highly regulated, and many regulations are complex, contradictory or unrealistic. The citrus industry has emerged, however, as a political entity with an influence on policy making at the local, state, and national levels that can alter some regulatory actions. Nevertheless, the industry must accept reasonable restrictions and resist primarily those that are genuinely contrary to common sense. The industry must cultivate broad public support and cultivate trust and respect. Many new regulations or restrictions are prompted by public concerns for consumer safety and environmental protection. Reliable experts must address these fears, and the public must be educated about requirements for a successful citrus business to operate.

The Florida citrus industry, as is all Florida agriculture, is also threatened by urban sprawl. As Florida's population continues to grow and residential and industrial development spreads into previously remote areas of the state, competition for land and water will increase. Conflicting needs and values will come into dispute. For example, most citrus growers have already converted from flood irrigation and high-volume sprinkler systems to low-volume irrigation equipment. This has somewhat relieved the competition for water, but drought conditions may still heighten tensions between the industry and the general public. In addition, residents object to heavy machinery moving in and around adjacent farmland and are apprehensive about their inadvertent exposure to agrichemicals in the soil, water, or air. Continued competition and conflict are inevitable, but creative urban and suburban planning may offer some resolution. Many cities and towns are working to control or limit growth and sprawl; their goal of maintaining their commercial core and promoting a sense of community may also serve to preserve some separation between residential areas and farmland. Growers will need to recognize the threat that urban sprawl represents to agriculture and support initiatives to control it. Growers should also embrace environmental initiatives to maintain and protect currently undeveloped areas surrounding farmland. These wild areas are intended primarily to preserve habitat for native flora and fauna, but they also serve to isolate farmland from suburban expansion (Figures 42.2a & b).

Finally, the public must be educated concerning the significance of Florida's citrus industry and its current threats. Consumers must realize that compliance with modern regulations is expensive for the citrus producer, and that, although prices for fresh and processed citrus may be increasing, grower returns are in decline. The industry must enlist public and governmental support as it struggles to reposition itself for long-term global viability. Florida citrus is a fundamental aspect of the state's economy, both directly and indirectly. In addition, Florida has long been identified with its citrus; lovely, fragrant citrus groves and delicious fresh fruit and juice delight residents and attract visitors. Working together with the public, Florida's citrus industry can and should remain viable into the foreseeable future providing jobs and contributing to the economy of the state.

Figures 42.2a & b Urban encroachment into citrus producing areas invariably leads to conflict between citrus growers, land developers and the public. Good zoning and community development plans are important to help all parties co-exist. The photographs above show the same area before and after urban development.

REFERENCES

Beattie, G. A. C., and D. Smith. 1997. Integrated pest management: sustainable pest control for the future based on the past? International Citrus Congress (8th 1996: Sun City, South Africa), Sun City, South Africa 1: 51-58.

Edwards, C. A., R. Lal, P. Madden, R. H. Miller, and G. House (eds.). 1990. *Sustainable Agricultural Systems.* Soil and Water Conservations Society, Ankeyy, Ia.

Ishii, T., S. Kirino, and K. Kadoya. 2003. *Sustainable citriculture using vesicular-arbuscular mycorrhizal fungi: Introduction of new soil management techniques.* International Society of Citriculture Congress (9th 2000: Orlando, Florida) Orlando, Florida USA 2: 1926-1029.

Porazinska, D. L., L. W. Duncan, R. McSorley, and J. H. Graham. 1999. *Nematode communities as indicators of status and processes of a soil ecosystem influenced by agricultural management practices.* Applied Soil Ecology 13: 69-86.

Schaffer, B. 1998. *Flooding responses and water-use efficiency of subtropical and tropical fruit trees in an environmentally-sensitive wetland.* Annals of Botany 81: 475-481.

Singh, M. and S. Tan. 1993. *Sustainable weed management in tree fruits.* Proceedings of the International Symposium Indian Society of Weed Science, Hisar. 1: 369-371.

Appendix: Additional Sources of Information

For more than 100 years, scientific research has been directed toward a better understanding of citrus production, improvements in citrus cultural practices, and solutions to short- and long-term problems facing the citrus industry. The results of this research are presented and analyzed in a vast array of publications. The scientific journals of many international, national, and state organizations of scientists and industry professionals provide very comprehensive, scholarly discussions of horticultural issues and current research. These journals include the Proceedings of the International Citrus Congress, the Proceedings of the American Society for Horticultural Science, and the Proceedings of the Florida State Horticultural Society. Industry trade magazines, including Citrus Industry, Citrus and Vegetable, and Florida Grower, target a broader audience and provide very accessible, popular reading. The University of Florida Cooperative Extension Service produces a wide selection of bulletins, circulars, fact sheets, and economic or statistical reports on specific topics of interest to the industry. Extension and research publications are available at IFAS research centers and county extension offices throughout the state and on the main University of Florida campus. Much of this information is available through an Internet accessible database (EDIS). In addition, various industry organizations and chemical manufacturers publish timely newsletters and reports on a variety of subjects. This information represents a tremendous resource that is currently underutilized by the Florida citrus industry.

The library at the University of Florida Citrus Research and Education Center at Lake Alfred is considered the preeminent repository of citrus publications in the world. The library, which is open to the public, currently maintains an impressive collection of more than 14,000 volumes and over 20,000 articles as reprints for the use of citrus researchers, producers, and others interested in the industry. The collection is organized for easy access, and the library staff is available to aid in information. Web pages have also been developed to help users locate information.

Abbreviations

AMS	Agricultural Marketing Service
APHIS	Animal Plant Health Inspection Service
ARS	Agricultural Research Service
CREC	Citrus Research and Education Center, Lake Alfred, Fla.
DEP	Department of Environmental Protection
EPA	Environmental Protection Agency
FASS	Florida Agricultural Statistics Service
FC	Frozen Concentrate
FCC	Florida Citrus Commission
FCM	Florida Citrus Mutual
FCP	Florida Citrus Packers
FCPRAC	Florida Citrus Production Research Advisory Council
FCPMA	Florida Citrus Production Managers Association
FDA	Food and Drug Administration
FDACS	Florida Department of Agriculture and Consumer Services
FDOC	Florida Department of Citrus
FFAA	Florida Fertilizer and Agrichemical Association
FFB	Florida Farm Bureau
FFVA	Florida Fruit and Vegetable Association
GCGA	Gulf Citrus Growers Association
GFSA	Gift Fruit Shippers Association
HCCGA	Highlands County Citrus Growers Association
ICM	Integrated Crop Management
IFAS	Institute of Food and Agricultural Sciences
IRCL	Indian River Citrus League
NFC	Not From Concentrate
OSHA	Occupational Safety and Health Administration
PRVCGA	Peace River Valley Citrus Growers Association
UF	University of Florida
USDA	U. S. Department of Agriculture
WMD	Water Management Districts

Contributors

L. Gene Albrigo
professor of horticulture at the University of Florida Citrus Research and Education Center, Lake Alfred

Brian J. Boman
professor of agricultural engineering at the University of Florida Indian River Research and Education Center, Ft. Pierce

Galen K. Brown
formerly a member of the Florida Department of Citrus, Lake Alfred

Jackie K. Burns
professor of horticulture at the University of Florida Citrus Research and Education Center, Lake Alfred

Frederick S. Davies
professor of horticulture at the University of Florida, Gainesville

Claire G. Erickson
formerly of the Monsanto Corporation, St. Louis, Missouri

Stephen H. Futch
multicounty extension agent at the University of Florida, Citrus Research and Education Center, Lake Alfred

Renee M. Goodrich
assistant professor of food science at the University of Florida, Citrus Research and Education Center, Lake Alfred

Tim R. Gottwald
research plant pathologist for the USDA Agricultural Research Service, Ft. Pierce, Florida

James T. Griffiths
formerly president of the Citrus Growers Associates, Lakeland, Florida

John L. Jackson, Jr.
formerly multicounty extension agent at the University of Florida, Tavares

Larry K. Jackson
professor emeritus of horticulture at the University of Florida, Citrus Research and Education Center, Lake Alfred

Michael C. Kesinger
chief of the Bureau of Citrus Budwood Registration at the Florida Department of Agriculture and Consumer Services, Winter Haven

Martin B. Main
associate professor of wildlife management at the University of Florida Southwest Florida Research and Education Center, Immokalee

William M. Miller
professor of agricultural engineering at the University of Florida Citrus Research and Education Center, Lake Alfred

Kelly T. Morgan
research manager at Citrus Research and Education Center, Lake Alfred, Florida

Ronald P. Muraro
professor of economics at the University of Florida Citrus Research and Education Center, Lake Alfred.

Thomas A. Obreza
professor of soil sciences at the University of Florida, Gainesville

Mickey E. Parish
professor and chair of the Nutrition and Food Science Department, University of Maryland, College Park

Lawrence R. Parsons
professor of horticulture at the University of Florida Citrus Research and Education Center, Lake Alfred

J. Steven Rogers
president of Ecostat, Inc., in Highland City and scientific coordinator at the Florida Citrus Production Research Advisory Council, Lake Alfred

Fritz M. Roka
associate professor of economics at the University of Florida Southwest Florida Research and Education Center, Immokalee

Ralph V. Russ
former bee keeper, Haines City, Florida

Ted J. Smith
professional geologist at B.C.I. Engineers and Scientists, Inc., Lakeland, Florida

Peter D. Spyke
president of Arapaho Citrus, Ft. Pierce, Florida

Edward W. Stover
curator and research leader at the Pacific West Area of the USDA Agricultural Research Service, National Clonal Germplasm Repository, Davis, California

James A. Stricker
formerly county extension director at the University of Florida, Bartow

J. P. Syvertsen
professor of plant physiology at the University of Florida Citrus Research and Education Center, Lake Alfred

Chet E. Townsend
senior account manager at Agrilink Florida Inc., Fort Denaud

David P.H. Tucker
professor emeritus of horticulture at the University of Florida Citrus Research and Education Center, Lake Alfred

T. Adair Wheaton
professor emeritus of horticulture at the University of Florida Citrus Research and Education Center, Lake Alfred

Michael R. Ziegler
president of Agricultural Resource Management, Vero Beach, Florida